QUANTITATIVE BIOLOGY

Biology at all scales has become a data-driven science, with large-scale datasets driving fields from population genomics to ecology. Practicing biologists have no choice but to use computational approaches, statistics, modeling, and other data science tools in their research. However, undergraduate biology education still primarily focuses on nonquantitative descriptions. This book provides students whose background is in biology with an introduction to modeling biological systems using mathematical, computational, and statistical tools. It is based on a series of hands-on analyses conducted with open-source tools that allow the students to discover for themselves emergent properties of biological systems that are not evident without using model-based approaches. The goal of this book is to provide a "turn-key" introductory quantitative biology course suitable for all biology students. The book provides the narrative for the analyses and discussions to be done in class, with support from the included website, slides, and test material.

Key Features

- Written in an accessible, narrative style
- Includes hands-on analyses with open-source tools
- Integrates biology across spatial and temporal scales
- Links to a course website with interactive tools
- Brings biological education into the "data science" era
- Each chapter includes a variety of exercises designed to actively engage the reader
- Lecture slides and animations to cover the key arguments and derivations in each chapter, as well as example exam questions, are available for qualified instructors.

Gavin Conant worked as a researcher in evolutionary and computational biology for more than 25 years and has authored or coauthored more than 80 peer-reviewed scholarly articles, as well as book chapters and articles for the popular press. His research spans bioinformatic algorithm development, data visualization, evolutionary biology, metabolic modeling, parallel computing, and microbial ecology.

QUANTITATIVE BIOLOGY
LIFE FROM THE NUMBERS

Gavin Conant

CRC Press
Taylor & Francis Group
Boca Raton New York London

CRC Press is an imprint of the
Taylor & Francis Group, an **informa** business

Designed cover image: Yongkiet Jitwattanatam/Shutterstock

First edition published 2026
by CRC Press
2385 NW Executive Center Drive, Suite 320, Boca Raton FL 33431

and by CRC Press
4 Park Square, Milton Park, Abingdon, Oxon, OX14 4RN

CRC Press is an imprint of Taylor & Francis Group, LLC

ISBN: 978-1-041-17016-7 (hbk)
ISBN: 978-1-041-17015-0 (pbk)
ISBN: 978-1-003-68750-4 (ebk)

DOI: 10.1201/9781003687504

Typeset in Utopia
by SPi Technologies India Pvt Ltd (Straive)

Contents

Preface

WHO IS THIS BOOK FOR?

Before you commit to reading this book, it is probably worth considering what I hoped to accomplish by writing it. (Whether I achieved this is, of course, a different matter.)

This book is first and foremost intended to *teach*. The reason I think it is worth noting this goal is that anyone writing a book of this type must decide whether it is primarily intended to help students learn about a topic or to serve as a reference to those who already know that topic to a degree. Thinking back to the classes I took as an undergraduate biology major, some of the textbooks we used were ideal as references on almost anything one might have wanted to know about—say, cell biology. However, trying to read them as a new student of cell biology was very difficult because you might read one sentence that gave a core concept followed by a sentence of explanation and then four sentences of exceptions and qualifiers. From that text, it was often hard to separate the essential from the more extraneous. At least for me, these textbooks did not help me *learn* these areas of biology, although now I find them quite useful when I want to refresh my memory. I have also now had the experience of reading science books to my two young sons, both old books from my childhood and new ones. The thing I find most striking about the new books on, say, dinosaurs is that while they are much more beautiful than my childhood books, they lack a core narrative. Instead, they are just a series of vignettes, pictures, and boxes. I find it very hard to learn from these books because the visual clutter overwhelms any story being told. Having experienced this issue, I have tried to make this book as narrative as possible: I have avoided sidebars, boxes, and other extra features that would distract from following the main argument through each chapter.

It therefore goes without saying that the book does not claim to exhaustively describe the research around any of the models used. Instead, what I have presented is, to the best of my knowledge and ability, correct. And, at least in some cases, what has been left out was left out intentionally, to focus attention on the central ideas I want to impart. In the same vein, the examples I have picked were chosen as much for their illustrative value as for their importance to modern biology. Hardly any practicing ecologists would use the Lotka–Volterra models as described in Chapter 3, but they are an excellent introduction to the problem of creating a mathematical model of a biological phenomenon.

BACKGROUND

No book can truly start from nothing. But a book that assumes the reader already knows a great deal is hardly a useful textbook. So, what background have I assumed from the reader here? The book assumes familiarity with differential calculus: integral calculus is a nice addition but not required. Some basic knowledge of biology, such as the concepts of DNA, inheritance, and evolution, will also be helpful. More complex mathematical and statistical concepts are covered, but my hope is that you can learn what you need to know about these concepts from the text itself.

A NOTE TO INSTRUCTORS

This book grew out of the course I teach in the Spring semester: it meets 2.5 hours per week. The classroom we use is a collaborative computing room, where the students are given laptops with the R and other course software installed, as well as whiteboards and presenting materials so that they can collaboratively discuss and solve the exercises in the book.

In a typical semester with 45–55 students in class, I can cover Chapters 1–3, 5–9, 11, and 13–15, though Chapters 14 and 15 often end up being somewhat rushed. I would say that the core chapters of the book are 1–3, 5–9, 11, and 13 because these chapters explicitly build on each other and cover the core idea of how mathematics, statistics, and computation apply to biological problems. Of course, I do not do *all* of the exercises in each chapter in class, but I hope I have provided enough choice among those exercises for different course delivery modes and time frames.

I have tried to have all of the exercises based on open-source data and software. Most of the beginning of the book can be done in R. I have created a website (http://qbio.statgen.ncsu.edu) with applets to cover those topics for which an accessible, free software tools is not available (especially flux-balance analysis in Chapter 6 and some simple statistical simulations for Chapters 8 and 9). In many cases, R and the other tools will allow students to replicate the figures in the book, although I have lightly edited those figures here for readability. One of my key goals for the book is for there to be nothing left hidden behind my back, so to speak: that the results shown can be understood and reproduced with the tools described and the material explained in the text.

I have also provided slides showing which topics in the book I emphasize when I lecture on this material. These slides include examples of which equation derivations I find important enough to go over in class, as well as some simple animations that help visualize some of the static diagrams in the printed book.

A NOTE ON AI

I have used artificial intelligence (AI) to generate a handful of the figures in this book. I have done so when I wanted to represent a common object like a sugar cube for illustrative purposes: none of the images of actual biological systems were created in this way. The figures with AI components are 1.3, 7.1, 8.2, 13.12, and 13.13.

Acknowledgments

One of the pleasures of writing this book has been realizing that I was only able to do so because I was remembering ideas I had been given by teachers and mentors over virtually my entire academic career. I am therefore going to indulge myself in what is perhaps a longer acknowledgment section than is warranted by the work that follows it.

The proximate cause of this book is a dinner I had with Jerry LeBlanc and Jane Lubisher at which they proposed creating a quantitative biology course, an idea that had never occurred to me. I therefore also appreciate Jerry's indulgence as department chair in allowing me to develop the course when I joined the NC State faculty.

I thank my sixth-grade math teacher Mr. Roland for showing me that math problems can be solved in more than one way and John England for illustrating how to make teaching science dynamic and interactive. I thank Ken Maier for teaching me how to program defensively and robustly. None of the code included here should be taken to follow his teaching, but whatever remnants of good programming practice I have are due to him. Ruth Roberts taught me to use commas and Eleanor Sullivan and Nancy Harris not to take myself quite so seriously in writing. (This second lesson did not take to any appreciable degree). Bill Neno showed me that biology was more than anatomy, but Karl Bitner showed me that even anatomy was fascinating, and Jeff Hay and Tony Davis got me my first science-adjacent job.

At university, Wayne Rickoll introduced me to the idea that biology could be quantitative, and Bob Beezer made me see the close linkages between computation and mathematics. Mark Ondrias convinced a budding biologist that statistical thermodynamics was not an esoteric discipline remote from living systems, while Robert Millar opened the door for me to working as a researcher in addition to teaching an outstanding cell biology class. Ed Bedrick and Aparna Huzurbazar gave me the confidence to apply the necessary statistical approaches in my research without needing to claim expertise in statistics, and Tim Thomas taught me how to actually use LINUX computers. Finally, a huge part of the genesis of this book was Stephanie Forrest's course in complex systems and the centrality of modeling for the scientific enterprise.

My career as a researcher owes an enormous debt to my four academic mentors: Paul Lewis, Andreas Wagner, Peter Stadler, and Ken Wolfe. Paul is the genesis of Chapter 13; indeed, many of the examples and explanations there are adapted from him. Andreas directly inspired the topics of Chapters 5, 6, and 14, but more importantly, his example gave me the confidence to keep my interests broad as my career advanced. In that sense, he is the genesis of my desire to write a book like this one. Peter helped me a great deal in thinking about computation and biology (Chapter 12); more importantly, he was my guide to what a good scientific group leader should be. Ideas that Ken helped me develop are found explicitly in Chapter 15, but my time with him also allowed me to develop the topics in Chapters 5, 6, and 14. He is also an amazing example of a person of both great knowledge and grace in sharing that knowledge.

I was supported in my doctoral work by the U.S. Department of Energy's Computational Sciences Graduate Fellowship (CSGF). In addition to the financial contribution, this fellowship opened my eyes to the range of science being done on computers, introduced me to amazing scientists from many different fields, let me work as a researcher at Sandia National Laboratories, and encouraged me to expand my horizons beyond biology. During that time, I was mentored by Steve Plimpton and Bruce Hendrickson, who also taught me practical parallel computing and algorithms. In a very real way, that fellowship was the prologue of this work.

I owe a great deal to a long list of other friends and researchers who have listened to my odd ideas and turned them into useful ones. These (unlucky?) sounding boards have included Brian Cusack, Carolin Frank, Jon Gordon, Grace Lyo, Matthew McIntyre, Axel Mosig, Jeff Mower, Åsa Perez-Bercoff, Devin Scannell, Marie Sémon, Suzie Shoup, Lydia Tapia, Sean Walston, and Meg Woolfit. I thank Mike Fuller for teaching me about ecology and allometric scaling, Mike Gilchrist for showing me how to be a developer of biological models, and Kevin Byrne for teaching me graphical web development and Irish (and American) politics. Annette Evangelisti introduced me to stochastic biochemical simulations, Amy Powell to the power of fungal genetics, and Nora Khaldi to how a mathematician views biology. The chapter on networks grew out of a joint lecture I developed with Frank Schmidt in 2016, and the chapter on cancer grew directly from lectures I attended sponsored by the Triangle Center for Evolutionary Medicine (TriCEM) that were organized by Jason Somarelli. David Hillis kindly provided his depiction of the tree of life for Chapter 13.

Since becoming a faculty member, whatever success I have had owes a very great deal to a wonderful group of mentors, collaborators, friends, and students. These include mentors like Jerry Taylor, Jim Birchler, and Matt Lucy and my students and postdocs Michaël Bekaert, Abbey Coppage, Amrit Dhillon, Yue Hue, Corey Hudson, Rocky Patil, Kathy Scienski, Mustafa Siddiqui, Huan Truong, Logan Williams, and Sara Wolff. I also thank my friends Francisco Aguilar, Lori Eggert, Fabio Gallazzi, Ali Siavosh Haghighi, Adam Hartstone-Rose, Dmitry Korkin, David Rasmussen, Rocío Rivera, Chi-Ren Shyu, and Jeff Thorne. I also am extremely grateful to Bill Lamberson and Chris Pires, who between them guided me through the perilous waters of being a new faculty member. Chris was also instrumental in helping me develop the material on understanding phylogenetic trees at the beginning of Chapter 13.

Finally, the simple chance to write this book comes from the love and support of my family: my grandfather Marion Duty, my mother and father, Ettajane and Henry, my sister Eleanore, and my sons Ian and Colin. And, more than anyone else, Michela Becchi has been on this scientific journey with me, reassuring me when things went badly and listening when I needed to work through frustrations or confusions.

On the Road
Dynamical Models of Infectious Discases and Physical Systems

1

"[T] o travel hopefully is a better thing than to arrive ..."
—Robert Louis Stevenson, *Virginibus Puerisque*

A PEEK INTO THE FUTURE: MODELING INFECTIONS

Imagine five friends from a small town of 5,000 people return home from a vacation all with some sort of flu-like illness. After 4 days they all recover, but in the meantime 10 other people in the town show symptoms of the same disease.

Is this the start of a horror film? Or is it just a week when the local pharmacy runs out of cough medicine? Let's keep our first example a relatively cheerful one and assume two things: first that individuals are fully recovered from the disease after the 4-day recovery period, and second that once a person has caught the disease, they cannot be reinfected with it again. Under those assumptions, what might happen in this town?

Figure 1.1 illustrates one way of answering this question. To understand the problem, we will first divide people in the town into three groups. The *susceptible* individuals (denoted S) are those who have not yet been in contact with an *infected* individual (here noted as I) but who might become sick if they do have such a contact. The *recovered* individuals (here R) will not become ill even if they are in contact with an infected individual because, under our assumption, they are immune to reinfection. When we break the town into these groups and make our assumption of an initial group of $I = 5$ individuals, we can run some mathematical analyses to track the course of the outbreak over a few weeks. *How* we will do that analysis is the topic of the next four chapters. Here we will just give the results: over the course of 75 days, many, but certainly not all, of the 5,000 people in the town catch the disease (Figure 1.1A).

An obvious question about these results is how different they might be if the characteristics of the disease were somewhat different. In Figure 1.1B, I have left all the features of the system the same except that I have assumed that the disease only lasts for 3 days. Now, as we will discuss in Chapter 2, we assume that the only way for a susceptible individual to become ill is to come into contact with an infected one. We might therefore expect that reducing the period of time that people have the disease would reduce the overall proportion of the town that catches the disease. And indeed, under these new conditions, Figure 1.1B shows that now only slightly more than half of the town catches the disease before the outbreak begins to subside.

The framework we have used here is called an *SIR* model of infectious disease [1, 2]. *Model* is a word that we will have a good deal more to say about over the course of this book: for the moment, we can think of a model as a way of representing some phenomenon of interest in such a way that we can study it through mathematics, statistics, computation, or some combination of the three.

This SIR model can actually tell us many interesting things: for the moment, we will pause and discuss only one. The duration of the illness has a surprisingly large effect on the spread of the epidemic. When we decrease the duration of the

DOI: 10.1201/9781003687504-1

A)

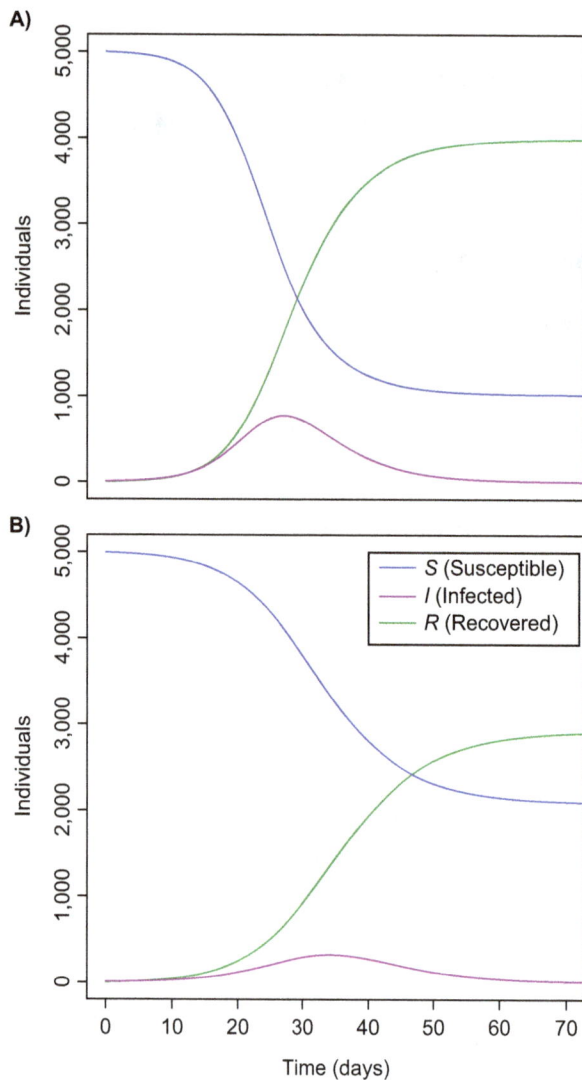

B)

Time (days)

Figure 1.1 Shown are two possible histories of an epidemic in a small town. Both start with 5 initially infected individuals returning to a town of 5,000. In **A**, each initially infected individual remains infectious for 4 days and infects, on average, two other people. In **B**, the "instantaneous" infectivity of the disease (the chance that an encounter between an infected and susceptible individual results in disease transmission) remains the same. However, the infection lasts only 3 days. As a result, those initially infected individuals infect on average only 1.5 other individuals before they recover and are no longer infectious.

infection by a quarter (4 to 3 days), we go from a situation where the large majority of the town gets the disease to one where only about half of the town does. Now, on one level, we can feel pleased that this change is exactly the one we predicted above (shorter duration means a smaller epidemic). On the other hand, this is a book about the value of quantitative approaches: there is a real difference between knowing only the direction of the expected change and knowing its size. In Chapter 2 we will revisit this model and discuss using it to inform public health strategies such as vaccination plans.

Before we can use this model, however, we need to understand a bit of the mathematics it is based on. To build that background, we will start with a simpler set of problems for which we already have some intuition. We will then build a set of tools and terminology that will allow us to solve, discuss, and understand models like the SIR one.

PHYSICAL EXAMPLE: POSITION AND VELOCITY

While the SIR model describes phenomena we are all familiar with, such as catching a cold from our family or acquaintances, the way those familiar events are converted into mathematics is a trifle less obvious. For other biological systems, such as the biochemistry of the cell, the connection between the mathematics and the biology are a bit more evident, but the biology itself is not as familiar. (Few of us have a true intuitive sense of what it would be like to stand inside a cell and watch the goings-on.)

So, with some reservations, I would like to take a brief detour from biology and introduce some of the mathematical "machinery" we will need using examples from physics. We will start with a quite simple example: a car with the cruise control set to maintain a constant speed of 100 km/h on a very straight and level road (perhaps Interstate 70 in Kansas). We can imagine watching this car through the course of its drive and recording its location over that period. Indeed, for simplicity, we can ignore the three-dimensional nature of actual driving and simply consider the car's location along the road. In this framework, we will define our starting point as distance 0 and all other locations in relation to that starting point (Figure 1.2).

In this scheme, we can also think about measuring the *change* in the car's position over an interval of time. We are so familiar with this idea, in fact, that we might not have stopped to think exactly what it means. When we say a car is traveling at 100 km/h (about 60 m/h for those of us in the United States), we do not mean to imply that the planned journey is exactly 100 km, nor that we will drive for exactly 1 hour [3]. Rather, by combining the distance and the time, we have created a new form of measurement that tells us the *rate of change* of our position.

A natural question now is: *When* are we traveling at 100 km/h? In most real-life situations, our speed along the road will vary quite a bit due to traffic, hills, and so forth. So we might think of two different answers to this question of "when." First, we might happen to look down at the speedometer at a particular moment and see that, right at that instant, our speed is 100 km/h. Or second, we might look back once we reach some destination and say, "Well, we traveled 100 km and it took 1 h, so our *average* speed was 100 km/h." In our example here, we will avoid this question altogether by setting the cruise control to 100 km/h, meaning that both our instantaneous speed and our average speed are the same 100 km/h.

Now, in an era when our cars and phones communicate instantly with satellites that can locate them within a meter anywhere on Earth, this idea of knowing, minute-to-minute, where you are seems quite simple. Given that position information, we can also compute the rate of change of that position. But thinking back even a few centuries, the problem of measuring relatively long distances (say, between cities) was a very difficult one: the local explorers' outfitting store would hardly carry tape measures in lengths of tens of kilometers.

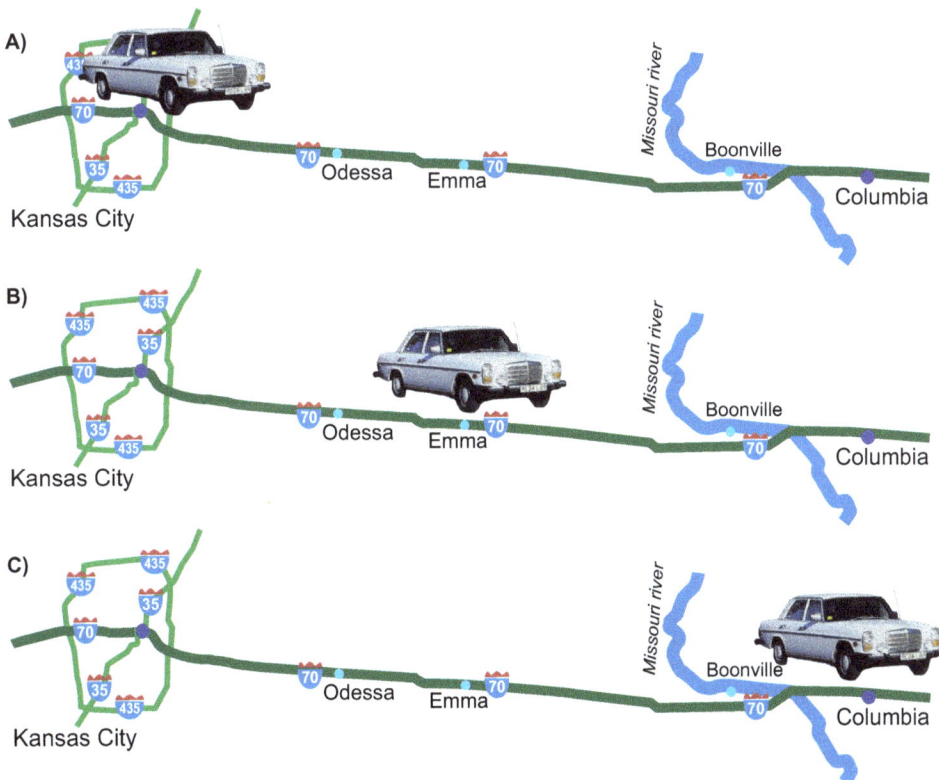

Figure 1.2 **An automobile trip through the US Midwest.** (**A**) We start in Kansas City, MO. (**B**) Pretending that we encounter no traffic, 1 h at 100 km/h brings us to Emma, MO. (**C**) A further hour then brings us to Columbia, MO. Notice that the complexity of the topography and the curving of the road are removed from this analysis: we treat the situation as a one-dimensional trip along the road.

The problem was far from merely a bookkeeping one, especially in the case of ships at sea. Sailors could fix their north/south position without great difficulties using the pole star. But to compute their east/west position, they were often reduced to using exactly the sort of speed/time calculation we have just described. When we recall that those ships had no continuous or fully accurate method for measuring that speed, the potential for disaster is obvious. After more than 2,000 British sailors and soldiers were killed in the simultaneous wreck of four Royal Navy ships in 1707, the British Parliament offered a prize of the equivalent of one million pounds today for any approach that could yield the longitude of a ship within one half of a degree [4]. The story of the spring-driven clocks that could accomplish this feat on the rolling deck of a ship is well worth a read if you are interested [4]. For our purposes, we can notice that the clock-based solution found avoided entirely the problem of determining the speed of the ship and directly located it based on the comparison of the local time to that at the prime meridian in Greenwich, England.

THE INVERSE PROBLEM

What does all of this have to do with infectious disease, or indeed biology more generally? As we will see in the upcoming chapters, information about many biological phenomena comes to us in this inverse form, where we know how fast a quantity is changing but not the quantity itself. In the case of the SIR model, we will have equations for the increase or decrease in I in time but not for the value of I itself. To get *that* value, we will use our knowledge of its rate of change, coupled with a known initial point, to compute the value of I at any time we desire. First, though, we need to find the mathematical and computational toolkit that can allow us to move from the knowledge of a quantity's rate of change to a knowledge of the quantity itself.

Functions

We will start by reminding ourselves of a language for representing quantities of interest in a mathematical form: the concept of a *function*. We can think of a function as a mathematical or computational machine that takes one or more numerical measurements or estimates and produces from them another numerical answer. We will talk more about functions in the abstract a bit later: for now, we will use our road trip as an example. Here, our measurement will be the time elapsed on our trip, which we will represent with t. We could use several different units of time in our measurements: for convenience, we will stick with hours. We would like to predict p, our position along the (one-dimensional) road, meaning our desired function is $p(t)$. Because of our assumption of a constant speed, computing $p(t)$ is quite simple:

$$p(t) = 100 \cdot t \tag{1.1}$$

Notice that we have omitted the units here: if we add them back underneath the main equation, we can see that they reinforce our sense that this function describes what we want it to:

$$
\begin{array}{ccc}
p(t) & = & 100 \cdot t \\
[\text{km}] & & \left[\dfrac{\text{km}}{\text{h}}\right] \cdot [\text{h}]
\end{array}
\tag{1.2}
$$

What then is the rate of change of our position in time? For this simple example, the answer is instinctive: 100 km/h. As a result, it is not very informative to illustrate the mathematics involved in computing it. Instead, we will create a slightly more complicated example where the rate of change of the position is not simply a constant. We will then circle back to this example after we understand that one.

Physical example: Adding Acceleration

For this new example we will again take a slightly artificial case, but one that is reasonably close to something we are all familiar with: a falling ball (**Figure 1.3**). We will simplify the situation by assuming that the ball is falling either in a vacuum or for a short enough distance that we can ignore the problem of air resistance slowing its fall.

If we think of pushing the ball off a shelf that is at some distance d above the ground, we actually have motion in two dimensions: motion horizontally along the ground due to our initial push, and motion vertically toward the ground due to the pull of gravity. The horizontal motion (p_x; again in a vacuum) follows our equations above for the moving car, and so we will ignore it for this example. Instead, we will focus on the vertical motion (p_y), particularly on discovering *when* the ball hits the ground and *how fast* it is traveling when it does so.

Thus, our first question requires a function $p_y(t)$ that gives the ball's position as a function of the time t after the push. Obviously in this case, units of hours would be quite cumbersome, and we will instead measure time in seconds (or fractions of one).

We can see at once that this problem is very much like the epidemic one we started with. We know the conditions of the system at the start $(t = 0)$ but do not instinctively know a function that gives the time-dependent behavior after $t = 0$. Unlike the car example, the ball's velocity in y is not constant. Instead, because of the pull of the Earth's gravity, its velocity is increasing as it falls. Notice that I have switched from writing about speed to velocity: the distinction is not particularly critical here, but because we are now confining our interest to change in position with respect to the ground, the concept of velocity (speed in a defined direction) is a bit more appropriate.

Ideally, we would work through the process of using our knowledge of the acceleration of gravity to work out $p_y(t)$. Unfortunately, we have not yet seen the mathematical tools we need to do so. Instead, we will start with the answer and work backward to understand how that answer was obtained. Having done so, we will be able use that approach in Chapter 2 to build the SIR model that was used to construct Figure 1.1.

The function that describes $p_y(t)$ is:

$$p_y(t) = d - \frac{9.81}{2} \cdot t^2 \tag{1.3}$$

Figure 1.3 A falling ball. The ball starts at a distance $y = d$ above the ground and an initial horizontal position $x = 0$. Because horizontal and vertical motion are independent, we can ignore the movement along x and focus on p_y: the ball's position in the air. (Image generated using ChatGPT).

Here d is the initial position of the ball in y, while time is measured in seconds and distance in meters. From Equation 1.3 we can compute when the ball strikes the ground by setting $p_y(t) = 0$, rearranging, and solving for t:

$$d = \frac{9.81}{2} \cdot t^2 \tag{1.4}$$

which yields:

$$\sqrt{\frac{2d}{9.81}} = t \tag{1.5}$$

With $d = 10$ m (the height of the highest diving platform in the Olympics), this equation implies the ball will hit the ground after 1.43 s.

How Fast Is the Ball Falling?

Our second question was how fast the ball is traveling when it hits the ground. But perhaps we should expand this question and ask: How fast is the ball traveling at each moment of its fall? Equation 1.3 suggests that this question is an interesting one: to see why, our first exercise (1.1) will be to tabulate the ball's position first at intervals of 0.1 s and then at intervals of 0.05 s. Table 1.1 gives the results for the interval size of 0.1 s—the second calculation is left to the exercise.

Given these positions, we can compute the ball's *average* speed over each interval as simply the distance traveled in that interval over the length of the interval. Glancing at the third column of Table 1.1, it is clear that the ball's velocity is increasing as it falls, just as we expected. As a result, these average velocities are exactly that: averages. In fact, the ball is traveling slower than that average at the beginning of each interval and faster at the end.

How might we correct for this error? If we reduce the length of the time interval, the difference in velocity between its beginning and its end will also be reduced. We can represent this idea by considering a short interval of time Δt. How far does the ball travel in this interval? We will define this distance to be Δp_y. From our function:

$$\Delta p_y = p_y(t + \Delta t) - p_y(t) \tag{1.6}$$

TABLE 1.1 POSITION AND VELOCITY OF A FALLING BALL

Time t (s)	Position p_y (m)	Average Velocity (m/s)
0	10	0
0.1	9.95	0.49
0.2	9.80	1.47
0.3	9.56	2.45
0.4	9.22	3.43
0.5	8.77	4.41
0.6	8.23	5.40
0.7	7.60	6.38
0.8	6.86	7.36
0.9	6.03	8.34
1	5.10	9.32
1.1	4.06	10.30
1.2	2.94	11.28
1.3	1.71	12.26
1.4	0.39	13.24

The velocity over this short interval is then simply:

$$\frac{\Delta p_y}{\Delta t} = \frac{p_y(t+\Delta t) - p_y(t)}{\Delta t} \tag{1.7}$$

In other words, it is the distance traveled over the time (here Δt). Now, using Equation 1.3, we can compute the p_y values to get the average speed for a short interval in our particular example. But our real goal is a more general understanding that would work with any distance function (and indeed any function at all). As a starting point, notice that while Equation 1.7 gives an average velocity, if we let the size of the time interval decrease, the difference between this average velocity and the instantaneous velocity at any point in the interval should also decrease.

INSTANTANEOUS VELOCITY: THE DERIVATIVE

This idea of reducing the interval size is the core idea of differential calculus, and it can be formalized in terms of taking a *limit*:

$$\lim_{\Delta t \to 0} \frac{\Delta p_y}{\Delta t} \tag{1.8}$$

We can expand Equation 1.7 by replacing its right side with the formula of Equation 1.3, which gives the following:

$$\lim_{\Delta t \to 0} \frac{\Delta p_y}{\Delta t} = \frac{d - \frac{9.81}{2} \cdot (t+\Delta t)^2 - \left(d - \frac{9.81}{2} \cdot t^2\right)}{\Delta t} \tag{1.9}$$

When we cancel the common d term and move the ½ × 9.81 term to the front, we get the less intimidating:

$$\lim_{\Delta t \to 0} \frac{\Delta p_y}{\Delta t} = \frac{9.81}{2} \frac{t^2 - (t+\Delta t)^2}{\Delta t} \tag{1.10}$$

If we now expand the $(t+\Delta t)^2$ term, we find:

$$\lim_{\Delta t \to 0} \frac{\Delta p_y}{\Delta t} = \frac{9.81}{2} \frac{t^2 - t^2 - 2t\Delta t - \Delta t^2}{\Delta t} \tag{1.11}$$

which, when we cancel positive and negative terms in t^2 and the Δt at the top and bottom of the right side of the equation, allows a further simplification to:

$$\lim_{\Delta t \to 0} \frac{\Delta p_y}{\Delta t} = -\frac{9.81}{2} \cdot (2t - \Delta t) \tag{1.12}$$

When we take this limit, the Δt approaches 0. As a result, this value grows closer and closer to $-9.81 \times t$. By convention, we represent the result of this limit with a new notation:

$$\lim_{\Delta t \to 0} \frac{\Delta p_y}{\Delta t} = \frac{dp_y}{dt} \tag{1.13}$$

and refer to this entire process as "taking the derivative" of the function $p_y(t)$:

$$\frac{dp_y}{dt} = -9.81 \cdot t \tag{1.14}$$

This result tells us a few things. The ball's velocity is a function of how long it has been falling: the longer it falls, the faster it goes. We describe such an object with an increasing velocity as *accelerating*. We also notice that the velocity is independent of the starting position (at least until the ball hits the ground!).

This problem of determining the rate of change in a function relative to the function's dependent variable (here t) is one that comes up over and over. Naturally, then, mathematicians have tabulated the general form of the derivatives of a very large number of functions. In **Table 1.2**, I give the derivatives of several common functions as examples [5]. Notice that another common notation for the derivative of a function $f(x)$ is $f'(x)$. One key case to watch is when we take derivative of a function of the form $f(x) = m \cdot x$: we find that $f'(x) = m$. If we look back to our original example of the moving car, we see that Equation 1.1 has exactly this form, with $m = 100$. Hence the derivative of Equation 1.1 is:

$$\frac{dp}{dt} = 100 \tag{1.15}$$

which corresponds to our intuitive solution for the speed of the car.

Graphical Interpretation of Derivatives

We can also consider this idea of a function and its derivative graphically. In **Figure 1.4**, I have plotted Equation 1.3 in red. I then show three possible intervals over which I might compute the average velocity, each centered at $t = 0.7$ s. To make the illustration a bit clearer, we compute the average velocity by first calculating $p_y(t - \frac{1}{2}\Delta t)$ and $p_y(t + \frac{1}{2}\Delta t)$, giving ourselves the average velocity in the interval that is centered around $t = 0.7$ s. The average velocity $\Delta p_y / \Delta t$ at t is then simply:

$$\frac{\Delta p_y}{\Delta t} = \frac{p_y\left(t + \frac{1}{2}\Delta t\right) - p_y\left(t - \frac{1}{2}\Delta t\right)}{\Delta t} \tag{1.16}$$

We will now draw a line that has a slope $\Delta p_y / \Delta t$ for $t = 0.7$. We will compute the y intercept of that line such that it intersects Equation 1.3 at both $p_y(t - \frac{1}{2}\Delta t)$ and at $p_y(t + \frac{1}{2}\Delta t)$. How we make that computation is not particularly important for this problem, but for completeness, notice that, with a fixed slope m given by Equation 1.16, we take the equation of the line $y = mt + b$, set $y = p_y(0.7 - \frac{1}{2}\Delta t)$ and solve for b.

We can see from Figure 1.4 that as the interval Δt shrinks, the line whose slope is given by $\Delta p_y / \Delta t$ in Equation 1.16 grows ever closer to a line that touches the curve of Equation 1.3 at a single point (namely $t = 0.7$). We can refer to this line touching at a single point as a *tangent* line. Its slope is given by Equation 1.14 (see Figure 1.4C). We can thus say that *the derivative of a function defines a line that is tangent to that function at every point along the curve for that function.*

TABLE 1.2 SOME COMMON DERIVATIVES

Function	Derivative
$f(x) = c$	$f'(x) = 0$
$f(x) = m \cdot x$	$f'(x) = m$
$f(x) = x^2$	$f'(x) = 2 \cdot x$
$f(x) = x^3$	$f'(x) = 3 \cdot x^2$
$f(x) = e^x$	$f'(x) = e^x$
$f(x) = \sin(x)$	$f'(x) = \cos(x)$
$f(x) = \cos(x)$	$f'(x) = -\sin(x)$

A)

B)

C)

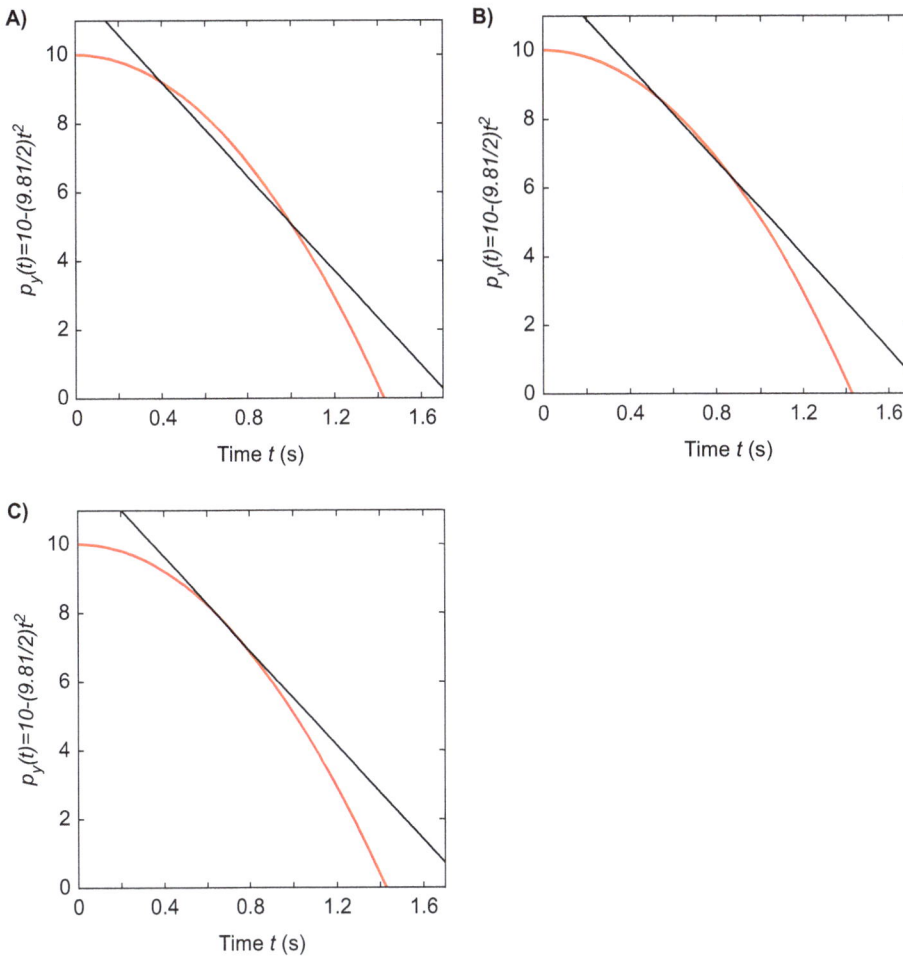

Figure 1.4 Average velocity lines (black) superimposed on the ball's position curve from Equation 1.3 (*red*) for three interval sizes centered at time *t* = 0.7. Because the initial curve is quadratic and we computed the average speed at $t \pm 1/2\,\Delta t$, all three average velocity lines have the same slope: when the intersection point is at $t = 0.7$, the line is tangent to the original curve at that point.

As another example, if we look back at Table 1.2, we can again see that the derivative of $f(x) = m \cdot x$ is $f'(x) = m$. The implication of this relationship is that the derivative defines a curve with the same slope (m) everywhere, a slope that is also the same as the slope of the original line. Now, since the tangent of a line is just the line itself, we can again see that the operation of taking a derivative describes the tangent of the original function at every point.

The Utility of the Derivative for the Inverse Problem

As we already mentioned, while computing the rate of change of a function at every point (e.g., computing its derivative) is quite a useful ability, many problems come to us framed in the opposite form: we *know* the rate of change but not necessarily the original function. A natural question is: *Can we use our understanding of derivatives to solve this inverse problem?*

Using our falling ball example, we will work through this idea, assuming we know the derivative of the function we desire but forgetting, for the moment, that we actually know the original function itself because it is written down in Equation 1.3. (We will also forget any of the integral calculus we may have picked up here and there.) Building off our discussions of tangent lines, we will first approach the question graphically.

As a proposal, we can consider approximating the continuous curve we are interested in (the red line of Figure 1.4), with a series of short line segments corresponding to the value of the derivative near that region of the curve. This idea may sound a bit odd, but a brief study of **Figure 1.5** should make it more obvious. We first decide the size of the segments we plan to use, which simply involves defining a time interval Δt and treating the velocity (e.g., the derivative of the position function) as constant across that interval. In Figure 1.5, Δt decreases from a quarter of a second in panel A to 1/32 of a second in panel D. We can make our estimate a bit more accurate in this case by computing the value of the derivative not at the beginning of each interval (e.g., 0 s, 0.25 s, etc. in A) but rather at the midpoint of each interval (0.125 s, 0.375 s, etc.). We then approximate the position function for the first interval using a line segment with a slope given by evaluating Equation 1.14 at $t = 0.125$. The line will start at $y = 10$ m. We call this point our *initial condition*, and we will explore initial conditions a good deal more later. The segment will end at:

$$10\,\text{m} + \left(-9.81\frac{\text{m}}{\text{s}^2} \cdot 0.125\,\text{s}\right) \cdot 0.25\,\text{s} \qquad (1.17)$$

This result corresponds to a velocity of $\left(-9.81\frac{\text{m}}{\text{s}^2} \cdot 0.125\,\text{s}\right) = -1.23\,\text{m}/\text{s}$. In 0.25 s, therefore, the ball will fall 0.31 m, meaning that the final position at $t = 0.25$ s is 9.69 m. Using that value as the initial position of the *next* interval, we simply repeat the process for the remaining intervals: once the position becomes negative, we can stop because our model is valid only for positive values of y. As you may know, this approach to finding an unknown function using its known derivative is referred to as *Euler's method* (6).

Figure 1.5 is interesting, because, at least for this example, we can make a rather accurate approximation of the position function even with relatively large intervals: in Figure 1.5C, we approximate the position using a total of 24 intervals over

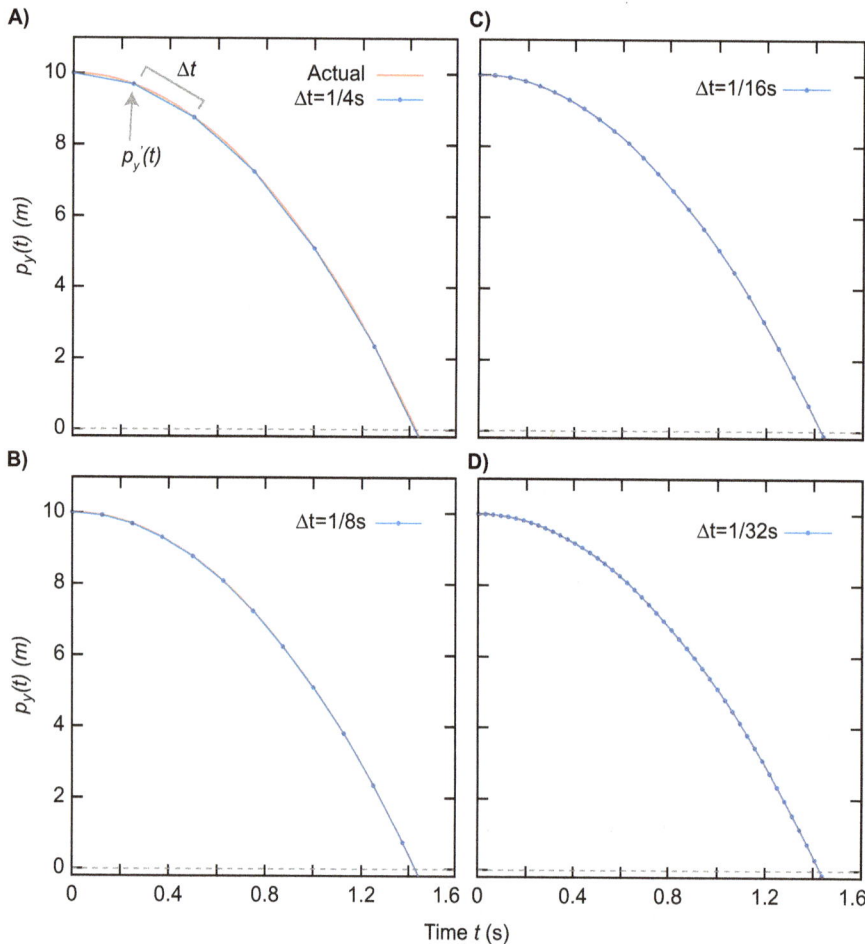

Figure 1.5 **Approximating the ball's position using its initial position (y = 10 m) and the value of the derivative over short time intervals.** The red curves are computed from the true position function of Equation 1.3. The blue lines correspond to computing the derivative at the interval midpoint with Equation 1.14: this value is our interval velocity. We then change in position from the end of the previous interval to the end of this interval by multiplying the interval velocity by the interval size Δt. (**A–D**) show the improvement in the approximation that results from decreasing Δt by successive factors of 2. In **A** I illustrate a different perspective on our computation, namely that we multiply the derivative value at time $t \left(p_y{'}(t) \right)$ by the interval size Δt. We further investigate this view of the calculation in Figure 1.6.

the course of the 1.43 s of the fall. Visually it is difficult to see large errors in that approximation. (Note that the example we have picked is one for which the approximation problem is particularly easy: if our position function were more complex, we would need smaller intervals for a good approximation.)

Now, if we change our perspective slightly, we can see that the computation that gave us Figure 1.5 can be viewed in another interesting way. We can break the computation into two parts: the initial conditions for each interval and the change along the interval. The initial conditions for $t = 0$ s is of course just the ball's starting position at $y = 10$ m. Similarly, the initial conditions for all the other intervals are just the endpoint of the previous intervals. What is more interesting, then, is the value of the change *over* an interval, which is given by:

$$\Delta y = p_y'\left(t\right) \cdot \Delta t \tag{1.18}$$

This idea is illustrated in Figure 1.5A, where each interval consists of the value of the derivative at time t multiplied by the interval length Δt. One slight complication is *when* we compute $p_y'\left(t\right)$: in Figure 1.5, we used the midpoint of the interval, but for the argument we are about to make that is an unnecessary complication. Instead we will compute $p_y'\left(t\right)$ at the *start* of each interval, with Δt still being our interval size.

What would it look like if we plotted Equation 1.18 together with the derivative function itself? **Figure 1.6** gives an answer. The first thing we should notice is that Equation 1.18 can be thought of as the area of a number of boxes, each of which is $p_y'\left(t\right)$ units tall and Δt units wide. I have therefore represented Equation 1.18 as a series of such boxes, each positioned to start at its respective value of t and having width Δt and a height $p_y'\left(t\right)$.

The next thing I have done is slightly surprising: I have also plotted $p_y'\left(t\right)$ on the same scale. If we consider for a moment, it makes sense that our boxes all touch this line, since the value of $p_y'\left(t\right)$ is given by the line's equation. But notice

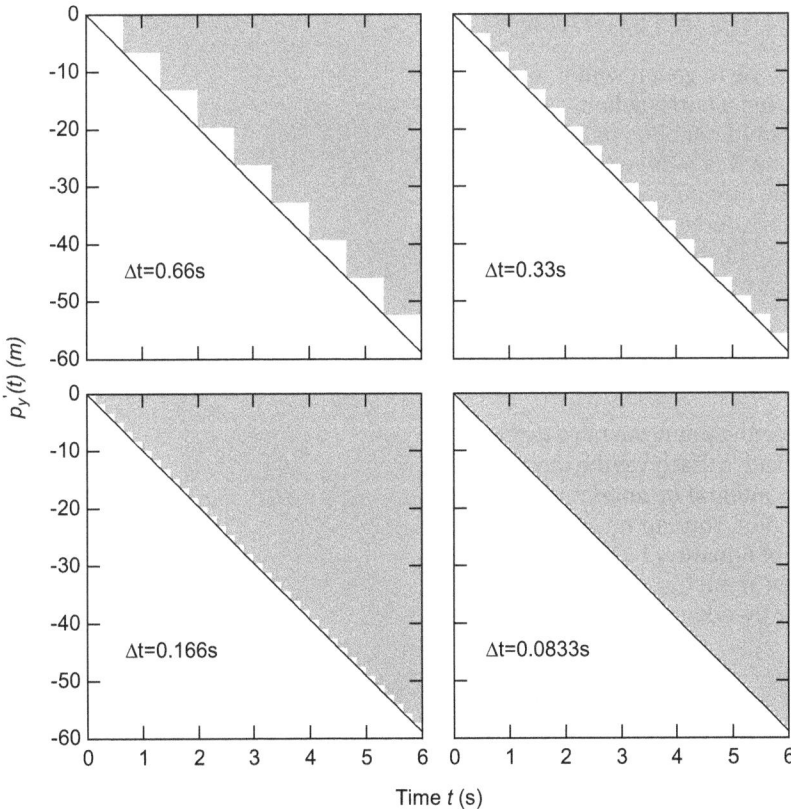

Figure 1.6 Plotting a slightly modified version of the interval changes from Figure 1.5 against the derivative (Equation 1.14). For simplicity, I have shown the value of $p_y'\left(t\right)$ at the *beginning* of each interval (not the midpoint as in Figure 1.5). As the interval $\left(\Delta t\right)$ shrinks, this procedure increasingly closely approximates the area between the derivative curve and the x-axis.

how the picture changes as we allow Δt to decrease. Recall that this reduction of Δt caused our approximation from Figure 1.5 to increasing closely resemble the true position of the ball. What we see here is that this same reduction in interval size causes the total area of our boxes to approximate ever more closely the area between the derivative curve and the x-axis. As unexpected as it may seem, the same process of increasing our approximation accuracy in Figure 1.5 is also the process of more accurately computing the area under the derivative curve in Figure 1.6. In other words, approximating a function with short intervals of its derivative and computing the area under (or over, in this case) a derivative curve are identical. And *that* observation leads toward the general solution of the inverse problem we have been looking for. First, let us write an equation for this area A as follows:

$$A = \sum_{i=0}^{n} \left[p_y{}' \left(i \left({}^{t_m}\!/\!_n \right) \right) \cdot {}^{t_m}\!/\!_n \right] \tag{1.19}$$

where

$$\Delta t = {}^{t_m}\!/\!_n \tag{1.20}$$

Such that n is the number of intervals and t_m is the end of our time interval (which for simplicity starts at $t = 0$). Now, if we take our approximation approach above and let the size of the interval Δt become infinitesimally small, the result is termed the *integral* of the function $p_y{}'(t)$:

$$\lim_{n \to \infty} \sum_{i=0}^{n} \left[p_y{}' \left(i(\Delta t) \right) \cdot \Delta t \right] = \int_0^{t_m} p_y^t(t) \cdot dt \tag{1.21}$$

THE INVERSE OF THE DIFFERENTIAL: INTEGRATION

Of course, many of you have been quietly asking me to get on with it for a couple of paragraphs now, because the preceding has been a fairly pedantic way to get to the idea of such integrals. But my obfuscation should not blind us to the remarkable nature of this idea: the process of summing up this infinite series of infinitely narrow boxes is the same as approximating some function with a set of infinitely short line segments computed from that function's derivative [7]. So, this process of integration in Equation 1.21 is *also* the inverse of the operation of taking the derivative of a function. Hence, for our example, we see that it must be true that:

$$\int -9.81 \cdot t \cdot dt = C - \frac{9.81}{2} \cdot t^2 \tag{1.22}$$

And if we replace C with our original 10 m starting point, we have regenerated Equation 1.3. Hence, just as a function has a derivative that gives the rate of change of that function at every point, we can take the integral or *antiderivative* of that derivative function and recover the original function. You can verify this argument yourself by taking the derivative of the right side of Equation 1.22 and find that you again arrive at Equation 1.3. (If you are wondering about how to take the derivative of the sum of two statements, it is done simply by taking the derivative of each piece and summing them.)

Analytic versus Computational Approaches to Integration

Table 1.3 gives some example *symbolic* integrals. It conveniently includes our case of the falling ball. However, we should not take too much comfort in that fact. Unfortunately, while integration is the inverse of derivation, it is not the case that

TABLE 1.3 SOME COMMON INTEGRALS

Function	Integral
$f'(x) = 0$	$f(x) = C$
$f'(x) = m$	$f(x) = m \cdot x + C$
$f'(x) = x$	$f(x) = x^2 / 2 + C$
$f'(x) = x^2$	$f(x) = x^3 / 3 + C$
$f'(x) = e^x$	$f(x) = e^x + C$
$f'(x) = \sin(x)$	$f(x) = -\cos(x) + C$
$f'(x) = \cos(x)$	$f(x) = \sin(x) + C$

the two operations are equally easy to carry out. Using a series of rules, some of which we will encounter in later chapters, any function you can write down is also one that you can take the derivative of. However, there are any number of functions (indeed, most functions) for which no symbolic form of the integral exists. In other words, one cannot write down a function in the form of Table 1.2 or 1.3 that has a derivative that is the original $f(x)$. A classic example is the function $f'(x) = e^x / x$: no symbolic form of the integral of this function is known:

$$\int \frac{e^x}{x} \cdot dx = ? \tag{1.23}$$

In a mathematics class, we would be quite concerned about this difficulty. However, we will not directly use integration in this book until we take up the topic of continuous random variables and probability density functions in Chapter 8. Instead, we have spent a fair amount of time with computationally approximating such integrals because doing so is a special case of computationally solving the inverse problem. Remember again that this problem involves knowing the rate of change of a system but not the values for the system itself. Our approach for treating the derivative as constant over a short interval and using each estimated endpoint as the starting point of the next interval will work also for the more complicated problems in the next chapters.

Again using the example of the problem of this chapter, we are arguing that if we do not have a symbolic integral available to us, we can approximate that integral using a computer as we did in Figure 1.5. The approach is straightforward, albeit potentially tedious. We employ increasingly small values of Δt and compute our approximate solution. When we find that decreasing Δt causes the change in our solution to be so small as to be irrelevant, we consider that solution to be sufficiently accurate for our needs, never having obtained a symbolic representation of our desired function.

REFLECTIONS AND PREVIEWS

In this chapter, we introduced the *inverse problem*: that of taking a quantity's rate of change and computing its value through time. We saw that there are both analytic and numeric approaches to this problem, but that the analytic approaches are not applicable to every problem. In fact, most of the biological systems we will be working with in the next chapters have no analytic solution. As a result, while we can write down descriptions of their rates of change, we cannot convert those rates of change into analytic equations. The equations themselves will also be more complex than those in Tables 1.2 and 1.3 because they will involve, say, interactions between infected and uninfected people in a population. However, they *are* susceptible to solution with the numerical approaches we have just seen: breaking the problem into a set of short time intervals and applying our rate of change functions in those intervals. In fact, this approach is so routine that there are many available software packages, such as R [8] or Mathematica [9], that will take as inputs the derivatives (rate of change functions) and handle this process of *numeric integration* for us. As a result, we will begin the next chapter where we started this one: with the time course of an infectious disease.

Exercise 1.1

Using a spreadsheet program and Equation 1.14, approximate the position of a falling ball using short intervals where the ball has a constant velocity given by the value of the derivative at the midpoint of that interval (Equation 1.14). Using an initial position of $p_y = 10\,\text{m}$, try four approximations, with $\Delta t = 1/4\,\text{s}$, $\Delta t = 1/8\,\text{s}$, $\Delta t = 1/16\,\text{s}$, and $\Delta t = 1/32\,\text{s}$. Plot your predictions and compare them to Figure 1.5.

REFERENCES

1. Kermack WO & McKendrick AG (1927) A contribution to the mathematical theory of epidemics. *Proceedings of the Royal Society of London. Series A, Containing Papers of a Mathematical and Physical Character* 115(772):700–721.

2. Vynnycky E & White R (2010) *An introduction to infectious disease modelling* (OUP Oxford).

3. Feynman RP, Leighton RB, & Sands M (2011) *The Feynman lectures on physics, Vol. I: The new millennium edition: mainly mechanics, radiation, and heat* (Basic Books).

4. Sobel D (2005) *Longitude: The true story of a lone genius who solved the greatest scientific problem of his time* (Macmillan).

5. Larson RE & Hostetler RP (1986) *Calculus with analytic geometry*, 3rd Edition (D. C. Heath and Company, Lexington, MA), p. 1013.

6. Boyce WE & DiPrima RC (1992) *Elementary differential equations and boundary value problems* (John Wiley and Sons, New York) p. 680.

7. Berlinski D (1997) *A tour of the calculus* (Vintage).

8. R Development Core Team (2008) *R: A language and environment for statistical computing* (R Foundation for Statistical Computing).

9. Wolfram S (1999) *The MATHEMATICA® book, version 4* (Cambridge University Press).

Outbreaks

Modeling an Infectious Disease Outbreak with Differential Equations

2

"[M]en have died from time to time and worms have eaten them, but not for love."
—*William Shakespeare*, As You Like It, *Act IV, Scene I*

THE SIR MODEL OF DISEASE SPREAD

With humanity having experienced in 2020 its first truly global pandemic since 1918, the questions of how fast an infectious disease will spread in a population and for how long come up at dinner tables as well as at academic seminars. In this chapter, we will work through the most basic of the mathematical models to treat this question, occasionally using the COVID-19 pandemic as an example.

Recall our assumptions from Chapter 1: a town of 5,000 people, an initial group of 5 infected individuals, a 4-day recovery period, and 10 new cases in those 4 days (**Figure 2.1**). How might we try to understand the future of this disease in the town?

One appealing approach might be to write a computer simulation of the interactions of the people in the town, trying to estimate how often they come into contact with each other and how often the disease is spread as a result of that contact. We might even be able to consider how differently the disease outbreak might progress due to chance events such as who meets whom at the supermarket.

The general name for this type of approach is "agent-based modeling," and it is called that because the computer simulation explicitly represents all 5,000 individuals ("agents") in the town and their interactions. The power of this approach is in its generality: virtually any phenomenon that you think might be important can be included in the model. Maybe the disease is only transmissible when the air temperature is below 32°C, so one can use historical weather data as part of the simulation. Or maybe contact must be maintained for more than 2 minutes for transmission to take place. Almost any rule you can imagine can be included in the simulations. We will return to this approach for a different problem in Chapter 11.

Unfortunately, the generality of the agent-based approach is also a weakness, for two key reasons. The first may come as something of a surprise: scientific understanding sometimes comes as much from the ideas we leave *out* as from those we include. In principle, a very large number of factors *might* influence a disease outbreak. But what if we find that we can predict real disease outbreaks without needing to include the air temperature, the time of the year, the altitude of the town, the number of supermarkets in it, or a dozen other variables? The implication is that those multitudinous variables are not the critical features of the process. Moreover, writing a simulation that not only includes all of these features but does so accurately will be very difficult (in fact, probably impossible with current technology and knowledge). The second weakness of the agent-based approach is that this generality requires considerable computing power. While it might appear that we live in a world of unlimited computing resources, we will see over the course of this book that this impression is an illusion. A simulation of a town of 5,000 is feasible: one of a planet of 7 billion is rather less so.

DOI: 10.1201/9781003687504-2

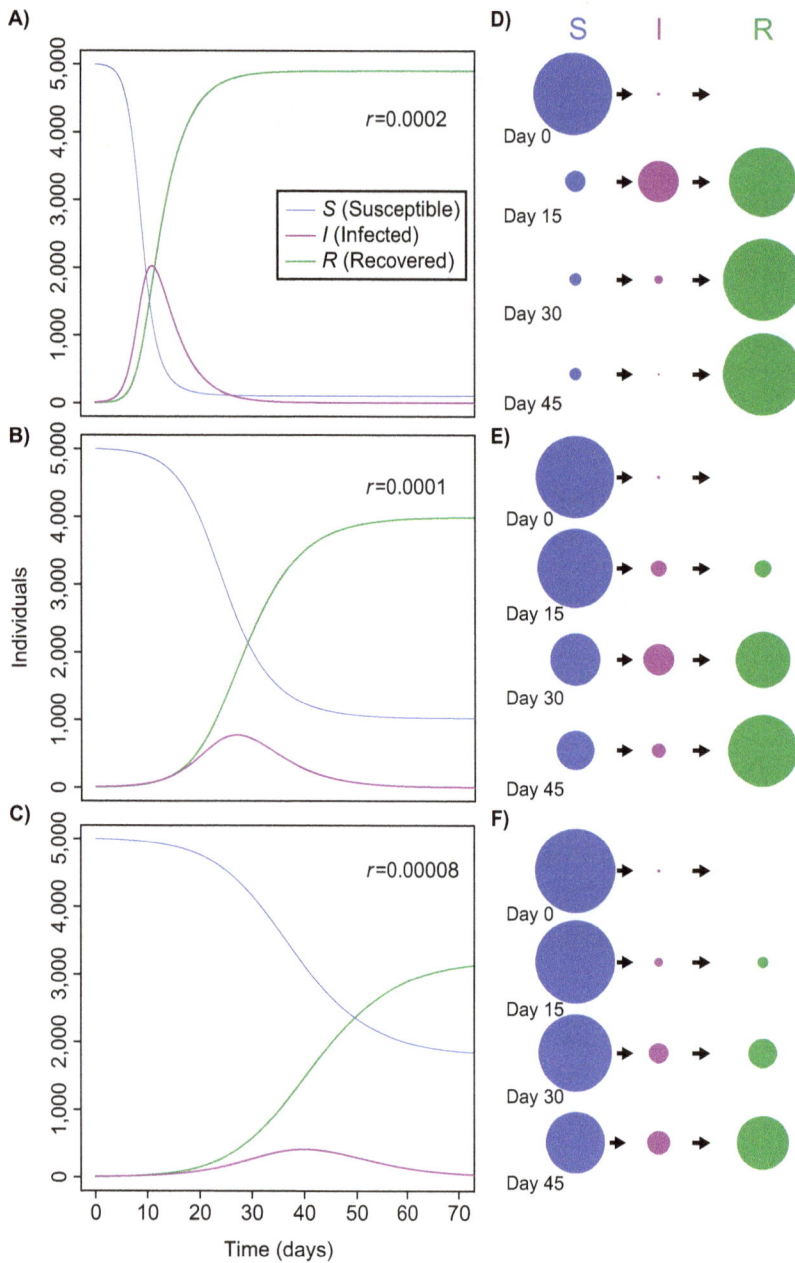

Figure 2.1 Time courses and diagrammatic views of three solutions to the SIR model in Equation 2.5 using different values of the infectivity parameter r (0.0002, 0.0001, and 0.00008 in (A), (B), and (C), respectively). The initial conditions used were $S(0) = 4,995$, $I(0) = 5$, and $R(0) = 0$. The recovery rate a was set to 0.25. Exercise 2.1 allows you to recreate these graphs using R. **(D–F)** give another view of these solutions, with the area of each circle proportional to the number of individuals in each of the three pools (solutions in D–F correspond to A–C). You can explore this system at http://qbio.statgen. ncsu.edu/SIR/.

What Is a Differential Equation? How Can We Use It?

As an alternative approach, we will explore a family of mathematical models that have a relatively light computational footprint but yield very useful insights. These models are based on *differential equations*. We have already seen such an equation in Chapter 1:

$$\frac{dp_y}{dt} = -9.81 \cdot t \qquad (2.1)$$

A differential equation is simply a description of a quantity of interest in terms of its rate of change. In the example in Chapter 1, the quantity of interest was the position of the falling ball. Under the assumptions we made, that position was dependent only on the time the ball had been falling. As a result, Equation 2.1 has only a single variable on the right side of the equation: t. It was therefore a differential equation that could be solved through integration, yielding:

$$\int -9.81 \cdot t \cdot dt = C - \frac{9.81}{2} \cdot t^2 \tag{2.2}$$

The constant C on the right side of the equation is important because it illustrates a key fact about differential equation models. The differential equations alone are not sufficient to model the system. We will always need to know the starting state (or *initial conditions*) of the system if we wish to model it with differential equations.

A falling ball is a simple case because it involves a single object. Most of the problems biologists are interested in involve several quantities of interest and hence are models that will involve *systems* of differential equations. In such systems, the different quantities will most often interact with each other.

THE SIR MODEL

The original version of our system of differential equations for describing a disease outbreak was developed by Kermack and McKendrick in the early 20th century [1]. Their approach breaks the population into three pools of individuals: the susceptibles (S) who can become infected, the infecteds (I) who currently have the disease, and those who have recovered (R). As a result, the model is referred to as an SIR model.

A core goal of the SIR model is to predict the number or proportion of individuals in each of these three classes as a function of time: $S(t)$, $I(t)$, and $R(t)$. These three quantities are also referred to as the *state variables* of the system, and the differential equations will give us the means to recover the values of these variables at different times.

We cannot, simply by intuition, guess functions that describe the three state variables. However, we can make some educated guesses about their rates of change. We will start with $R(t)$, since it will be the most straightforward. In our simple view of disease, the only way someone can recover from the disease is by being infected with it. In this version of the model, all recovered individuals are also resistant to new infections. We will add the idea of resistance due to vaccination in a few pages.

So, the rate of change of $R(t)$ will depend on the number of infected individuals: the more infected individuals there are, the faster $R(t)$ will grow in time. But not everyone who is infected will recover instantly: the disease must last for some period of time. We will call this recovery rate a. Without worrying just yet about its units, we can describe the rate of change of $R(t)$ as:

$$\frac{dR}{dt} = a \cdot I \tag{2.3}$$

Although this equation might at first look similar to Equation 2.1, notice that the rate of change of R depends on I, and we do not yet know what I is or how it changes. Thus, we cannot solve this equation with simple integration as we did for Equation 2.1.

So, what about $I(t)$? We already know part of its rate of change: all the $a \cdot I$ individuals that recover from the disease and enter state R are lost from the pool of individuals in I. As a result, the same $a \cdot I$ term that appears as a positive value in the differential equation for dR/dt will appear as a *negative* value in the equation for dI/dt. Now, if recovery were the only process occurring, the infection would quickly die out, because the number of infected individuals would only ever decrease. Where do *new* infected individuals come from? Well, by our assumptions, infection occurs through the contact between an infected individual and a susceptible one. Now, this idea raises several important questions. Does every contact between individuals in pools I and S result in an infection? How often do such contacts happen? Our model could include both these values (or indeed even more detail). But if we reflect for a moment, it becomes clear that if we were to double the contact rate and halve the chance of infection at each contact, the model would have the same behavior in both cases. For the moment then, we can imagine that the growth rate of the I population will depend on the frequency of contact between infected and susceptible individuals, which we will represent with $S \cdot I$. We will then scale this interaction term with an infection rate constant r:

$$\frac{dI}{dt} = r \cdot S \cdot I - a \cdot I \tag{2.4}$$

The fact that dI/dt depends on the product of I and S might make us pause for a moment, but it is quite reasonable upon further reflection. We can imagine individuals wandering effectively at random on some landscape. This landscape need not be the real physical landscape of our town: it is instead a conceptual landscape that describes the places people meet and the relative frequency with which they travel to each. But regardless of the form of the landscape, the chance of an I individual encountering someone from S will be proportional to the density of both I and S individuals on that landscape. The value of r then allows us to model the frequency of contact and the chance of infection at that contact.

With these two ideas of infection and recovery built into the model, we can write the full system of differential equations fairly easily:

$$\frac{dS}{dt} = -r \cdot S \cdot I$$
$$\frac{dI}{dt} = r \cdot S \cdot I - a \cdot I \tag{2.5}$$
$$\frac{dR}{dt} = a \cdot I$$

Using the same logic we used for R and I, we can see that any new infected individuals are necessarily lost to the S pool, meaning that the $r \cdot S \cdot I$ term in dI/dt reappears as a negative quantity in dS/dt. Since for the moment no one is being born in our town, nor is anyone immigrating in, dS/dt can only ever decrease in this form of the model.

SOLVING DIFFERENTIAL EQUATIONS COMPUTATIONALLY

What do we do with these equations? From the perspective of a mathematician, we would like to know if we could write down a set of three state functions that describe the system's behavior in time. In other words, could we find three functions $S(t)$, $I(t)$, and $R(t)$ that, when differentiated, gave us back the system in Equation 2.5? As it happens, the system in Equation 2.5 does not have such an *analytic* solution [1]. So, as far as mathematicians know right now, we *cannot* write down three equations in any conventional way that we can differentiate back to Equation 2.5. In fact, most of the models we study in this book will not have analytic solutions. Therefore, we will not devote much time to the study of the analytic solutions of differential equations. Instead, we will recognize that we are fortunate that computers are now ubiquitous and that there is a variety of high-quality software tools that can computationally solve systems of differential equations. The exercises in this book use the R [2] package deSolve [3] because it is free and because we will also use R elsewhere in the book. But everything we do could also be accomplished with other tools such as Mathematica [4] or indeed your own computer code and standard computer libraries for differential equations [5].

In a way, depending on such software will feel like cheating. However, we have already seen in Chapter 1 that we conceptually understand how such programs work. They treat the function as being made up of many, many short linear pieces computed from the derivatives, as we saw in Figure 2.1. For concreteness, in Figure 2.2, I give an example of solving this SIR system numerically and compare it to the deSolve solution. The approach is quite simple but effective. We take our initial conditions (see discussion that follows) and solve Equation 2.5 using those

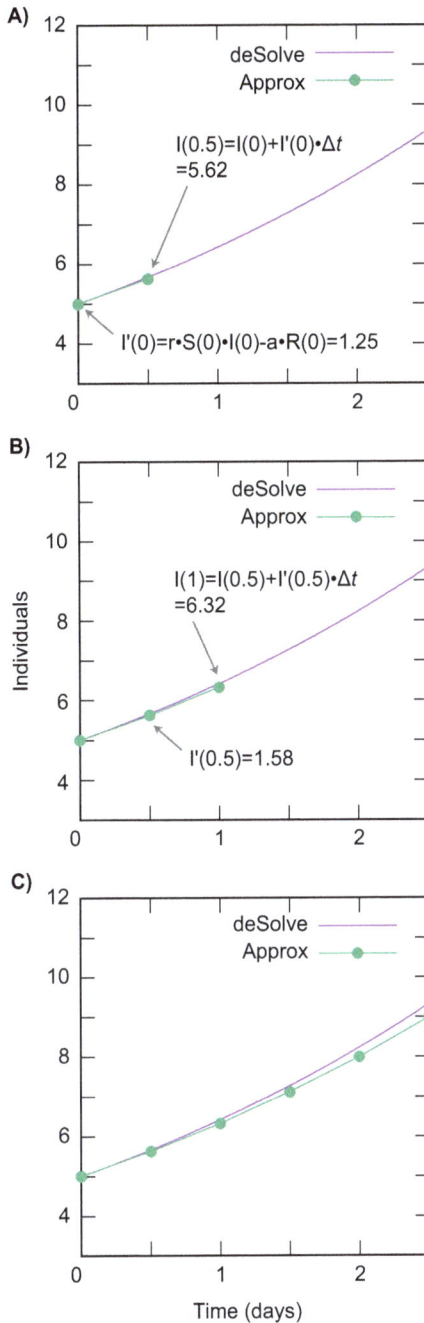

A)

B)

C)

Time (days)

Individuals

Figure 2.2 Numerically solving a system of differential equations using a spreadsheet. (**A**) Starting with the initial conditions $S(0)=4{,}995$, $I(0)=5$, $R(0)=0$, as well as $r=0.0001$ and $a=0.25$, we solve Equation 2.5 for $t=0$, which gives us derivative values for dS/dt, dI/dt, and dR/dt at $t=0$. We denote these three values as $S'(0)$, $I'(0)$, and $R'(0)$. We then pick a time-step size (here $\Delta t=0.5$) and extrapolate $I(t+\Delta t)$ using the computed $I(t)$ and $I'(t)$. Specifically, $I(t+\Delta t)=I(t)+I'(t)\cdot\Delta t$. (**B**) The next approximation step simply uses A as its starting point and repeats the procedure. (**C**) Over several timesteps, our approximation diverges slightly from the solution found with R and deSolve. Notice that although $S(t+\Delta t)$ and $R(t+\Delta t)$ are not shown, we need to approximate all three values to use Equation 2.5 to compute derivatives for the next timestep.

values. We then treat those derivatives from Equation 2.5 as constant over a short time interval Δt and compute new values of $S, I,$ and R, which we in turn use to compute new derivatives. We then repeat this process for as long as we wish. Of course, the real software tools are more clever than the approach in Figure 2.2, but the basic principles we used there still apply. In particular, when a more accurate solution is required, this can be achieved by reducing the time step size. Indeed, even the rather large Δt of 0.5 days does not give a terrible result in Figure 2.2.

Figure 2.1 shows deSolve's solution for Equation 2.5, the results we also saw at the beginning of Chapter 1 (Exercise 2.1). Now, however, it is time to take a little more care in understanding how such a solution is obtained. The first point to note again is that the equations comprising Equation 2.5 are incomplete: they alone cannot describe the course of the infection. We also need to know the initial state of the system. For instance, suppose we start with no infected individuals: $I(0)=0$ —what could we expect? Well, since dS/dt depends on I, when $I=0$, the rate of change of S also becomes 0, meaning no susceptible individuals become infected.

Exercise 2.1

Solve the SIR model using http://qbio.statgen.ncsu.edu/SIR or using the R package's differential equation solver. Using R, you can:

1. Load the solver:
   - `library(deSolve)`
2. Define the parameters:
   - `parameters<-c(r=0.0001, a=0.25)`
3. Define the initial conditions:
   - `state<-c(S=4995,I=5,R=0)`
4. Define the model:
   - `SIR_model<-function(t,state, parameters)`
     `{with(as.list(c(state,parameters)), {`
   - `dS <- -r*S*I`
   - `dI <- r*S*I-a*I`
   - `dR <- a*I`
   - `list(c(dS,dI,dR))})}`
5. Set the integration timescale:
   - `times <- seq(0, 100, by = 0.01)`
6. Solve the equations:
   - `out <- ode(y=state, times=times, func=SIR_model,`
     `parms = parameters)`
7. Plot the result:
   - `plot(out[,"time"], out[,"R"], type="l",`
     `xlab="time", ylab="Individuals", col="green",`
     `xlim=c(0,70),ylim=c(0,5000))`
   - `points(out[,"time"], out[,"S"], type="l",`
     `xlab="time", ylab="Individuals", col="blue")`
   - `points(out[,"time"], out[,"I"], type="l",`
     `xlab="time", ylab="Individuals", col="red")`

In other words, once the pool of infected individuals is completely depleted, the system freezes, with the remaining susceptible individuals remaining susceptible. In our case, we assumed instead that $S(0) = N-5$, $I(0) = 5$, and $R(0) = 0$. For the moment, $N = 5,000$—the entire population of the town. With these values we can solve the system, but before we do so, we need to stop and consider how we have represented both the infection rate r and the recovery rate a, which are the *parameters* of our model.

PARAMETERS IN BIOLOGICAL MODELS

We will return to the idea of a parameter repeatedly in this book, so it is worth spending a moment to define what one is. Essentially, a parameter is a quantity that relates a model to a specific instance of the system under study. Hence, unlike the number of individuals in the pools $S, I,$ and R, the model parameters are not functions of time and do not change over the system's lifetime. However, they will differ from disease to disease.

Table 2.1 gives the estimated value of R_0, which combines a and r, for a number of diseases. The table illustrates how parameters differ across diseases, although we will understand these differences better once we explore the meaning of R_0 in a couple of pages. For the moment, we should just notice that the power of having such parameters is significant: by simply changing a and r, the same SIR equations can be used to model a wide range of diseases.

So, what values of these parameters should we use? The a parameter is relatively easy to understand: it is the proportion of the infected population that recovers from the infection in a single time unit (which we will assume is 1 day). It is therefore measured in units of 1/days. In Figure 2.1, $a = 0.25 \cdot 1/$ days, meaning that a quarter of the members of I recover each day, making the mean infection time 4 days. In Chapter 1, we already explored the effects of making a smaller: if the infection lasts only 3 days

TABLE 2.1 R_0 VALUES FOR SELECTED DISEASES		
Disease	R_0	*Reference*
Diphtheria	6–7	[6]
Malaria	5–100	[6]
Measles	12–18	[6]
Mumps	4–7	[6]
Pertussis	12–17	[6]
Polio	5–7	[6]
Rubella	6–7	[6]
Smallpox	5–7	[6]
1918 Influenza pandemic	2–3	[7]
SARS	≈3.5	[8]
Ebola	1.7–2.0	[9]
COVID-19 alpha	≈3.3	[10]

($a = 0.33 \cdot 1 / $ days), the size of the epidemic is smaller and the number of remaining susceptibles at its end is correspondingly larger. This result makes intuitive sense: a shorter infective period corresponds to less time to infect others. As a result, if we keep the rate an infected person meets and transmits the infection to a susceptible one constant, a shorter infection period corresponds to fewer total infections.

We should also recognize the value of including units in our models. Our three differential equations in Equation 2.5 have units of individuals/day: they are rates of change in pools of individuals. If a has units of 1/days, when we multiply it by I in the equation for dR/dt, we also get a unit of individuals/day, as we should.

The r parameter is more complex, as it models the interaction of susceptible and infected individuals. Its units are $1/(individuals \cdot days)$: in other words, larger values of r imply either more contacts per day per individual or more infections resulting per contact per day. In Chapter 1, we saw the effect of a change in a: here Figure 2.1 shows the effect of changing r. As expected, higher infectivity rates increase the size of the peak epidemic in red: less obviously, they also shorten its duration, as more of the available susceptible individuals are infected and recover in a short interval.

FLATTENING THE CURVE

What are the implications of a faster infection wave? There are two points to notice. Intuitively, the total number of people who end up being infected is greater for higher infection rates (bigger r). We can prove this point by noticing that since every individual in the model who is infected passes through I and ends up in R, the size of R at the end of the epidemic gives the total number of people infected over the whole epidemic. Seeing that the relative height of the R curve decreases as r decreases from Figure 2.1A to 2.1C tells us that fewer total infections occur when r is small. For larger values of r, however, we also see the height of the I peak becoming both higher and occurring earlier in time. If you remember the initial news coverage of the COVID-19 pandemic, there was a good deal of discussion about "flattening the curve." We are now able to understand this phrase in the context of our model.

Interventions Acting on r

As we have seen, as the value of r decreases, both the epidemic's size and its time span change. When we decrease the epidemic *size*, we also increase its "span": the epidemic peak comes later in time than is the case with a large value of r. Roughly speaking, because fewer contacts between infected and susceptible individuals result in disease transmission, the epidemic grows more slowly. How might we as a society alter r? Two interventions that you may remember from the pandemic are the use of masks and social distancing: asking people to go out less and to remain further apart from each other when they do go out. How do social distancing and masks act from the perspective of our SIR model? They both act directly on the value of r: social distancing reduces the number of contacts, and masks reduce the chances that a contact results in an infection. Effectively, these behaviors reduce the apparent infectivity of the disease. More generally, we can also notice how the r parameter combines many real-world factors into a single number, which is both a blessing and a curse for models like this one.

Because reducing r extends the time horizon of the epidemic and lowers the epidemic peak, we can see that social distancing and masks do indeed, in principle, flatten the curve by lowering r. And here is where mathematical biology gives us potentially concrete advice regarding our reaction to an epidemic. Flattening the curve buys us two valuable things. First, it gives us more time to study the epidemic and learn how to treat it. Possibly more importantly, flattening the curve reduces the size of the epidemic's peak. This second point is rather important: if the number of doctors, hospital beds, and ventilators is a fixed quantity, reducing r reduces the chances of running out of treatment options for sick people at the outbreak's peak, potentially saving lives even beyond the reduction in R that we have seen comes with a smaller value of r.

Exploring the SIR Model: Conservation, Birth and Death, and Periodicity

For all the utility we have found in the SIR model so far, we should not be lulled into believing that we have learned all we can from it. For instance, what happens if we add the three equations in Equation 2.5 together?

$$\frac{dS}{dt} + \frac{dI}{dt} + \frac{dR}{dt} = -r \cdot S \cdot I + r \cdot S \cdot I - a \cdot I + a \cdot I = 0 \tag{2.6}$$

What does it mean that the sum of these rates of change is zero? Without quite seeing that we had done it, we have introduced *conservation* into our model: individuals in the model may move between pools $S, I,$ and R, but they cannot leave the model altogether. This constraint is useful: it will be harder to understand the system if the total number of people in it is also increasing or decreasing in time. But it could also appear unrealistic: What if the disease is sometimes fatal? In fact, the R pool need not only contain recovered individuals: any individual who dies is also in the R pool from the model's perspective because she or he can no longer infect others. If this lumping of the recovered and the deceased troubles us, we could add a fourth pool to the model (D) that contains those individuals the disease has killed. We would then need to add another model parameter, d, giving the relative frequency of recovery or death (Exercise 2.2).

Exercise 2.2

Develop a system of *four* differential equations for an SIR model that includes death. Assume that a gives the recovery rate for individuals in I and d gives the death rate. Solve the model in R and compare to Figure 2.1.

SIR MODELS OVER LONGER TIME PERIODS

Another way we could modify this conservation assumption is to allow births and deaths to occur in the model. For simplicity, we will assume that the birth and death rates are equal, so that the total population size (N) remains constant [11]. Doing so adds a new birth/death rate parameter u to the model:

$$\frac{dS}{dt} = -r \cdot S \cdot I + u \cdot N - u \cdot S$$

$$\frac{dI}{dt} = r \cdot S \cdot I - a \cdot I - u \cdot I \qquad\qquad (2.7)$$

$$\frac{dR}{dt} = a \cdot I - u \cdot R$$

In this new form of the model, all individuals are born into the susceptible pool and die in equal proportions from all three pools: the disease itself is nonfatal. The units of u are the same as those of a: 1/days. Numerically, however, u is much smaller because the rate of births and deaths in the population is on the order of years, while a is on the order of days. Notice that despite the addition of births and deaths, the sum of the three equations in Equation 2.7 is still zero: the population size remains constant.

When we solve the system in Equation 2.7 with some reasonable values of the parameters and a large population ($u = 1/70$ years; $N = 10^6$ individuals), we see something interesting. As **Figure 2.3** shows, there is not a single epidemic, but a series of them, with each successive one being somewhat smaller than the previous one. The process starts because the initial number of susceptible individuals is enough to spark an epidemic. As before, the epidemic burns through the population, leaving a pool of susceptible individuals too small to sustain that epidemic. However, new susceptible individuals are now being born all the time, meaning that S begins to slowly increase again. At some point it will reach a level large enough to spark a new epidemic. Because the pool of individuals in S is smaller than for the first epidemic, this new epidemic will be smaller too. And this process continues, with each succeeding epidemic smaller than the last. Eventually, the system approaches a stable state where the appearance of new susceptible individuals is balanced by the rate of their infection and progression to state R, meaning that the population suffers a continuous low level of infection, at which point we could call it an endemic disease [12].

Having a model that predicts this sort of periodic disease outbreaks is all very well. But does that model relate to how diseases behave in real populations? Our version of the SIR model is of course quite simplified and would not produce useful quantitative predictions about outbreak timing and scale. But it is interesting that exactly these sorts of damped epidemics *were* observed in the series of plague outbreaks we now refer to as the Black Death. These outbreaks occurred in Europe starting in the middle of the 14th century. Genetic work has confirmed that they were caused by the bacterium *Yersinia pestis* [13, 14]. In the initial outbreak of 1348–1349, the death toll was close to catastrophic: perhaps on the order of 30% of the European population died [15]. In the wake of this wave of death came a variety of social changes and upheavals, partly due to the suddenly increased value of manual labor in a society that had placed too little value on such work [16].

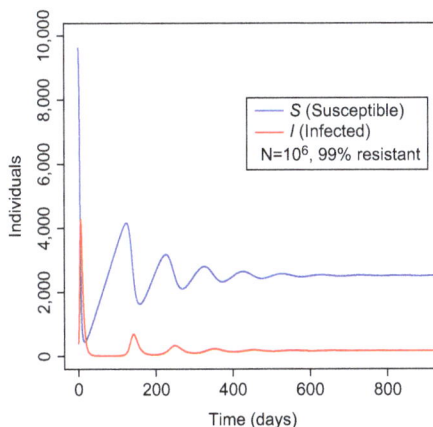

Figure 2.3 An infectious disease time course showing the behavior of a population experiencing births and deaths in the presence of that disease. For argument's sake, we assume that 99% of the individuals in the population are in the R state when we start our analysis and that 0.04% were actively infected. The values of r and a are the same as used in Figure 2.1, but the population is now 10^6 individuals. Individuals are born and die at a rate $u = 1/(70 \cdot 365)$, corresponding to an average lifespan of 70 years.

Beyond its overwhelming immediate impact, however, one of the most disheartening aspects of this plague was that it returned many times. Worse, in these new outbreaks, mortality was concentrated among the young. The reason for this pattern was not understood at the time, but it is obvious to us due to our understanding of acquired immunity [17]. Any person born after the initial outbreak had no acquired immunity to the disease [15]. Such individuals were therefore more likely to contract the disease and more likely to die if they contracted it, concentrating the plague's effect among young people.

For our purposes, this parallel to real diseases is intriguing, but the model in Equation 2.7 is only partly biologically realistic [11]. For instance, notice that because we treat the populations as continuous (see discussion that follows), the value of I never actually reaches 0. In a real village, a disease like the plague may in fact die out completely: the epidemic waves seen in medieval Europe were also driven by the movement of infected people around the continent, something our current SIR model does not include. As a result, we will explore this idea of *periodic* behavior more carefully in the next chapter with a different model. Instead, we now will explore another value of the model: understanding how to eliminate an infectious disease. But before we do that, there are some details of the model we need to explore.

EPIDEMIC SIZE AND THE BASIC REPRODUCTION RATE

Are epidemics always a feature of SIR models? In the examples presented so far, the number of infected individuals has increased through time for a period and then fallen off once there were not enough remaining susceptible individuals to maintain the epidemic. But can there be combinations of the initial number of susceptible individuals, recovery rates, and infection rates that do not give rise to further cases?

While this question might at first seem difficult to answer, it actually can be explored using the equations comprising Equation 2.5 without even needing to solve them for $S(t)$, $I(t)$, and $R(t)$. In particular, we can ask what it means for an epidemic to occur. Looking back at Equations 2.4 and 2.5, we have:

$$\frac{dI}{dt} = r \cdot S \cdot I - a \cdot I$$

An epidemic implies that the number of infected individuals must be, at least at some point, increasing in time. Another way to say this is that dI/dt must be positive. We can then define the boundary of an epidemic as the case when $dI/dt = 0$, which gives us:

$$0 = r \cdot S \cdot I - a \cdot I \tag{2.8}$$

Or, by rearranging and canceling the common I term:

$$a = r \cdot S \tag{2.9}$$

This equation is not completely helpful because it includes the unknown term $S(t)$. But referring back to Equation 2.5, we notice that dS/dt is never greater than zero. The reason for this is that no new susceptible people are entering the population but some are being removed by becoming infected. As a result, we know that the number of susceptible individuals in the population will never exceed the initial number of susceptibles, which we will write as $S(0) = S_i$. We can now define a new parameter R_0:

$$R_0 = \frac{r \cdot S_i}{a} \tag{2.10}$$

The parameter R_0 is often referred to as the *basic reproduction rate* of the disease [11, 18]. A more formal definition is: "[T]he number of secondary infections

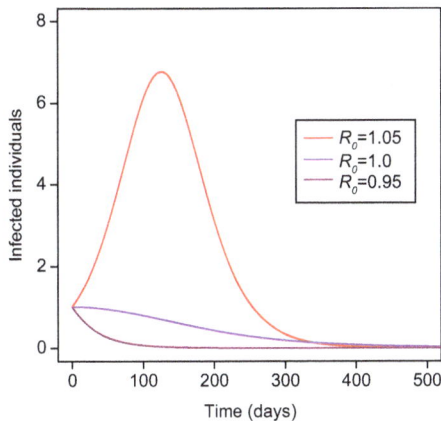

Figure 2.4 Size of the infected population I(t) for three time courses with differing values of the basic reproduction rate R_0. We start with a single infected individual $(I(0)=1)$, and only when $R_0 > 1.0$ does the size of the infected pool grow after that initial infection. Here $N = 5,000$, $r = 0.0001$, and a is computed from $S(0)$ and r with Equation 2.10 to give the desired value of R_0.

produced by one primary infection in a wholly susceptible population" [18]. We should also notice that R_0 is *not* the initial number of resistant individuals, $(R(0) = R_i)$.

If we solve for a in Equation 2.10 and plug the result back into the original dI/dt equation, we get:

$$\frac{dI}{dt} = r \cdot S \cdot I - \frac{r \cdot S_i}{R_0} \cdot I \qquad (2.11)$$

If we now let $R_0 = 1.0$ and plug that into Equation 2.11 at $t = 0$ (i.e., $S = S_i$), we have $dI/dt = 0$. We can see therefore that $R_0 = 1$ is the threshold between an epidemic and the lack of one. In Exercise 2.3, you can explore the effects of different values of R_0 on the size of the infected pool. **Figure 2.4** shows some results that should be very similar to what you find. Notice that near $R_0 = 1$, the rate of the infection spreading is quite a bit smaller than what we saw in Figure 2.1: as a result, the timescale of this figure is much longer than that of Figure 2.1. Nonetheless, at or below $R_0 = 1$, the value of dI/dt is always less than 0, and hence the size of the infected population is always decreasing.

Exercise 2.3

Explore the behavior of the SIR model near $R_0 = 1.0$. Using the commands from Exercise 2.1 or the website, compare the number of recovered individuals, $R(t)$, when $R_0 = 0.95$, 1.0, and 1.05. How do you account for your results?

R_0 VALUES FOR DIFFERENT DISEASES

By defining R_0 we can see one of the useful features developing models like the SIR one. It is possible, albeit challenging, to infer the value of R_0 for different diseases (under some assumptions). In Table 2.1 I list estimated values of R_0 for several different diseases, including the initial COVID-19 variant. We can now view R_0 as telling us to what degree a particular disease is likely to spread. Table 2.1 may take us by surprise in some cases. Would we have expected the R_0 value of COVID-19 to be so small? In fact, probably yes. As dire as the COVID-19 epidemic has been, the initial spread was not terribly rapid, and tools such as masks appear to have significantly reduced that spread [19–21].

LETHALITY AND THE SIR MODEL

One topic that we have skirted several times in this chapter is that of the lethality (or lack of lethality) of a disease. The reason we have done so is that while parameters like r and a are not untethered to that lethality, the relationships between them are complex. Thus, when we look at Table 2.1, we find diseases like SARS and Ebola with quite different lethality but rather similar R_0 values.

The easiest way to grasp this complexity is to examine a and r from the perspective of the infectious agent's evolutionary success [22, 23]. The infectious agent (e.g., the virus) will generally seek to maximize its R_0 value as it evolves, since R_0 can be thought of as the long-term reproductive success of the disease agent. To understand why it is R_0 that has this critical role, realize that it does not matter how many copies of itself a virus can make in a single host. At some point, that host will either recover or die, and all those viral copies will be lost. Instead, to be evolutionarily successful, the virus must spread well from one host to the next. Because disease agents mutate very quickly and are subject to strong evolutionary pressures from the host's immune system, they generally evolve very rapidly to achieve efficient spread [24–26]. We can therefore usually assume that viruses that have been infecting humans for a long time are in some sense at their evolutionary optimum infective behavior (i.e., have an R_0 value that maximizes their evolutionary success). Now, we all know of viruses like the common cold that have a high R_0 while killing almost none of their hosts, presumably because their optimal strategy for spreading does not involve fatalities. In fact, we could argue that since dead hosts do not transmit diseases well, we could generally expect viruses to evolve so as *not* to kill their hosts.

What then do we make of diseases that do at least occasionally kill? In those cases we would need to argue that that lethality represents some unavoidable cost to the host of the disease agent achieving its optimal R_0 [23]. If this hypothesis is correct, we would expect that when a disease agent and its host have evolved together for some time, the lethality should remain basically constant.

Data on the lethality of various diseases over time can be very difficult to come by, particularly as human populations, social structures, and medical practices do not remain constant. But it is at least intriguing that while estimates of the fatality rate of smallpox differ among outbreaks, a level of 15–30% for unvaccinated individuals has been seen across many centuries, even up to some of the last outbreaks in the United States and the United Kingdom [27, 28]. Moreover, looking at the known smallpox deaths in London (where the historical records are unusually complete) in the pre-vaccine era does not show any apparent global trend in death rates that is independent of population growth [29].

Newly Introduced Diseases and Mortality

The situation becomes more complex for diseases that have been newly introduced to a population, usually as a result of the "leap" of the disease agent from another animal species. In that case, lethality or other harm to the host might be an undesirable side effect of infection of the new host and might therefore decrease as the virus evolves to maximize R_0 [30]. Consistent with this hypothesis, there are known examples of where lethality has dropped after a viral introduction. For instance, the myxoma virus was introduced into Australia to control the population of introduced rabbits. The introduced virus initially had high virulence, killing many infected rabbits. Over a few decades, that virulence dropped [31]. However, that reduction in virulence did not continue indefinitely; rather, it appeared to reach a stable intermediate level. Mathematical biologists have made more complex models that include *both* disease agent evolution and SIR-type host dynamics. When they do so, they find that indeed an intermediate and stable level of lethality is one of the possible dynamic outcomes to the evolutionary struggle between diseases and hosts [31].

COVID-19 notwithstanding, from a global health perspective, the true danger is the escape of a much more lethal virus such as Ebola into the human population at large. Since Ebola is tens of times more lethal than COVID-19, its escape could produce a death toll on the order of the Black Death. On the other hand, the arguments above would at least give the hope that Ebola, were it to spread throughout the human population, would see its lethality decrease somewhat. Remarkably, in the Ebola outbreaks that have occurred so far, it does appear that the death rate has fallen at least somewhat over the course of the longer/larger outbreaks [32].

That observation would be consistent with the argument above: the initial lethality was partly a side effect of the jump of Ebola into human hosts. During the outbreak, the virus then evolved to improve its transmissibility in ways that reduced lethality.

CONTINUOUS VERSUS DISCRETE POPULATIONS

Looking at Figure 2.4, there is something you might find a bit odd. The number of infected individuals is not always (or even often) a whole number. What does it mean biologically to have 1.83 infected individuals? Would it not be better to use models where the number of individuals is always an integer? As it happens, there is a branch of mathematics that deals with *difference equations*, which are the discrete counterpart to the continuous models we have used so far [33]. In Chapter 3 we will explore discrete approaches to population modeling. However, we will also see in Chapter 7 that the apparent improvement in precision a discrete model would give is probably illusory. The reason is that when the size of a population is small, stochastic effects become important and make any deterministic model, difference *or* differential equation-based, incomplete. Since the mathematical "machinery" of difference equations is also more difficult to work with, for the moment we will continue with our continuous model.

FURTHER UTILITY OF DISEASE MODELING: ESTIMATING VACCINE EFFECTIVENESS

We are now ready to return to the question of whether the SIR model can help us think about the problem of eliminating a disease entirely. To do so, we will use its predictions to plan a vaccination program. In so doing, we will reveal one of the most important powers of modeling in science: the discovery of the unexpected.

So far, we have not inquired too deeply into the assumption that once an individual has recovered from a disease and entered state R, they remain there and cannot be reinfected with the disease. Our own experience might seem to suggest that resistance is in fact only temporary: after all, many of us catch a cold nearly every year. But in fact, mammalian immune systems do learn to respond to infections, with populations of antibody-producing B cells and infection-detecting T cells kept in memory and poised to respond to another attempted infection that is sufficiently similar to the original one [17]. As a result, if a human is infected with and recovers from most of the diseases in Table 2.1, she or he is generally immune to that disease for life.

Vaccination, then, involves creating a compound or compounds that can train the immune system to recognize a disease without the individual having to undergo the danger and unpleasantness of contracting the disease itself. The nature of vaccines varies quite a bit, from live but weakened versions of the infectious agent (generally a virus) to single proteins such as toxins that the agent produces as part of the infection process [17]. For our purposes, we will just view a vaccine as a treatment that moves some fraction of the population from state S to R without passing through I.

A full model of this idea would include features such as the age at which individuals become susceptible to the disease (many infants are protected from a number of diseases by antibodies in their mothers' breast milk), the age at which the vaccine is administered, and the effectiveness of that vaccine (what proportion of people to whom it is administered actually become resistant) [34]. Omitting such complexities, we will explore a simplified model [11] that elaborates on Equation 2.7:

$$\frac{dS}{dt} = -r \cdot S \cdot I + u \cdot N (1-p) - u \cdot S$$

$$\frac{dI}{dt} = r \cdot S \cdot I - a \cdot I - u \cdot I \qquad (2.12)$$

$$\frac{dR}{dt} = a \cdot I - u \cdot R + u \cdot N \cdot p$$

The only change we have made to Equation 2.7 is to add the parameter p. It gives the proportion of new individuals in the population who are effectively vaccinated against the disease at birth by being born into group R. Notice that when $p = 0$, these equations devolve back to the original versions of Equation 2.7.

Figure 2.5 shows the behavior of this system when we use the same initial conditions as those of Figure 2.3. It is important not to take too much away from this version of the model. We have somewhat unreasonably modeled what would happen if a vaccine were introduced to a previously unvaccinated population at $t = 0$ but with only newborn individuals vaccinated. We have also used a set of initial conditions that are not the long-term *steady state* of the model. These two assumptions drive some of the form of the trajectories shown.

However, two larger points are apparent. First, increasing the proportion of vaccinated individuals makes the time between epidemics grow longer because it takes more time to build up the necessary number of susceptibles to spark a

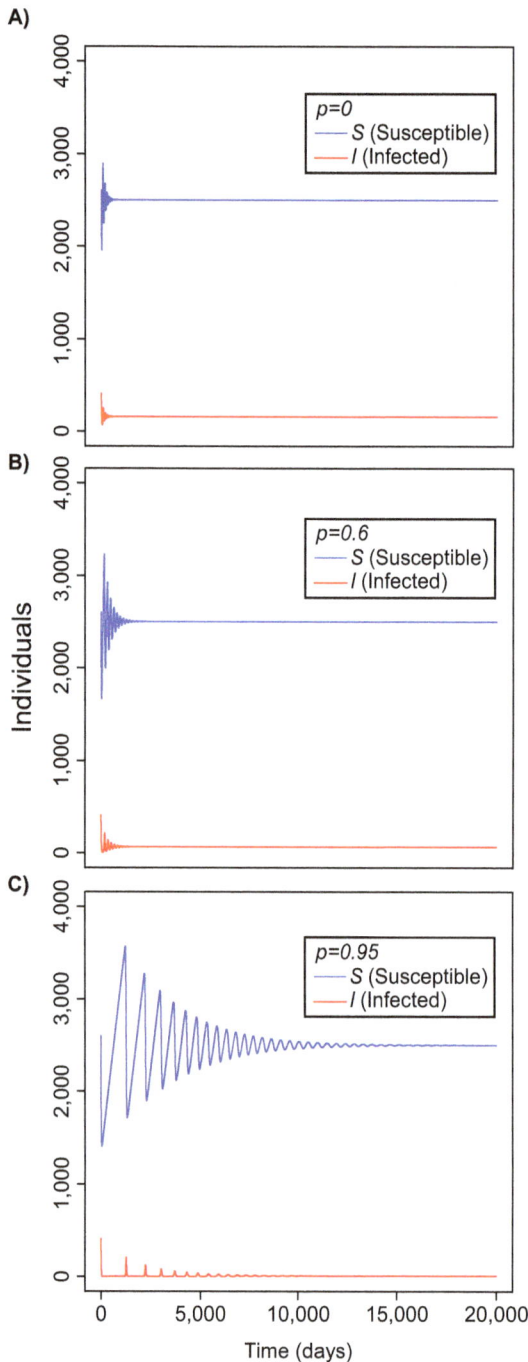

Figure 2.5 **Effects of the vaccination of a proportion p of new individuals on the time course of the disease from Figure 2.3. (A)** This panel uses parameters identical to Figure 2.3: **(B, C)** These show how increasing the proportion of vaccinated individuals lengthens the periods between the oscillation waves of infection and decreases the end-state number of infected individuals.

new epidemic. Second, as the proportion of individuals vaccinated increases, the number of infecteds left at the end of the modeling run decreases.

HERD IMMUNITY

But how much does this number of infected individuals decrease over time? In **Figure 2.6** I have taken a slightly different set of model parameters and initial conditions and focused only on the number of infected individuals. When the proportion of vaccinated individuals is low, we see that we eventually achieve a long-term balance of susceptibles and infecteds: in other words, there is a small but constant proportion of the population infected at all times. Of course, were we to set $p = 1.0$ (100% successful vaccination rate), the number of infecteds would drop to 0. What is striking about Figure 2.6 is that we actually achieve this goal of 0 infected individuals with a vaccination rate of *less than* 100% (namely $p = 0.97$, Exercise 2.4). How is this possible?

Exercise 2.4

Explore the behavior of the SIR model with vaccination for proportions of vaccinated individuals equal to 95%, 96%, and 97%, using the initial conditions and parameters of Figure 2.6. You can use either R or the course website. You will need to extend the time steps of your solutions to use the new model.

What we have discovered is a phenomenon called *herd immunity*: If enough individuals in a population are vaccinated, the fraction of the population that is susceptible to the disease is too small to sustain the disease in that population. Herd immunity is the mechanism by which diseases can be eradicated: smallpox was driven to extinction not by vaccinating every last human being against it, but by vaccinating enough that there were not enough susceptible humans left to maintain a chain of smallpox infections. Then, once the last infected person died or recovered, because the virus has no other natural hosts, it was extinct.

Can we compute the proportion p_h of individuals we need to vaccinate to achieve herd immunity? To do so, we will first return to the R_0 parameter. As we saw, we can compute R_0 by setting $I(0) = 1$ and $dI/dt = 0$ in Equation 2.12, giving the following equation for R_0 in this model:

$$R_0 = \frac{r \cdot S_T}{a + u} \tag{2.13}$$

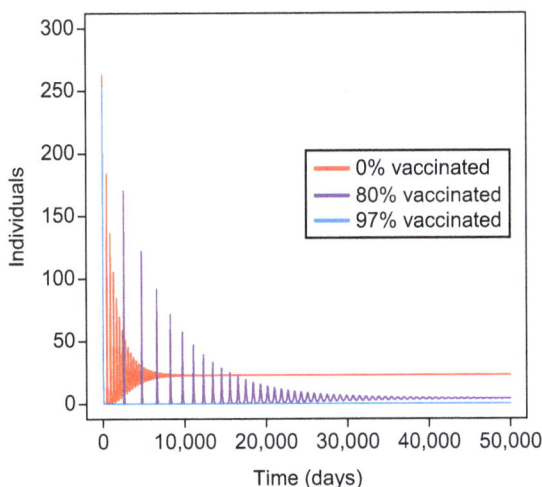

Figure 2.6 The trajectory of the number of infected individuals in the population for three values of the percentage of individuals vaccinated at birth: 0%, 80%, and 97% (Equation 2.12). We assumed a constant population of 200,000 individuals, $r = 0.00005$, $a = 0.3333$, and $u = 1/(70 \cdot 365)$. An initial population of 7,840 susceptible and 160 infected individuals was used. Notice that when the proportion of vaccinated individuals is 97%, the number of infected individuals drops to 0 and remains there. This phenomenon is known as *herd immunity*.

Because our equations now allow the number of susceptibles to grow, we have replaced the S_0 value in Equation 2.10 with S_T. S_T will be the threshold value at which the infection can be sustained. Setting $R_0 = 1$ and rearranging, we can solve for this S_T, yielding:

$$S_T = \frac{a+u}{r} \tag{2.14}$$

If the number of suceptibles drops below S_T and remains there, the disease becomes extinct. When might such a drop occur? If we now look at the dS/dt portion of Equation 2.12, we can see that it takes on its maximal value when $I(t) = 0$. At that point, the rate of change of $S(t)$ is determined by the balance of birth and death. Figures 2.5 and 2.6 show us that this model tends (at least in these examples) to converge to a steady state. What does such a steady look like in the equations comprising Equation 2.12? Steady here just means unchanging, which is another way of saying that the derivative is 0. Hence, a steady state is one where all the equations in Equation 2.12 have a value of 0. If we set $dS/dt = 0$ under the assumption of $I(t) = 0$, we find:

$$N \cdot (1-p) \cdot u = S \cdot u \tag{2.15}$$

If we cancel the common u term and substitute our expression for S_T from Equation 2.14 for S, we get:

$$N \cdot (1-p) = \frac{a+u}{r} \tag{2.16}$$

Now we can rearrange and solve for $p = p_T$: the proportion of individuals needing to be vaccinated to have the steady-state number of susceptibles be exactly equal to S_T:

$$p_T = 1 - \frac{a+u}{r \cdot N} \tag{2.17}$$

Recall that at S_T the number of infecteds does not increase, meaning that our neglect of the terms in I in Equation 2.12 will not be misleading. If we solve this equation for the model in Figure 2.6, we get $p_T = 0.966$. In other words, for this particular disease, if we vaccinate 96.6% or more of the population, we can drive the disease to extinction.

More generally, we can see from Equation 2.17 that p_T depends on a and r, or alternatively on R_0. The larger R_0 is, the higher p_T will need to be to achieve herd immunity. While our models here are too simplified to be used for building a vaccination program, the same conceptual approach is used for those programs. In **Table 2.2**, I give estimated p_T values for some of the diseases in Table 2.1. As expected, diseases with larger R_0 values, like measles, require a more intensive

TABLE 2.2 VACCINATION LEVELS (p_t) REQUIRED FOR HERD IMMUNITY IN DIFFERENT DISEASES		
Disease	P_t	*Reference*
Diphtheria	85%	[6]
Malaria	80–99%	[6]
Measles	83–94%	[6]
Mumps	75–86%	[6]
Pertussis	92–94%	[6]
Polio	80–86%	[6]
Rubella	83–85%	[6]
Smallpox	80–85%	[6]

vaccination program to achieve eradication, since for models such as those used here there is a direct relationship between R_0 and p_t.

Reflections and Previews

In this chapter we have seen our first biological models, created as part of a quest to understand the outbreak of an infectious disease. Although the SIR model has no analytic solution, we had little difficulty in using computers to solve it. However, we realized that for it to predict the dynamics of a real epidemic, we would need to measure at least two *parameters* for the model. At the same time, even this simplified model gave us insights into ideas such a "flattening the curve."

We also saw the power of being able to elaborate on our models. By adding birth, death, and vaccination to our model we found our first *emergent property*. An emergent property of a system is one that arises from the behavior of the system's individual elements but, in some sense, is not obviously predictable from that individual-level behavior. Herd immunity is just such a property, and we will find several more emergent features of life over the course of this book.

More broadly, now that we have explored a mathematical model and seen some of its uses, we need to do a better job of understanding how scientists go about creating or discovering such models. That is the topic of the next chapter.

REFERENCES

1. Kermack WO & McKendrick AG (1927) A contribution to the mathematical theory of epidemics. *Proceedings of the Royal Society of London. Series A, Containing Papers of a Mathematical and Physical Character* 115(772):700–721.

2. R Development Core Team (2008) *R: A language and environment for statistical computing* (R Foundation for Statistical Computing).

3. Soetaert KE, Petzoldt T, & Setzer RW (2010) Solving differential equations in R: package deSolve. *Journal of Statistical Software* 33:1–25.

4. Wolfram S (1999) *The MATHEMATICA® book, version 4* (Cambridge University Press, Cambridge).

5. Galassi M, Davies J, Theiler J, Gough B, Jungman G, Alken P, Booth M, Rossi F, & Ulerich R (2002) *GNU scientific library* (Network Theory Limited, Godalming, UK).

6. Fine PE (1993) Herd immunity: history, theory, practice. *Epidemiologic Reviews* 15(2):265–302.

7. Mills CE, Robins JM, & Lipsitch M (2004) Transmissibility of 1918 pandemic influenza. *Nature* 432(7019):904–906.

8. Wallinga J & Teunis P (2004) Different epidemic curves for severe acute respiratory syndrome reveal similar impacts of control measures. *American Journal of Epidemiology* 160(6):509–516.

9. Team WER (2014) Ebola virus disease in West Africa—the first 9 months of the epidemic and forward projections. *The New England Journal of Medicine* 2014(371):1481–1495.

10. Liu Y, Gayle AA, Wilder-Smith A, & Rocklöv J (2020) The reproductive number of COVID-19 is higher compared to SARS coronavirus. *Journal of Travel Medicine* 27(2):taaa021.

11. Anderson RM & May RM (1982) Directly transmitted infections diseases: control by vaccination. *Science* 215(4536):1053–1060.

12. Vynnycky E & White R (2010) *An introduction to infectious disease modelling* (OUP Oxford, Oxford).

13. Bos KI, Schuenemann VJ, Golding GB, Burbano HA, Waglechner N, Coombes BK, McPhee JB, DeWitte SN, Meyer M, & Schmedes S (2011) A draft genome of Yersinia pestis from victims of the black death. *Nature* 478(7370):506–510.

14. Haensch S, Bianucci R, Signoli M, Rajerison M, Schultz M, Kacki S, Vermunt M, Weston DA, Hurst D, & Achtman M (2010) Distinct clones of Yersinia pestis caused the black death. *PLoS pathogens* 6(10):e1001134.

15. Tuchman BW (1977) *A distant mirror: the calamitous 14th century* (Alfred A. Knopf, New York), 677 pp.

16. Cohn S (2007) After the black death: labour legislation and attitudes towards labour in late-medieval western Europe. *The Economic History Review* 60(3):457–485.

17. Roitt I, Brostoff J, & Male D (1998) *Immunology*. 5th Edition (Mosby International Ltd., London), 423 pp.

18. Murray JD (2001) *Mathematical biology, I: an introduction*. 3rd Edition (Springer, New York).

19. Lyu W & Wehby GL (2020) Community use of face masks and COVID-19: evidence from a natural experiment of state mandates in the US: study examines impact on COVID-19 growth rates associated with state government mandates requiring face mask use in public. *Health Affairs* 39(8):1419–1425.

20. Lerner AM, Folkers GK, & Fauci AS (2020) Preventing the spread of SARS-CoV-2 with masks and other "low-tech" interventions. *JAMA* 324(19):1935–1936.

21. Meyerowitz EA, Richterman A, Gandhi RT, & Sax PE Transmission of SARS-CoV-2: a review of viral, host, and environmental factors. *Annals of Internal Medicine* 174(1):69–79.

22. Nowak MA & May RM (1994) Superinfection and the evolution of parasite virulence. *Proceedings of the Royal Society of London. Series B: Biological Sciences* 255(1342):81–89.

23. Futuyma DJ (1998) *Evolutionary biology*. 3rd Edition (Sinauer Associates, Inc, Sunderland, MA).

24. Harrigan PR, Bloor S, & Larder BA (1998) Relative replicative fitness of zidovudine-resistant human immunodeficiency virus type 1 isolates in vitro. *Journal of Virology* 72(5):3773–3778.

25. Duffy S, Shackelton LA, & Holmes EC (2008) Rates of evolutionary change in viruses: patterns and determinants. *Nature Reviews Genetics* 9(4):267.

26. Lieberman TD, Michel J-B, Aingaran M, Potter-Bynoe G, Roux D, Davis Jr MR, Skurnik D, Leiby N, LiPuma JJ, & Goldberg JB (2011) Parallel bacterial evolution within multiple patients identifies candidate pathogenicity genes. *Nature genetics* 43(12):1275.

27. Brabin B (2020) An analysis of the United States and United Kingdom smallpox epidemics (1901–5)–The special relationship that tested public health strategies for disease control. *Medical History* 64(1):1–31.

28. Berche P (2022) Life and death of smallpox. *La Presse Médicale* 51(3):104117.

29. Krylova O & Earn DJ (2020) Patterns of smallpox mortality in London, England, over three centuries. *PLoS biology* 18(12): e3000506.

30. Visher E, Evensen C, Guth S, Lai E, Norfolk M, Rozins C, Sokolov NA, Sui M, & Boots M (2021) The three Ts of virulence evolution during zoonotic emergence. *Proceedings of the Royal Society B* 288(1956):20210900.

31. May RM & Anderson RM (1983) Epidemiology and genetics in the coevolution of parasites and hosts. *Proceedings of the Royal society of London. Series B. Biological Sciences* 219(1216): 281–313.

32. Lefebvre A, Fiet C, Belpois-Duchamp C, Tiv M, Astruc K, & Glélé LA (2014) Case fatality rates of Ebola virus diseases: a meta-analysis of world health organization data. *Médecine et Maladies Infectieuses* 44(9):412–416.

33. Boyce WE & DiPrima RC (1992) *Elementary differential equations and boundary value problems* (John Wiley & Sons, New York), 680 pp.

34. Anderson RM & May RM (1985) Vaccination and herd immunity to infectious diseases. *Nature* 318(6044):323–329.

Building a Better Cat
Building a Predator–Prey Model and Chaos

3

"The best material model of a cat is another, or preferably the same, cat."
—Rosenblueth and Wiener [1]

"Since all models are wrong the scientist must be alert to what is importantly wrong. It is inappropriate to be concerned about mice when there are tigers abroad."
—George Box [2]

"The underlying physical laws necessary for the mathematical theory of a large part of physics and the whole of chemistry are thus completely known, and the difficulty is only that the exact solution of these laws leads to equations much too complicated to be soluble."
—Paul Dirac [3]

WHAT ARE MODELS?

The word "model" is a slippery one in science: most of us have a sense that we know what a model is and have seen some examples of how to use them. At the same time, neither the word nor the concept is heavily emphasized in many introductory science courses. As a result, the degree to which models underlie most or all scientific disciplines is often not made obvious to us as new students.

The quotations that open this chapter are, collectively, an excellent implicit definition of a model. I cannot be as pithy, but if I had to be concrete, I would write something along the lines of "a simplified, yet accurate, description of some phenomenon." There are three important parts to this statement.

First, all scientific statements about the world involve some simplification of the actual situation. Whatever your object of study, be it an atom, a steel beam, a rat cell, a forest, or a supernova, it will be, of course, most accurately represented by the thing itself. But aside from the practical difficulties of studying a supernova in a laboratory setting, the goal of scientific inquiry is some combination of understanding and prediction or manipulation. Since none of those three goals are possible using the study object alone, we as scientists use simplified abstractions of our study subjects in our work. Second, models are "accurate." Now, these first two statements are, of course, a contradiction: no simplification can be a fully accurate representation. So perhaps a better way to write this part of the definition would "an accurate description of the features of the system we are interested in." If we are concerned with the force necessary to break a steel beam, our model could easily be wrong about the color the beam is painted, but that inaccuracy is irrelevant for our study. (Of course, it might be quite relevant for the engineer tasked with cooling the building, since dark colors absorb more heat.) Third and finally, a model seeks to represent some other object, phenomenon, or process. This relationship is the test of the model: Is the accuracy sufficient to understand or predict behavior "in the real world"?

DOI: 10.1201/9781003687504-3

In my admittedly personal and idiosyncratic view, essentially every scientific statement, theory, or hypothesis is a model. If this view is even partly correct, then many scientific ideas and concepts for which we would not generally use the word model are, in fact, models. **Figure 3.1** gives an (incomplete) taxonomy of models,

Verbal/Descriptive

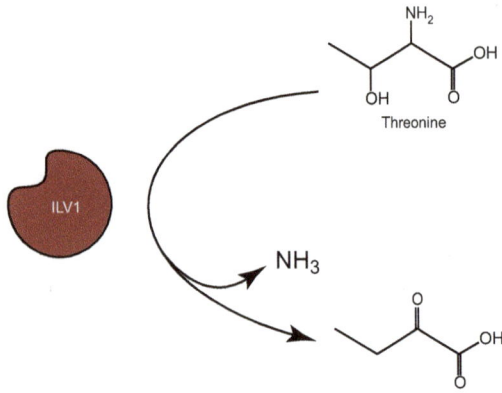

Statistical

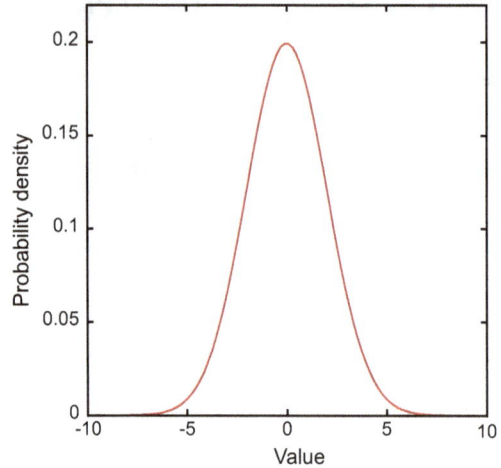

Mathematical

$$F = G\,\frac{m_1 m_2}{d^2}$$

Network

Computational

```
#!/usr/bin/python

import random
import sys


if (len(sys.argv)<3):
    print "Usage sim_genetic_drift.py population_size
    sys.exit(-1)
else:
    population_size=int(sys.argv[1])
    initial_freq_A=float(sys.argv[2])
```

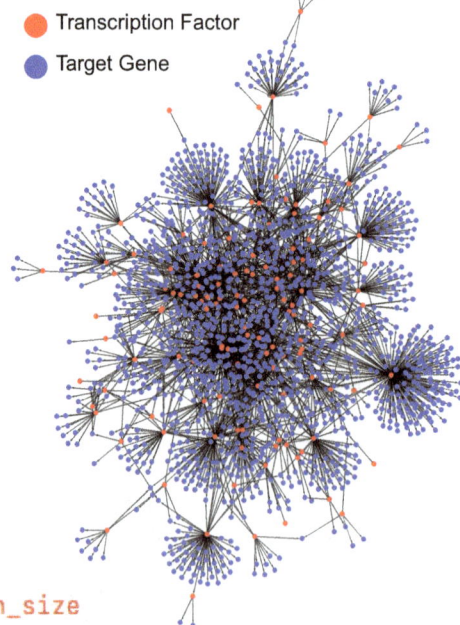

Figure 3.1 A few of the different kinds of models we will explore in this book. Verbal models can be very complex and rich but tend to hide some of their assumptions. For instance, the chemical reaction shown could proceed very slowly, but this fact would not be apparent from the diagram given. Much of the first half of this book is given over to mathematical models: we will primarily use models based on differential equations, but in Chapter 6 we will see a powerful type of model based on linear equations. Statistical models are closely related to mathematical models but allow for uncertainty or randomness in the model's behavior. Computational models are sometimes used when it is difficult to express a phenomenon in terms of common statistical or mathematical approaches. We will implement a computational model of genetic drift in Chapter 11. Network models are also mathematical in origins but allow researchers to study large systems with relatively sparse data (Chapter 14).

taken from the examples we will explore in this book. And, in what follows, I offer some further examples of this generalized notion of a model to suggest that the idea of science as a modeling enterprise is at least more plausible than it sounds. However, you need not accept this personal view of science to find the reminder of this book useful.

As an example of this broad reach of models, in my own research I use computation to study the large-scale patterns of which proteins bind to which other proteins in the cell (see Chapter 14). The data that underlie my studies can often be traced back to a scientific paper with the statement "A binds to B." I do not mean to oversimplify these papers: the authors are well aware of the many qualifications that accompany that bald statement. "A binds to B" is a *verbal* model, and it is not a useless one. For instance, suppose you have a genetic disease where you know that protein B is involved in, say, the underproduction of some needed compound. If you then discover that the individuals with the disease have mutated copies of protein A, the statement "A binds to B" provides a potential link between the genetics and the disease. If you alternatively knew with high confidence that "A will not bind to B," you would need to structure your investigations of the disease very differently.

Biology is full of verbal models: Darwin expressed the concept of natural selection as a verbal model (see Chapter 4). The concept of a species as a "group of potentially interbreeding individuals that can produce fertile offspring with each other" is likewise a core verbal model in evolution and ecology.

Some Examples of Mathematical Models

We are all more used to thinking of mathematical models in science. The following are equations for two rather well-known models and one you have may not have seen before:

$$F = G\frac{m_1 m_2}{d^2} \tag{3.1}$$

$$E = mc^2 \tag{3.2}$$

$$Q = \frac{N_e s}{N} \cdot \frac{1}{1 - e^{-2N_e s}} \tag{3.3}$$

Equation 3.1 gives Newton's model of the force of the gravitational attraction between two objects with masses m_1 and m_2 that are a distance d apart. Here, G is a universal constant describing the nature of gravity [4]. As such, this equation was the basis for the falling ball equation in Chapter 1, where the mass of the Earth (m_1) is so much larger than that of the ball that we can approximate the acceleration due to gravity with the constant 9.81 m/s². If we look up at the night sky and want to predict where the moon will be in 3 days, Equation 3.1 would be the model that allows us to do this.

Equation 3.2 gives Einstein's famous relationship between an object's mass m and the energy inherent in its atoms (E), which are related by the speed of light (c) squared. We are all very familiar with this equation because it describes the destructive power of atomic weapons, explaining how a tiny mass of fissile material can yield an enormous explosion.

Equation 3.3 is likely unfamiliar to you: it is from the field of population genetics and evolution. It describes how different versions a gene in a population change in their relative frequencies in time. To understand Equation 3.3, we will assume for the moment that there is a gene in the population for which all the members of the population have an identical version. This assumption is not as unreasonable as it might sound: any two individual humans differ from each other at only about one DNA base in a thousand [5].

Now, if one individual in the population experiences a mutation in that gene, Equation 3.3 describes how likely it is that that new mutation will actually spread through the population over evolutionary time such that it eventually replaces the original version of the gene in every member of that population. (As surprising as this replacement might seem, we will explore this process in Chapter 4 and then more carefully in Chapters 11 and 12.) To compute this replacement probability, we need to know several things, including the population size N [6]. We will also need to know s: the reproductive advantage $(s > 0)$ or disadvantage $(s < 0)$ that the mutation confers on its possessors relative to the alterative original version of the gene, which is often called the *wild-type* version. Meanwhile N_e is called the *effective population size*. It scales the population size N value by a variety of complicated factors relating to the amount of inbreeding in the population, its sex ratios, and the degree of nonrandom mating [7]. You should not worry if the details of Equation 3.3 are not intuitive yet, since we will return to it in Chapter 11. But do notice one of its interesting features: sometimes mutations that confer a disadvantage on their possessors are still fixed in a population, and sometimes mutations that confer an advantage still fail to replace the wild-type gene. Thus, Equation 3.3 modifies Darwin's verbal model of natural selection with more precise predictions about how genomes change in time.

The Power of More Precise Models

As we proceed in this book, we will see how models based on mathematics, statistics, and computation (see Figure 3.1) can improve on verbal models by (1) being more explicit about the modeling assumptions used and (2) providing greater precision in their predictions and hence a better framework for testing the model against what is observed in the laboratory or the field. We will pay particular attention to assumptions, because, as noted in our definition of models, *all* models have assumptions. Scientists new to model-based thinking tend to treat all assumptions as equally undesirable: by the end of this book, you should be able to explain why it is important to note your assumptions in any scientific work and to distinguish between appropriate and inappropriate assumptions for different problems.

Where Do Models Come From?

Science textbooks are full of equations and models, and as a student, it is easy to believe that only the great geniuses of science could have created such intricate intellectual structures. While it is certainly fair to think that Newton and Einstein might have been a bit more clever than you or I, we should also remember that science is an iterative process. We saw examples of this process in the previous chapter, where we started with an SIR model with only two parameters and five terms and then expanded it with parameters for birth, death, and vaccination, allowing us to explore new questions. In other science classes, the models covered are often very complex for the obvious reason that they most closely reflect our current understanding of a complex world. Yet often these complex models represent a series of refinements that were made over time to an original model that is not so far beyond what we could create ourselves. If we think back to the equation describing the actions of natural selection on gene variants in Equation 3.3, we can imagine starting with Darwin's insight that, on average, heritable variants that allow for increased numbers of offspring in the next generation will tend to be more common in that generation. We could then refine that idea as our understanding of genetics and statistics improves, until it was possible to write down Equation 3.3. To see how we might do this, we will take a different problem as our example: modeling the interactions of predators and prey in an ecosystem.

BUILDING A MODEL: AN EXAMPLE USING THE INTERACTIONS OF PREY SPECIES AND THEIR PREDATORS

Figure 3.2 outlines the ecological system for which we will develop a model. We have deliberately chosen an example that is oversimplified because our goal is to understand the process of model building, and a simple system makes this goal more approachable. Of course, we are not the first researchers to consider this

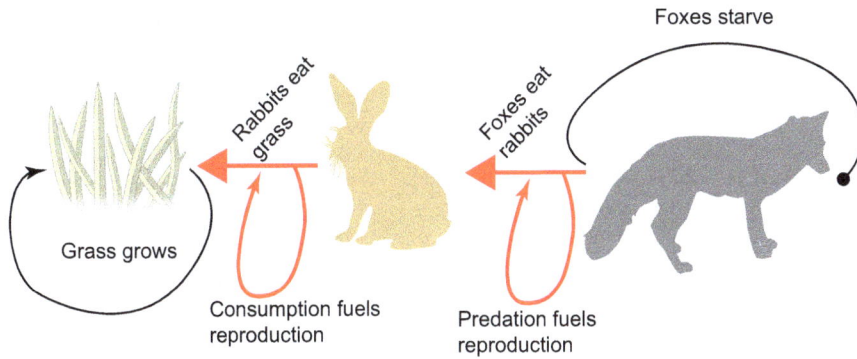

Figure 3.2 A schematic of the ecological system we propose to model. The scheme is highly simplified: as a result, the model we develop will be more of use for gaining general intuition about ecological systems than for modeling real ones [8–10]. (Grass, hare, and fox art courtesy of NIAID.)

problem: the model we create will be the classic Lotka–Volterra model [11, 12], the history of which I will discuss a bit later in the chapter.

As diagramed, the system has three components: grass, rabbits, and foxes. The first simplification we are going to make is to omit the grass. We can do this by assuming that the size of the grass plot is so large that the rabbits cannot meaningfully alter it through grazing. Now, as we will see shortly, were the rabbits allowed to reproduce without limit, this assumption would be very quickly violated. However, we are going to hope that the addition of the foxes to our model will keep the rabbit population enough in check that indeed the grass area is only modestly altered.

With this simplification, we find that we need a model of four processes:

1. The growth of the rabbit population in the absence of foxes
2. The effect of fox predation on the rabbit population
3. The ability of the fox population to grow as a result of rabbit predation
4. The decline of the fox population if the supply of rabbits is insufficient.

We will start with the growth of the rabbit population. Assuming grass is present in abundance, what determines the growth rate of the rabbit population? The answer is probably obvious to you, but we will nonetheless consider a simpler case: bacterial cells.

Modeling Population Growth

The most common bacterial species used in microbiology research is *Escherichia coli*. Under optimal conditions in the laboratory, cells of this species can divide about once every 20 min [13, 14]. In Exercise 3.1 you will compute the number of *E. coli* cells in every generation over the course of 1 day, assuming a generation time of 20 min and an initial population of a single cell. In **Figure 3.3A** I show the results of this analysis. We can express this idea of a population of cells doubling every 20 min with what is termed a *recurrence relation*. That relation gives the value of the system at time t as a function of time $t-1$. You might notice that in this framework *time* is measured in discrete units of generations rather than minutes or days:

$$N(t) = 2 \cdot N(t-1) \tag{3.4}$$

Exercise 3.1

Model the growth of an *E. coli* culture using a spreadsheet. Assume that at $t = 0$ min you have a single *E. coli* cell and that every cell in the culture divides every 20 min (hence, the number of cells doubles every 20 min). Continue your computation until you reach $t = 1{,}440$ min (1,440 min / 60 min/h = 24 h). In addition to the number of cells at each interval, compute the mass of the culture, assuming that one cell has a mass of 6.25×10^{-10} grams. Plot your results and compare to Figure 3.3A.

A)

B)

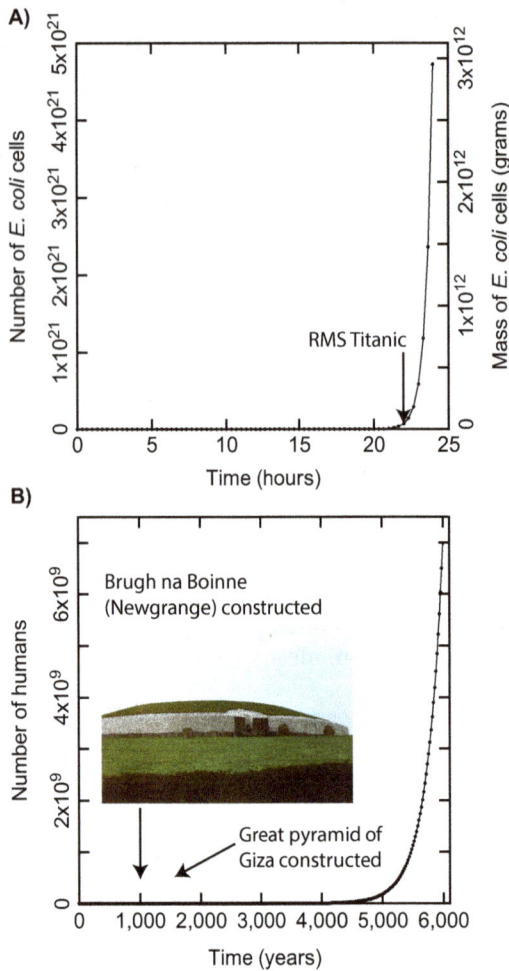

Figure 3.3 Population growth in bacteria and humans. (A) In this panel, an initial population of a single *E. coli* cell is allowed to double every 20 min for 1 day. By about 22 hours, the mass of the resulting cells would weigh more than the *Titanic*. **(B)** In this panel, an initial human population of two individuals is assumed to increase by 5% every generation of 20 years. In 6,000 years, this growth rate yields something approximating the modern human population. Pictured is the Neolithic tomb Brugh na Boinne (Newgrange) in Ireland, which dates from a few hundred years prior to the Great Pyramid, both of which must have been built (in this inaccurate model) by the handful of people alive at the time.

This equation is effectively the discrete analog of a differential equation, and, like a differential equation, it can be solved so long as we have a value of the initial condition of the system, which in this case is $N(0) = 1$. We will not discuss solving such equations here, but, for the initial condition given, it can be shown that the solution of Equation 3.4 is given by [15]:

$$N = N_0 \cdot 2^t \tag{3.5}$$

Here, $N_0 = N(0)$ (e.g., our initial population of one cell). Again, recall that the time t is measured in units of (integer) generations.

It is worth pausing for another look at Figure 3.3A: starting with a single cell, dividing every 20 minutes, if we provided enough food, we could have *E. coli* cells with a total weight exceeding that of the *RMS Titanic* in less than a day. Of course, such an absurd result would not occur in practice. Not only would we run out of food and space, but some of the cells would die during that day, decreasing the growth rate at least somewhat. However, because of our human cognitive biases toward linear thinking, we need to remind ourselves that exponential growth can get out of hand in a hurry. We should, therefore, probably keep the power of exponentials in mind, given that we are about to start talking about rabbits.

What about other populations? A human population generally does not double every generation, but a 5% increase every generation of 20 years is not unreasonable. That trend would be governed by the equation:

$$N = N_0 \cdot 1.05^t \tag{3.6}$$

A plot of this model is shown in **Figure 3.3B**. Occasionally, you see an analysis of the form of Figure 3.3B being used to suggest that the human race is not very old, because you can get from two individuals to something like the modern 9 billion or so in only about 6,000 years. (The number of initial humans and the time span of 6,000 years are not arbitrary, because the people making this argument are looking for mathematical "proof" of the literal truth of the account of Earth's creation in the biblical book of Genesis.)

A closer look at Figure 3.3B belies this naïve idea: the model predicts that in 2500 BCE there were fewer than a thousand people on the entire planet. Given that we know from well-attested historical sources [16] that this was the century in which the Great Pyramid at Giza was constructed, accepting this model leaves us with the essentially unanswerable question of how this handful of people managed such a construction project. This difficulty is especially troubling given that this same small group had, a few hundred years earlier, completed the megalithic monument Brú na Boinne (Newgrange; see Figure 3.3B) thousands of miles away in Ireland [17].

Transition to Continuous Time

Equations 3.5 and 3.6 are in some sense satisfying because they only allow integer values, meaning the number of cells or people is always a whole number. However, there is less to this apparent precision than it appears. Even with *E. coli* cells, after a few rounds of cell division, the cells lose *synchronicity*, meaning that they do not all divide simultaneously after 20 minutes. Of course, human generations are self-evidently *not* synchronized, so a model that assumes that there are x humans up to the end of one generation and $1.05x$ individuals at the beginning of the next suggests a precision that does not exist in actual populations.

Hence, for the reasons of mathematical and computational convenience mentioned in Chapter 2, it is worth looking for a model of population growth where both time and population are measured continuously. Such a model would still need to assume that the growth rate of the population is proportional to its size. How might we write a differential equation in continuous time to represent this idea? For a population of size N, its growth rate can be expressed as a function of that size itself times a constant a:

$$\frac{dN}{dt} = a \cdot N \tag{3.7}$$

This equation is simply telling us that the speed at which the population is increasing is proportional to the size of the population itself.

As it happens, Equation 3.7 is a differential equation for which an analytic solution *does* exist. To prove this fact to ourselves, we can first let $a = 1$, meaning that the rate of change of N is equal to itself. Referring back to Table 1.2 in Chapter 1, we see that if $N(t) = e^t$, then dN / dt is also e^t, in agreement with Equation 3.7. What about when $a \neq 1$? You may recall the *chain rule* for differentiation from your calculus class. This rule states that if we can break a complex function F into two simpler functions f and g, such that $F(x) = f(g(x))$, then we can write the derivative of $F(x)$ with:

$$\frac{d}{dx}(F(x)) = \frac{d}{dx}(f(g(x))) \tag{3.8}$$

Here the notation d / dt refers to the operation of taking the derivative of the quantity enclosed in parentheses. The chain rule then states that:

$$\frac{d}{dx}(f(g(x))) = f'(g(x)) \cdot g'(x) \tag{3.9}$$

where

$$\frac{d}{d\chi}(f(\chi)) = f'(\chi) \tag{3.10}$$

for some arbitrary variable χ.

Let us see if we can make use of this relationship. We will propose that there is some function $N(t) = e^{g(t)}$. If we were to take the derivative of this function, we would have:

$$\frac{d}{dt}\big(N(t)\big) = \frac{d}{dt}\Big(e^{g(t)}\Big) \tag{3.11}$$

By the chain rule, we can rewrite the right-hand side of the above as:

$$\frac{d}{dt}\big(N(t)\big) = \frac{d}{dg(t)}\Big(e^{g(t)}\Big) \cdot \frac{d}{dt}\big(g(t)\big) \tag{3.12}$$

For the first term on the right-hand side we are taking the derivative with respect to $g(t)$, meaning that the derivative of $e^{g(t)}$ is just $e^{g(t)}$ (see Table 1.1 in Chapter 1). Hence, to match Equation 3.7, we need a function that, when differentiated with respect to t, gives a constant. The function $g(t) = a \cdot t$ satisfies this condition:

$$\frac{d}{dt}\big(N(t)\big) = e^{at} \cdot \frac{d}{dt}(a \cdot t) \tag{3.13}$$

Which, when simplified, gives us Equation 3.7 again, since $\frac{d}{dt}(a \cdot t) = a$. As a result, we conclude $N(t) = e^{at}$ is a solution to Equation 3.7.

One thing missing from our treatment so far is initial conditions: the number of cells or people we start with. We will denote our initial population as N_0; the question then becomes how we incorporate N_0 into our new solution for Equation 3.7.

Notice that $e^0 = 1$, regardless of the value of a. So perhaps we could write the full solution for Equation 3.7 as:

$$N(t) = N_0 \cdot e^{at} \tag{3.14}$$

We can see that indeed for $t = 0$, $N(0) = N_0$, which is encouraging. We can also use another tool from calculus, called the product rule, to take the derivative of Equation 3.14 and check if it agrees with Equation 3.7 (Exercise 3.2).

The product rule states that:

$$\frac{d}{dx}\big(f(x) \cdot g(x)\big) = f'(x) \cdot g(x) + f(x) \cdot g'(x) \tag{3.15}$$

Applying it to Equation 3.14 with $f = N_0$ and $g = e^{at}$ gives us:

$$\frac{d}{dt}\big(N(t)\big) = \frac{d}{dt}(N_0) \cdot e^{at} + N_0 \cdot \frac{d}{dt}\Big(e^{at}\Big) \tag{3.16}$$

Since the derivative of a constant (N_0) is zero, we are back to the equation we already worked with, except that we have added the extra constant N_0:

$$\frac{d}{dt}\big(N(t)\big) = N_0 \cdot \frac{d}{dt}\Big(e^{at}\Big) \tag{3.17}$$

Which just gives:

$$\frac{dN}{dt} = a \cdot N_0 \cdot e^{at} \tag{3.18}$$

And since a and N_0 are both constants, Equation 3.7 is still satisfied, making Equation 3.14 a general solution for Equation 3.7 for any arbitrary initial population size N_0.

Building the Predator–Prey Model

We have now developed a model for the rabbits exclusive of foxes. If we look back at Figure 3.2, we can see that the behavior of foxes in the absence of rabbits is effectively the mirror image of the rabbit equation. What I mean by that statement is that the rate of *death* of foxes is simply proportional to the number of foxes. Another way of expressing that idea is that a fixed proportion of all the living foxes die at a particular instant. We could express the process as a differential equation:

$$\frac{dF}{dt} = -d \cdot F \tag{3.19}$$

Notice that we are now using F as our variable to remind ourselves that we are considering the foxes. Equation 3.19 differs only from Equation 3.7 in that we have a $-d$ rather than a positive a. However, since nothing we did above required that $a > 0$, we should be able to take all the same steps assuming $d < 0$. If we do, our new fox equation would simply be:

$$F(t) = F_0 \cdot e^{-dt} \tag{3.20}$$

In Exercise 3.3, you can check that differentiating Equation 3.20 satisfies Equation 3.19.

Exercise 3.2

Confirm that taking the derivative of Equation 3.14 with respect to t gives Equation 3.7.

Exercise 3.3

Confirm that taking the derivative of Equation 3.20 with respect to t gives Equation 3.19.

Modeling the Interaction of Foxes and Rabbits

We have already made a great deal of progress in our modeling efforts: the remaining pieces of the model we need are the terms for how the number of foxes affects the number of rabbits, and vice versa. An obvious possible answer comes from Chapter 2: perhaps the frequency of predation behaves in a manner similar to the spread of an infectious disease from infecteds to susceptibles. As with the SIR model, we can imagine that a large number of factors might influence the chance of a fox encountering and consuming a rabbit. However, most of those factors will act like the constant r in the SIR model: they will be independent of the actual number of foxes and rabbits. So, we can imagine a model with two constants b and c. Here b will represent the rate at which foxes meet and remove rabbits from the landscape, and c will represent the rate at which rabbit predation allows the foxes to reproduce. We can then write our differential equation model as:

$$\frac{dR}{dt} = a \cdot R - b \cdot R \cdot F$$
$$\frac{dF}{dt} = c \cdot R \cdot F - d \cdot F \tag{3.21}$$

You can immediately see the rabbit-only and fox-only terms that we derived above in the two equations, as well as the two new interaction terms. Another thing to notice about the equations comprising Equation 3.21 is that, unlike the SIR model, they do not sum to 0. As a result, we do not have conservation of individuals: one fox might need to consume several rabbits to reproduce, meaning that the total number of individuals in the ecosystem would drop. Biologically this assumption will sound trivial, but if we had unwisely set $b = c$, it would not have been allowed.

THE LOTKA–VOLTERRA MODEL

The equations comprising Equation 3.21 are known as the Lotka–Volterra model, named after the two mathematicians who developed them independently in the 1920s [11, 12]. Interestingly, Lotka initially proposed the equations to describe a chemical system rather than an ecological one [11]. However, he later applied more complex forms of the model to many biological problems, including ecological ones [18]. We will return to this surprising association of ecology and chemistry in Chapter 5. Likewise, the fact that independent researchers happened on the same equations reminds us that model-based thinking does not require us to have blinding and unique flashes of insight to proceed (although such flashes never hurt).

The major problem with the Lotka-Volterra model is the parameters: unlike the SIR model, where one could at least argue that r and a are constant for a given disease throughout the world, a, b, c, and d probably differ from forest to forest or from year to year, as the animals encounter each other more or less frequently, are better camouflaged at different times of the year, and so forth. Indeed, even the idea that a predator species has only a single target prey species is quite unrealistic. We will nonetheless explore this model because it allows us to probe our understanding of ecology in several ways.

The most apparent thing about **Figure 3.4A** is that the system is *periodic*: it cycles through the same sets of rabbit and fox populations again and again. With the SIR model with birth and death (see Figure 2.2), we saw oscillations, but they were *damped*. In other words, no matter the starting conditions, the system converged over time to a single combination of values of $S(t)$, $I(t)$, and $R(t)$. The predator–prey model does not always converge to a steady state in this way. However, **Figure 3.4B** suggests that this model may at least possess a steady state. We will first attempt to derive its value and then to explore how the steady state and the initial conditions are related.

Finding the Steady State

For our system to be in steady state, the values of $R(t)$ and $F(t)$ should be unchanging in time, at least after some initial transient oscillations. Of course, we do not have analytic expressions for the functions $R(t)$ and $F(t)$, so one might wonder how we might determine if such a steady state exists. But claiming that a function is unchanging in time is simply another way of saying that the rate of change of that function, or its derivative, is 0. So, we can take the differential equations comprising Equation 3.21 and set them equal to 0:

$$0 = a \cdot R - b \cdot R \cdot F$$
$$0 = c \cdot R \cdot F - d \cdot F$$

(3.22)

Now we need to solve for values of R and F that simultaneously satisfy both of these equations. Because we can cancel the R in the first equation and the F in the second, it is easy to see that both foxes and rabbits have an unchanging population size when:

$$F(t) = \frac{a}{b}$$
$$R(t) = \frac{d}{c}$$

(3.23)

Using the values in Equation 3.23 as initial conditions, we can solve the Lotka–Volterra model to produce Figure 3.4B, where $R(t)$ and $F(t)$ are indeed constant.

Finding this steady state is encouraging. However, we might wonder what happens if the system starts out near to but not at the steady state. As a reminder, in the

A)

B)

C)

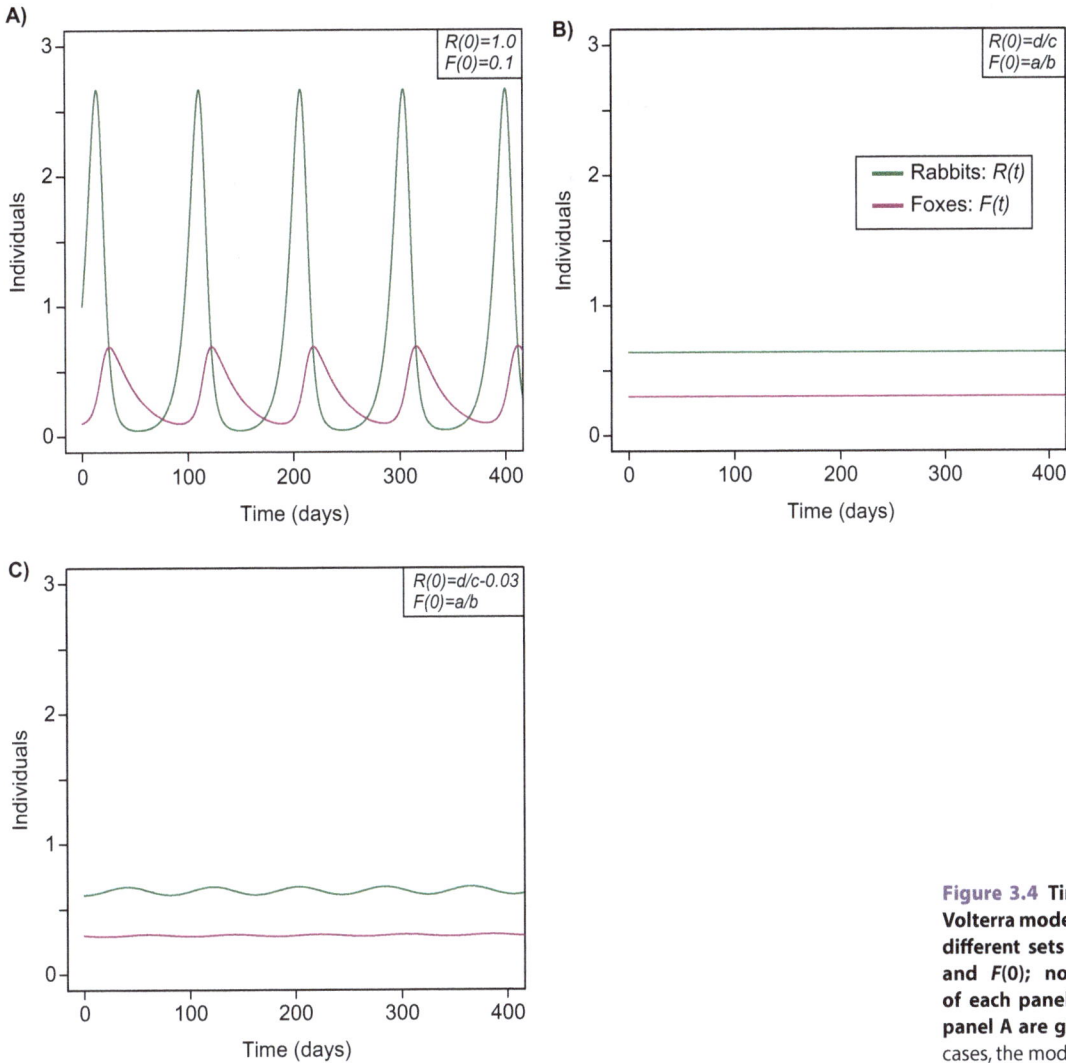

Figure 3.4 Time courses of the Lotka–Volterra model (Equation 3.21) for three different sets of initial conditions ($R(0)$ and $F(0)$); noted at the upper right of each panel. Commands to generate panel A are given in Exercise 3.4). In all cases, the model parameters were $a = 0.15$, $b = 0.5$, $c = b / 8$, $d = 0.04$.

SIR model, the oscillations would in fact decay into the steady state if we waited long enough. To explore the behavior of the Lotka–Volterra model, in Figure 3.4C I have used values of R and F that are near to but not exactly those of Equation 3.23. What we find is that initial conditions slightly different from those in Equation 3.23 will *not* converge to the steady state but will remain periodic. Why is this true? One way to answer this question is to take a graphical view of the system. For the system to reach the steady state, there must be some value of t at which the system simultaneously achieves the value of R and the value of F satisfying Equation 3.23. However, by definition, a periodic system must pass through all of its points repeatedly. For this principle, then, the only possibly periodic system that can repeatedly pass through the points of the steady state is then necessarily the one that remains at those points. If this claim is not intuitive, do not be alarmed. In the following section we are going to develop a new graphical approach to the problem that will make it much more obvious. To simplify matters, we will at the same time do something about the excessive number of parameters in Equation 3.21.

REPARAMETERIZING THE LOTKA–VOLTERRA MODEL

As we just mentioned, the parameters of the Lotka–Volterra model are a problem both because there are a relatively large number of them (four) and because they are difficult to measure for real biological systems. Hence, a figure like Figure 3.4 is only valid for a particular set of the four parameters. You could easily be excused for wondering how different the picture would be if those parameters changed slightly.

We will address this situation with two surprisingly simple tricks: changing the meaning of *time* and adjusting some constants. In our current model we have been very loose with our units, but it makes sense that our time unit would be days. In that case, the units of the parameter a are 1/days: the per-day rate of the rabbit population's growth. In somewhat familiar terms, in the absence of foxes, the doubling time t_h of the rabbit population is:

$$t_h = \frac{\ln(2)}{a} \tag{3.24}$$

After t_h days, we will have twice as many rabbits. What we will do now is use a as our new timescale, replacing t from the equations comprising Equation 3.21 with a new time unit τ, where $\tau = a \cdot t$. We will call our new functions of rabbits and foxes $r(\tau)$ and $f(\tau)$, respectively. Because it will be useful later, we will assume that:

$$\begin{aligned} R(t) &= \mu_1 \cdot r(\tau) \\ F(t) &= \mu_2 \cdot f(\tau) \end{aligned} \tag{3.25}$$

where μ_1 and μ_2 are constants. It should not be immediately apparent *why* I have proposed these two constants: we will see in a bit how they are useful.

Now, of course, we do not know the functions $R(t)$ and $F(t)$, so we also do not know $r(\tau)$ and $f(\tau)$. However, we do know the derivatives of $R(t)$ and $F(t)$, so we can ask if we can compute the derivatives of $r(\tau)$ and $f(\tau)$. We do so by taking the derivative with respect to t of both sides of Equation 3.25:

$$\begin{aligned} \frac{d}{dt}\big[R(t)\big] &= \frac{d}{dt}\big[\mu_1 \cdot r(\tau)\big] \\ \frac{d}{dt}\big[F(t)\big] &= \frac{d}{dt}\big[\mu_2 \cdot f(\tau)\big] \end{aligned} \tag{3.26}$$

The derivatives of $R(t)$ and $F(t)$ with respect to t are just dR/dt and dF/dt in our existing notation. If we move the constants outside of the derivative operator on the right side of Equation 3.26, we can compute the derivatives of $r(\tau)$ and $f(\tau)$ using the chain rule:

$$\begin{aligned} \frac{dR}{dt} &= \mu_1 \cdot \frac{dr}{d\tau} \cdot \frac{d\tau}{dt} \\ \frac{dF}{dt} &= \mu_2 \cdot \frac{df}{d\tau} \cdot \frac{d\tau}{dt} \end{aligned} \tag{3.27}$$

But since $\tau = a \cdot t$, $d\tau/dt$ is simply a, giving us:

$$\begin{aligned} \frac{dR}{dt} &= \mu_1 \cdot a \cdot \frac{dr}{d\tau} \\ \frac{dF}{dt} &= \mu_2 \cdot a \cdot \frac{df}{d\tau} \end{aligned} \tag{3.28}$$

If we plug our definitions of $r(\tau)$ and $f(\tau)$ and the values of their derivatives from Equation 3.28 back into our original Lotka–Volterra equations from Equation 3.21, we obtain:

$$\begin{aligned} a \cdot \mu_1 \cdot \frac{dr}{d\tau} &= a \cdot \mu_1 \cdot r(\tau) - b \cdot \mu_1 \cdot r(\tau) \cdot \mu_2 \cdot f(\tau) \\ a \cdot \mu_2 \cdot \frac{df}{d\tau} &= c \cdot \mu_1 \cdot r(\tau) \cdot \mu_2 \cdot f(\tau) - d \cdot \mu_2 \cdot f(\tau) \end{aligned} \tag{3.29}$$

If we make the useful substitution $\mu_2 = a/b$, we can rewrite the $dr/d\tau$ equation as:

$$\frac{dr}{d\tau} = r(\tau) \cdot (1 - f(\tau)) \tag{3.30}$$

Next, if we cancel the common μ_2 in $df / d\tau$, we can rewrite it as:

$$\frac{df}{d\tau} = \frac{1}{a} (c \cdot \mu_1 \cdot r(\tau) \cdot f(\tau) - d \cdot f(\tau)) \tag{3.31}$$

Now it appears that it might be useful to substitute $\mu_1 = d / c$, allowing us to write:

$$\frac{df}{d\tau} = \frac{d}{a} \cdot f(\tau) \cdot (r(\tau) - 1) \tag{3.32}$$

Our system of equations in four parameters can now be written as a system in a single parameter if we make one final substitution, namely $\alpha = d / a$:

$$\frac{dr}{d\tau} = r(\tau) \cdot (1 - f(\tau))$$
$$\frac{df}{d\tau} = \alpha \cdot f(\tau) \cdot (r(\tau) - 1) \tag{3.33}$$

This transformation of the model is referred to as *nondimensionalization* [19], because the original units or dimensions of the system, as set by the four parameters, have been removed. Nondimensionalization allows us to explore the more general nature of the system without worrying about the values of the parameters.

Phase Plots

Figure 3.5 illustrates the periodicity of the nondimensionalized model for one set of initial conditions. However, as mentioned, it would be nice to visualize the system in such a way as to understand its behavior across a range of initial conditions. **Figure 3.6** shows an approach to this problem known as a *phase plot*. In this plot, the number of rabbits is plotted along the x-axis and the number of foxes along the y-axis. In such a plot, we do not explicitly include time: instead, we mark little arrows showing in what direction the values of x and y change in time. Because this system is periodic, the phase plot of it forms a series of closed loops, each corresponding to a different set of initial conditions. Hence, each panel of Figure 3.6

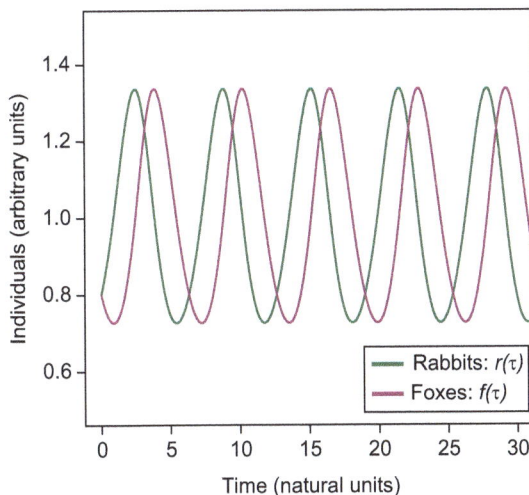

Figure 3.5 A time course of the non-dimensionalized Lotka–Volterra model (Equation 3.33) for initial conditions $r(0) = f(0) = 0.8$ and $\alpha = 1.0$.

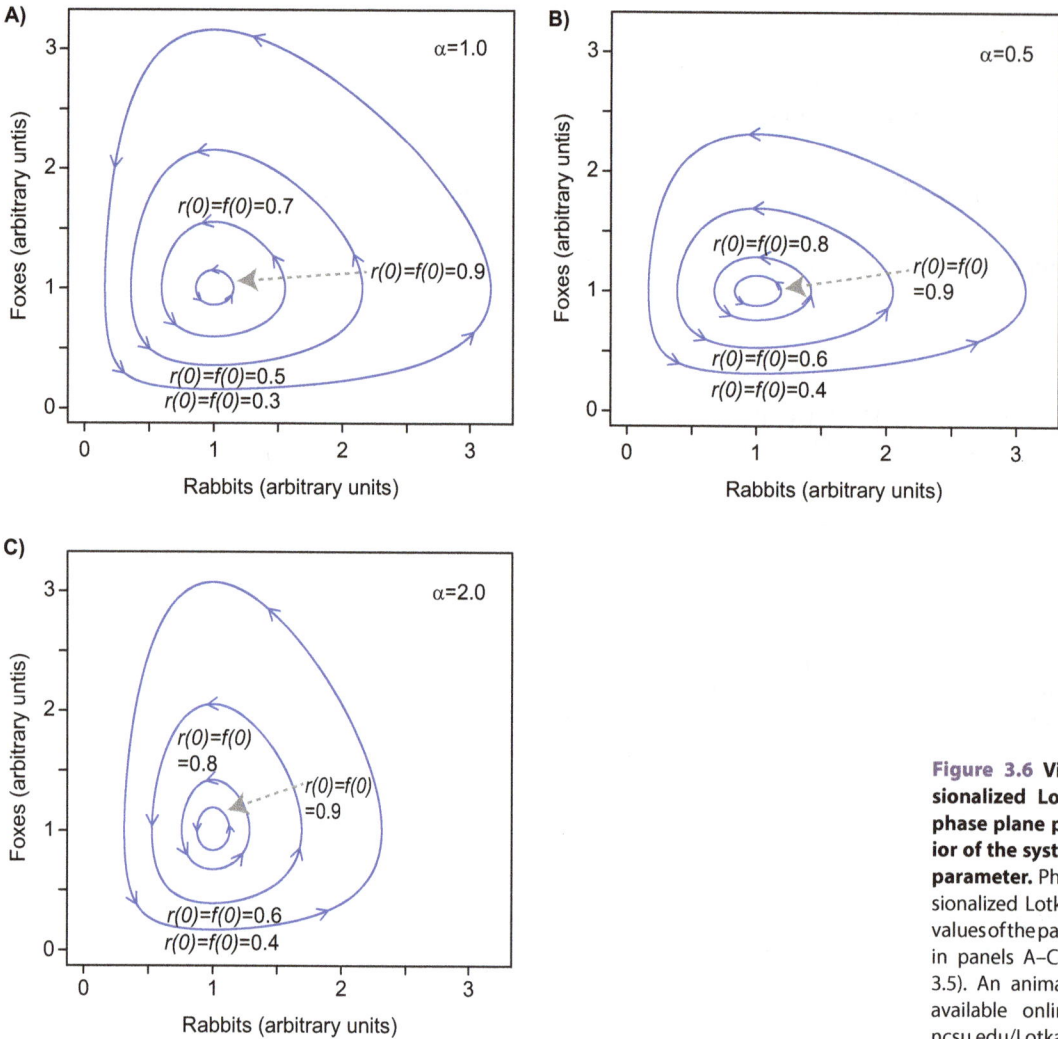

Figure 3.6 Visualizing the nondimensionalized Lotka–Volterra model with phase plane plots. Shown is the behavior of the system for three values of the parameter. Phase plots of the nondimensionalized Lotka–Volterra model for three values of the parameter α ($\alpha = 1, 0.5$ and 2.0 in panels A–C, respectively; see Exercise 3.5). An animated version of panel A is available online at: http://qbio.statgen.ncsu.edu/LotkaVolterra.html.

can show the behavior of the system across a large range of initial conditions, giving a much better sense of the system's possibilities. The panels themselves differ in the value of α. As can be seen, the α parameter changes the shape of these curves. Here, larger values of α indicate shorter-lived foxes relative to rabbit births: a large d implies a high per-unit time death rate of foxes. Hence, we see a larger relative "swing" in the fox population than in the rabbits (y-axis for foxes and x-axis for rabbits; see Figure 3.6C) in that case. Smaller values of α correspond to the opposite case, with the rapidly reproducing rabbits experiencing wider swings than the longer-lived foxes (see Figure 3.6B). From Figure 3.6 we can therefore essentially understand most of what we would like to about this two-species model of ecological interactions.

Thoughts on Initial Conditions

Figure 3.7 shows another set of phase-plane results from nondimensionalized Lotka–Volterra model, where I have intentionally selected initial conditions with many rabbits and relatively few foxes. What is striking about this plot is that the three traces come very close to each other at some times and are quite far apart at others. If we thus consider the problem of measuring the abundances of foxes and rabbits and then trying to fit a Lotka–Volterra model to those measurements, we can see that even a small amount of experimental error in our measurements could, in some circumstances, give rise to rather large errors in

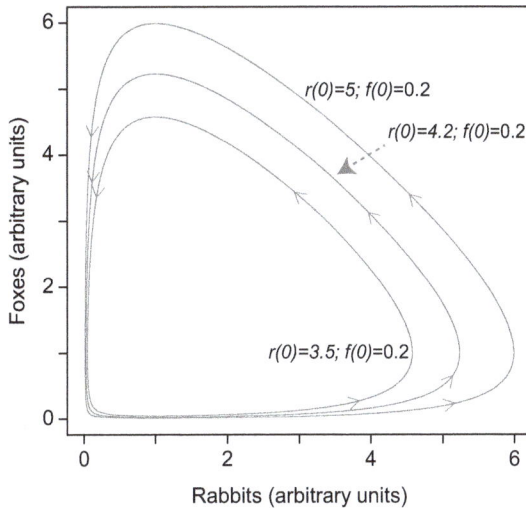

our predictions. In fact, this question of noise in measurements has quite large consequences for modeling natural systems with differential equations, as we will see in the next section.

INTRODUCTION TO CHAOS: THREE-SPECIES PREDATOR–PREY MODELS

It is conceptually easy to see that models of the form in Equation 3.21 could be extended to allow for three or more species. We will not explore this idea exhaustively, but there is a very interesting example identified by Areneodo, Coullet, and Tresser [20] that we can consider. This three-species model has nine model parameters but can be tuned such that only a single one, α, gives systems of quite different behavior. Hence, the other eight parameters are fixed at the numerical values given in the equations:

$$\frac{dx_1}{dt} = x_1(t) - \frac{1}{2} \cdot x_1(t)^2 - \frac{1}{2} \cdot x_1(t) \cdot x_2(t) - 0.1 \cdot x_1(t) \cdot x_3(t)$$

$$\frac{dx_2}{dt} = -\frac{1}{2} \cdot x_2(t) + 0.1 \cdot x_2(t)^2 + \frac{1}{2} \cdot x_1(t) \cdot x_2(t) - 0.1 \cdot x_2(t) \cdot x_3(t) \qquad (3.34)$$

$$\frac{dx_3}{dt} = (\alpha + 0.2) \cdot x_3(t) - 0.1 \cdot x_3(t)^2 - \alpha \cdot x_1(t) \cdot x_3(t) - 0.1 \cdot x_2(t) \cdot x_3(t)$$

These equations are very similar to the ones we have been using, save for the fact that they each include an additional *damping* term of the form $c \cdot x_i^2$. These terms prevent the unlimited growth of rabbits in the absence of foxes that was a slightly unrealistic property of the model we used above [21].

Figure 3.8 shows the behavior of the system in Equation 3.34 for three different values of the parameter α. In Figure 3.8A, the system behaves similarly to the two-species case, with a cycling of the populations of the three species. Figure 3.8B shows the type of convergence to a steady state we saw for some versions of the SIR model in Chapter 2. But Figure 3.8C shows something we have not seen before. In fact, looking at the three-species time course makes it difficult to see any sort of pattern in the dynamics.

Moving to three-dimensional phase plots (**Figure 3.9**) makes the situation clearer. Figures 3.9A and 3.9B look much as we would expect, namely periodic and converging. But looking at the $\alpha = 1.5$ case, we can see something rather different. The system certainly does not converge. But while it is not periodic in the conventional sense of the word, it does not move through space in an arbitrary way either: it remains relatively confined to a general area, without ever revisiting the same point in three-dimensional space. Mathematicians have termed this behavior *chaotic* [21].

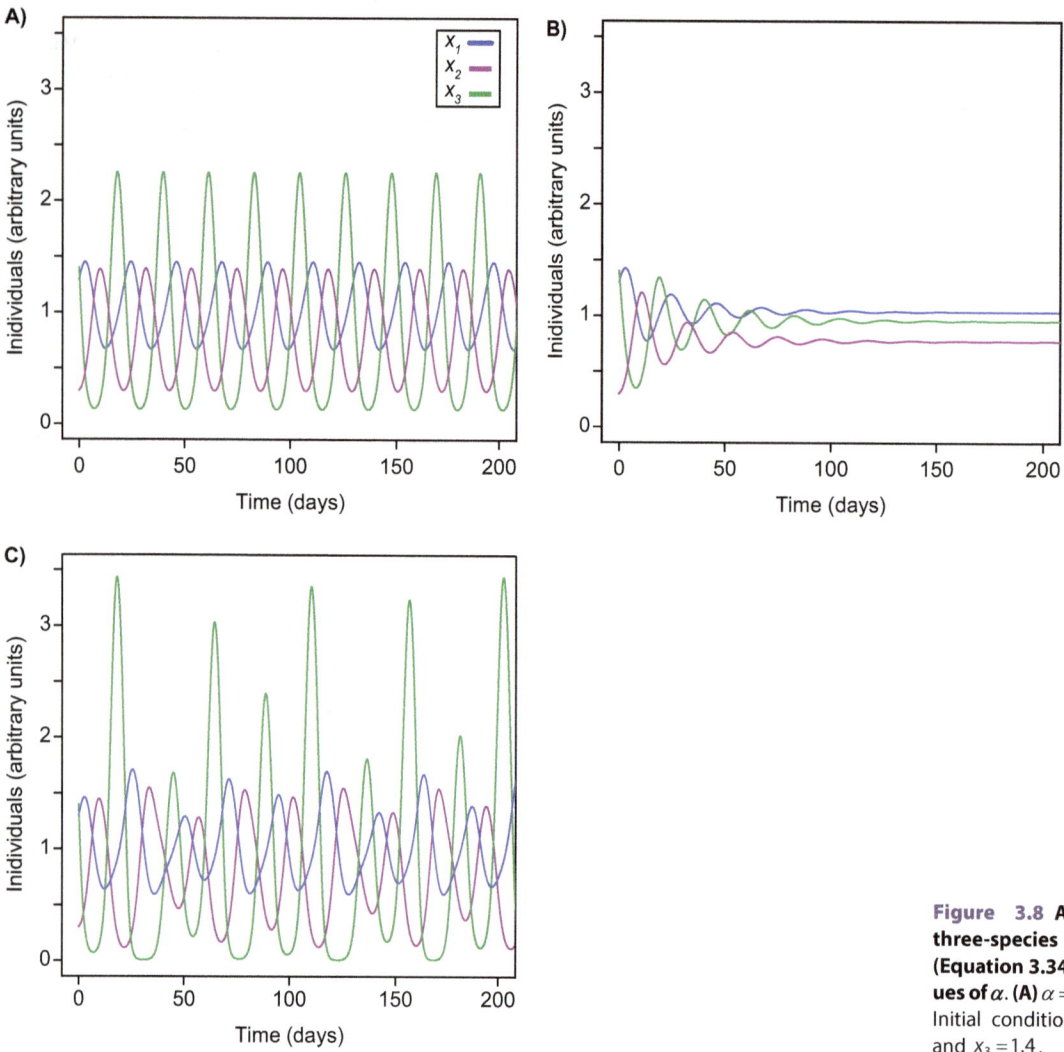

Figure 3.8 A time course of the three-species Lotka–Volterra model (Equation 3.34) with three different values of α. **(A)** $\alpha = 1.2$. **(B)** $\alpha = 0.75$. **(C)** $\alpha = 1.5$. Initial conditions were $x_1 = 1.3$, $x_2 = 0.3$, and $x_3 = 1.4$.

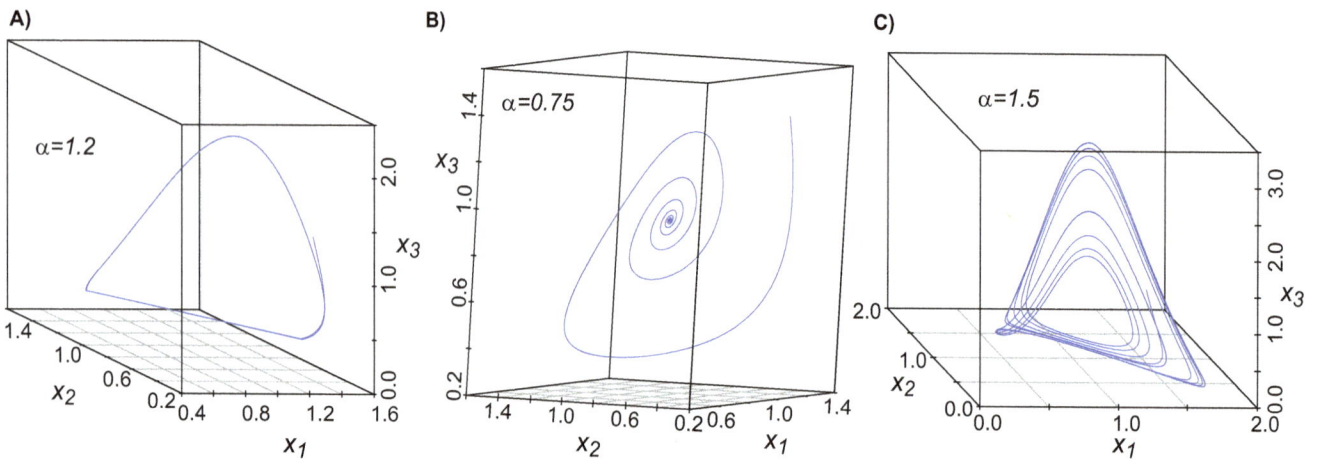

Figure 3.9 Phase plots of the three-species Lotka–Volterra model for three values of the parameter α. (See Equation 3.34). **(A)** $\alpha = 1.2$. **(B)** $\alpha = 0.75$. **(C)** $\alpha = 1.5$. Initial conditions were $x_1 = 1.3$, $x_2 = 0.3$, and $x_3 = 1.4$. An animated version of panel C is available at: http://qbio.statgen.ncsu.edu/LotkaVolterra.html.

The first point to recognize about a chaotic system is that it is not a *random* system: if the initial conditions are known precisely and one has access to a computing machine of infinite precision, then the state of the system at any time point can be computed exactly. We will return to the problematic nature of the assumptions about initial conditions and computation in a moment. But to finish our description of chaos, notice two other very interesting features. First, while the system's state may approach a point it passed through previously, it will never pass *exactly* through a prior point. Hence, the system is not just one with a long and complex period. This rule also explains why we saw chaos in a three-dimensional model but not in a two-dimensional one. In two dimensions, there is no way to draw a curve that continuously revisits the same neighborhood of points but never crosses itself, since crossing itself means to revisit exactly the same point [21]. However, in three dimensions, such noncrossing paths do exist. Second, though the system does not repeat itself, it also does not take on arbitrarily large values. Systems such as the one displayed in Figure 3.9C are called *strange attractors* because they appear bounded but nonrepeating [22].

Chaos in Ecological Systems

One might ask whether real systems of interacting species display the chaotic behaviors shown in Figure 3.9C. This question is deceptively hard to answer, as real ecosystems are also subject to many extrinsic variations (weather, seasons, human activity) that might disrupt periodic patterns of the type shown in Figure 3.4. However, researchers have created experimental communities where the external conditions are kept constant and found that those communities do display several characteristics of chaotic systems [23]. For instance, models of them give good predictions for periods of a few weeks but not over longer periods of time. Analyses of one such system have also suggested that the chaotic dynamics are driven more by resource competition between the prey species than by the actions of predators [24]. Notice that all these ideas require the *comparison* of observation to models: in the absence of such models, if one observed an empirical pattern like that in Figure 3.8C, it would be difficult to describe it without recourse to terms like "random."

Chaos and the Mathematics of the Real World

The existence of dynamical systems with chaotic behavior was unsuspected by mathematicians until the second half of the 20th century [21, 22], and their presence has some very interesting implications for how science models the world. One of the most important of these is how predictable such systems are. As mentioned, under the assumption of exact knowledge of initial conditions and arbitrary precision computing [25], chaotic systems are deterministic. However, real measurement instruments have limited accuracy, and real computers have limited precision (for instance, we cannot perfectly represent one-third using the 1s and 0s of a computer; see Chapter 10). Hence, as hinted by Figure 3.7, the existence of chaos means that there are biological phenomena for which, even if we can develop high-quality mathematical models, those models will not give us the ability to predict system states at arbitrary points in the future.

That, at least, is the pessimistic view. Perhaps the more optimistic view is that chaos provides a mathematical window into the great majority of things in the world that we are interested in that are *not* periodic or constant. Animal and plant populations, weather, prices in an economy, and, as we will see in Chapter 7, compounds in a cell exhibit behaviors that are somewhere in between "simple" (e.g., periodic or constant) and truly random and unpredictable. Chaos might be a useful metaphor for such systems.

REFLECTIONS AND PREVIEWS

This chapter presented the basics of constructing a differential equation model of a biological system. Although the model we developed was, in some sense, too simple to be a useful predictor of real biological systems, it still provided several insights. To start with, we encountered our first cycling, or *periodic*, system. We will consider a second oscillating system, the cell, in the next chapter. We also found an approach to dealing with a model with excess parameters when we nondimensionalized the Lotka–Volterra model. We saw with the phase-plane plots that visualization can be tremendously helpful in understanding biological systems, so we should always be on the lookout for new and improved visualization approaches. Finally, we found another emergent property when we extended our system to three dimensions: *chaotic* dynamics.

In the next chapter we will apply these ecological models of competition among organisms to one of the most fundamental problems in biology: evolution by natural selection. Having done so, we will then continue our search for a larger order inside an apparently simple system in Chapter 5. There we will consider the behavior of one of the smallest parts of a living thing: the metabolism of a cell.

Exercise 3.4

Solve the Lotka–Volterra model for the parameter values and initial conditions of Figure 3.4A.

```
• library(deSolve)
• a = 0.15
• b = 0.5
• c = b/8
• d = 0.04
• parameters<-c(a,b,c,d)
• state<-c(R = 1.0, F = 0.1)
• LV_model<-function(t,state,paramters)
  {with(as.list(c(state,parameters)), {
• dR <- a*R+-b*R*F
• dF <- c*R*F -d*F
• list(c(dR,dF))})}
• times <- seq(0, 500, by = 0.01)
• out <- ode(y=state, times=times, func=LV_model,
  parms = parameters)
• plot(out[,"time"], out[,"R"], type="l", xlab="-
  time", ylab="Individuals", col="green",
  xlim=c(0,500),ylim=c(0,4))
• points(out[,"time"], out[,"F"], type="l", xlab="-
  time", ylab="Individuals", col="red")
• legend("topright",legend=c("Rabbits", "Foxes"),
  fill=c("green", "red"), col=c("green", "red"))
```

Exercise 3.5

Solve the nondimensionalized Lotka–Volterra model from Equation 3.33 with $\alpha > 1.0$ and at least three different initial conditions and similarly for $\alpha < 1.0$. Make a phase plot of the results for each value of α. Example R commands:

```
• library(deSolve)
• parameters<-c(alpha=1.0)
• LV_nd_model<-function(t,state,paramters)
  {with(as.list(c(state,parameters)), {
• dr <- r - r*f
• df <- alpha*f*r-alpha*f
• list(c(dr,df))})}
• times <- seq(0, 200, by = 0.01)
```

```
• state<-c(r=0.3,f=0.3)
• out <- ode(y=state, times=times, func=LV_nd_model,
  parms = parameters)
• plot(out[,"r"], out[,"f"], type="l", xlab="Rab-
  bits", ylab="Foxes", col="blue",
  xlim=c(0,3.2),ylim=c(0,3.2))
• state<-c(r=0.5,f=0.5)
• out2 <- ode(y=state, times=times, func=LV_nd_model,
  parms = parameters)
• points(out2[,"r"], out2[,"f"], type="l",
  xlab="Rabbits", ylab="Foxes", col="blue")
• state<-c(r=0.7,f=0.7)
• out3 <- ode(y=state, times=times, func=LV_nd_model,
  parms = parameters)
• points(out3[,"r"], out3[,"f"], type="l",
  xlab="Rabbits", ylab="Foxes", col="blue")
• state<-c(r=0.9,f=0.9)
• out4 <- ode(y=state, times=times, func=LV_nd_model,
  parms = parameters)
• arrows(out[2,"r"], out-
  [2,"f"],out[5,"r"],out[5,"f"], length=0.15,
  col="blue")
• arrows(out[150,"r"], out[150,"f"],out[153,"r"],out[
  153,"f"], length=0.15, col="blue")
```

REFERENCES

1. Rosenblueth A & Wiener N (1945) The role of models in science. *Philosophy of Science* 12(4):316–321.
2. Box GE (1976) Science and statistics. *Journal of the American Statistical Association* 71(356):791–799.
3. Dirac PAM (1929) Quantum mechanics of many-electron systems. *Proceedings of the Royal Society of London, Series A* 123:714–733.
4. Feynman RP, Leighton RB, & Sands M (2011) *The Feynman lectures on physics, Vol. I: The new millennium edition: mainly mechanics, radiation, and heat* (Basic Books).
5. The International HapMap Consortium (2003) The International HapMap Project. *Nature* 426(6968):789–796.
6. Charlesworth B (2009) Effective population size and patterns of molecular evolution and variation. *Nature Reviews Genetics* 10(3):195–205.
7. Hartl DL & Clark AG (1997) *Principles of population genetics*. 3rd Edition (Sinauer Associates, Sunderland MA).
8. NIAID Visual and Medical Arts (9/16/2024) Hare Silhouette. in *NIAID BIOART Source*. https://bioart.niaid.nih.gov/bioart/195
9. NIAID Visual and Medical Arts (9/5/2024) Fox Silhouette. in *NIAID BIOART Source*. https://bioart.niaid.nih.gov/bioart/164
10. NIAID Visual and Medical Arts (9/10/2024) Grass. in *NIAID BIOART Source*. https://bioart.niaid.nih.gov/bioart/181
11. Lotka AJ (1920) Undamped oscillations derived from the law of mass action. *Journal of the American Chemical Society* 42(8):1595–1599.
12. Volterra V (1926) Variazioni e fluttuazioni del numero d'individui in specie animali conviventi. *Memoria della Reale Accademia Nazionale dei Lincei* 2:31–113.
13. Liang ST, Ehrenberg M, Dennis P, & Bremer H (1999) Decay of rplN and lacZ mRNA in Escherichia coli. *Journal of Molecular Biology* 288(4):521–538.
14. Chandler M, Bird R, & Caro L (1975) The replication time of the Escherichia coli K12 chromosome as a function of cell doubling time. *Journal of Molecular Biology* 94(1):127–132.
15. Boyce WE & DiPrima RC (1992) *Elementary differential equations and boundary value problems* (John Wiley and Sons, New York), p. 680.
16. Wilkinson T (2013) *The rise and fall of ancient Egypt* (Random House Trade Paperbacks).
17. Biggs F & Killeen R (2021) *A pocket guide to newgrange and the boyne valley* (Gil Books, Dublin, Ireland), p. 256.
18. Lotka AJ (1956) *Elements of mathematical biology* (Dover Publications, New York).
19. Murray JD (2001) *Mathematical biology, I: An introduction*. 3rd Edition (Springer, New York).
20. Arneodo A, Coullet P, & Tresser C (1980) Occurence of strange attractors in three-dimensional Volterra equations. *Physics Letters A* 79(4):259–263.
21. Flake GW (1998) *The computational beauty of nature: computer explorations of fractals, chaos, complex systems, and adaptation* (The MIT Press, Cambridge, MA).
22. Gleick J (1988) *Chaos: making a new science* (Penguin Books, New York), p. 352.
23. Benincà E, Huisman J, Heerkloss R, Jöhnk KD, Branco P, Van Nes EH, Scheffer M, & Ellner SP (2008) Chaos in a long-term experiment with a plankton community. *Nature* 451(7180): 822–825.
24. Benincà E, Jöhnk KD, Heerkloss R, & Huisman J (2009) Coupled predator–prey oscillations in a chaotic food web. *Ecology Letters* 12(12):1367–1378.
25. Ghazi KR, Lefèvre V, Théveny P, & Zimmermann P (2010) Why and how to use arbitrary precision. *Computing in Science & Engineering* 12(1–3):5–5.

Survival of the Fastest

Modeling Competition between Species and between Cells

4

"The old order changeth, yielding place to new,
And God fulfils himself in many ways,
Lest one good custom should corrupt the world"
—Alfred, Lord Tennyson, Idylls of the King

NATURAL SELECTION: ANOTHER USE OF THE PREDATOR–PREY MODEL

When we explored the Lotka–Volterra model in the previous chapter, we noted that the rabbit population would grow without limit in the absence of foxes. Since we were only interested in cases where the number of foxes was not zero, this weakness did not cause us any particular difficulties. However, there are versions of the predator–prey model that surmount this problem [1], and in this chapter we are going to use one of them. Our interest in this new model will not be for its value in understanding predator–prey dynamics. Instead, it will illuminate two other areas of biology that at first blush seem unrelated both to each other and to predator–prey dynamics: evolutionary biology and the treatment of cancer.

Carrying Capacity

To solve the problem of infinite population growth, we will add a *carrying capacity* to the model. This capacity is simply a constant K that describes the largest possible number of individuals that can be supported by that environment. When the population is much smaller than K, it will experience effectively the same exponential growth that we saw in the last chapter. However, as the population approaches its maximum size, that growth rate will fall to zero.

Developmental Stages

For reasons that will not be immediately obvious, we are also going to introduce another elaboration to the rabbit model of the previous chapter. We will split the rabbits into two *developmental stages*: immature rabbits (R_y) that have not yet reached reproductive maturity and mature rabbits (R) that can reproduce. The young rabbits will grow out of their immaturity at a rate g: however, depending on how close the population is to the carrying capacity, not all of these young rabbits will become reproductive adults. We will assume that the units of g are 1/years, so that $g = 0.5$ implies an immature period of 2 years. Just as with the foxes in the previous chapter, g then will act as a decay parameter, giving something like the time it takes for young rabbit to grow into an adult. The adult rabbits will also have a finite lifespan, with a death rate d describing their half-life.

DOI: 10.1201/9781003687504-4

Finally, we will have a reproductive rate r: the rate at which mature rabbits are giving birth to young ones. Putting all these parameters and variables together gives the following pair of differential equations:

$$\frac{dR_y}{dt} = -g \cdot R_y + r \cdot R$$

$$\frac{dR}{dt} = g \cdot R_y \cdot \left(1 - \frac{R}{K}\right) - d \cdot R$$

(4.1)

The $1 - R/K$ term in Equation 4.1 gives the chance of a young rabbit reaching sexual maturity at differing population levels: it approaches 1 for small population sizes and is near 0 as we get close to the carrying capacity.

We could solve Equation 4.1 immediately, but it is actually more instructive to ask first if it has any steady-state solutions. Recall from the previous chapter that we can find such solutions by setting the (known) derivatives of our (unknown) state functions to zero. In the terms of Equation 4.1, we have therefore $dR_y / dt = 0$ and $dR / dt = 0$:

$$0 = -g \cdot R_y + r \cdot R$$

$$0 = g \cdot R_y \cdot \left(1 - \frac{R}{K}\right) - d \cdot R$$

(4.2)

Using the first equation, we can solve for R, finding:

$$R = \frac{g}{r} \cdot R_y$$

(4.3)

Replacing R in the second equation of Equation 4.2, we find:

$$0 = g \cdot R_y \cdot \left(1 - \frac{1}{K} \cdot \frac{g}{r} \cdot R_y\right) - d \cdot \frac{g}{r} \cdot R_y$$

(4.4)

If we place all of the terms on the right side over a common $1 / K \cdot r$, we have:

$$0 = \frac{K \cdot r\left(g \cdot R_y\right) - g^2 \cdot R_y{}^2 - dKg \cdot R_y}{K \cdot r}$$

(4.5)

We can pull out the common $g \cdot R_y / K \cdot r$ term to get:

$$0 = R_y \frac{g}{r \cdot K}\left(r \cdot K - g \cdot R_y - d \cdot K\right)$$

(4.6)

One solution to this equation is obviously $R_y = 0$, which also implies $R = 0$. However, if we assume $R_y \neq 0$, we can solve for the expression in the paratheses:

$$R_y = \frac{K}{g}\left(r - d\right)$$

(4.7)

Which, using Equation 4.3, gives a value of R of:

$$R = \frac{K}{r}\left(r - d\right)$$

(4.8)

The first thing we should notice in Equation 4.8 is that if $r < d$, then R is negative, which is not biologically reasonable. On reflection, this relationship is sensible: if the reproduction rate is less than the death rate, the population will decay to 0. On the other hand, if $r \gg d$, then Equation 4.8 and the steady-state number of rabbits approaches K. In other words, the population is producing new offspring much faster than individuals are dying, and the population approaches the carrying capacity K.

General Solutions of Rabbit Growth with Carrying Capacity

Moving beyond the steady-state solution, we can return to the general model of Equation 4.1. **Figure 4.1** shows solutions to these equations, using the initial conditions and parameter values of Exercise 4.1. A few points are worth noticing. As the reproductive rate increases, the steady-state population approaches K. However, this point is partly a result of the way we have represented birth and death in the model and will not generalize to other forms of the model. We could in fact make another version of the model where the carrying capacity is modeled in a more intuitive way, but for the purposes of this chapter, the form of Equation 4.1 will be sufficient.

Exercise 4.1

Using R, solve Equation 4.1 for an initial condition of $R_y = 5$, $R = 50$, $K = 500$, $g = 0.5$, $d = 0.1$, and $r = 0.3$. Repeat your analysis with the same parameters except that $r = 0.6$. Plot your results for both cases. Example R commands:

```
• times <- seq(0, 60, by = 0.1)
• g=0.5
• K=500
• d=0.1
• r=0.3
• SingModel<-function(t,state,paramters)
  {with(as.list(c(state,parameters)), {
  o dRy <- -g*Ry +r*R
  o dR <- g*Ry*(1-R/K) - d *R
  o list(c(dRy,dR))})}
• state<-c(Ry=5, R=50)
• out <- ode(y=state, times=times, func=SingModel,
  parms = parameters)
• plot(out[,"time"], out[,"R"], type="l", xlab="-
  time", ylab="Individuals", col="green",
  xlim=c(0,60),ylim=c(0,500))
• points(out[,"time"], out[,"Ry"], type="l",
  xlab="time", ylab="Individuals", col="purple")
```

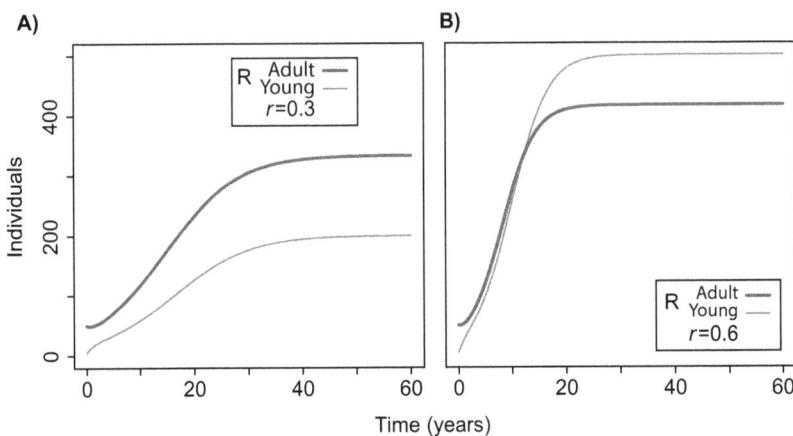

A) **B)**

Figure 4.1 **Comparing populations that differ in their reproductive rates r (A versus B) but otherwise have identical growth rates as they approach the carrying capacity of their environment (Equation 4.1).** Following Exercise 4.1, the initial conditions are $R_y = 5$ (young) and $R = 50$ (adult), and parameter values are $K = 500$, $g = 0.5$, $d = 0.1$. For A, $r = 0.3$, and for B, $r = 0.6$.

More interesting is the behavior of R_y: as r increases, the population of young rabbits exceeds that of adults, implying that, near the carrying capacity, most of these young rabbits do not survive to become reproductive. And finally, of course, as r increases, the rate at which R approaches the carrying capacity increases.

Two Population Models

The previous discussion probably strikes you as a trifle abbreviated: this is because we are not all that interested in the behavior of a single population or in the meaning of the various model parameters. Instead, we want to use this model to consider two populations that are in competition with each other in an environment. We will assume that these two populations occupy identical ecological niches, such that the carrying capacity of the environment is dictated by the sum of the two populations. We can therefore elaborate on Equation 4.1 to write:

$$\frac{dR1_y}{dt} = -g \cdot R1_y + r_1 \cdot R1$$

$$\frac{dR2_y}{dt} = -g \cdot R2_y + r_2 \cdot R2$$

$$\frac{dR1}{dt} = g \cdot R1_y \cdot \left(1 - \frac{(R1+R2)}{K}\right) - d \cdot R1 \tag{4.9}$$

$$\frac{dR2}{dt} = g \cdot R2_y \cdot \left(1 - \frac{(R1+R2)}{K}\right) - d \cdot R2$$

We have assumed that the population dynamics of the two populations are identical except that we will have $r_2 > r_1$: population 2 reproduces more effectively. We can solve Equation 4.9 for similar initial conditions and parameters we used in Figure 4.1. Doing so produces **Figure 4.2** (Exercise 4.2).

Exercise 4.2

Using R, solve Equation 4.9 for an initial condition of $R1_y = R2_y = 5$, $R1 = R2 = 50$, $K = 500$, $g = 0.5$, $d = 0.1$, $r_1 = 0.3$, and $r_2 = 0.303$. Repeat your analysis with the same parameters except that $r_2 = 0.33$. Plot your results for both cases. Example R commands:

```
• times <- seq(0, 300, by = 0.1)
• state<-c(R1y=5, R2y=5, R1=50, R2=50)
• g=0.5
• K=500
• d=0.1
• r1=0.3
• r2=0.303
• CompYModel<-function(t,state,paramters)
  {with(as.list(c(state,parameters)), {
  o dR1y <- -g*R1y +r1*R1
  o dR2y<- -g*R2y + r2*R2
  o dR1 <- g*R1y*(1-((R1+R2)/K)) -d*R1
  o dR2 <- g*R2y*(1-((R1+R2)/K)) -d*R2
  o list(c(dR1y, dR2y,dR1,dR2))})}
• out <- ode(y=state, times=times, func=CompYModel,
  parms = parameters)
• plot(out[,"time"], out[,"R1"], type="l", xlab="-
  time", ylab="Individuals", col="green",
  xlim=c(0,300),ylim=c(0,400))
• points(out[,"time"], out[,"R2"], type="l",
  xlab="time", ylab="Individuals", col="blue")
```

```
• points(out[,"time"], out[,"R1y"], type="l",
  xlab="time", ylab="Individuals", col="purple")
• points(out[,"time"], out[,"R2y"], type="l",
  xlab="time", ylab="Individuals", col="red")
```

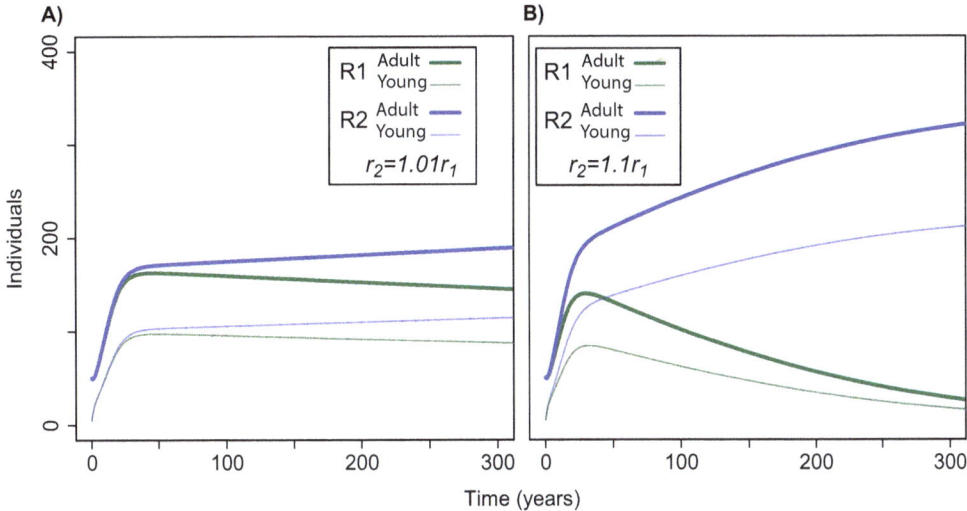

Figure 4.2 Competition between two populations with identical demographics save for a difference in reproductive rates r_1 and r_2 (Equation 4.9). Following Exercise 4.2, the initial conditions are $R1_y = R2_y = 5$ (young) and $R1 = R2 = 50$ (adult). The parameter values are $K = 500$, $g = 0.5$, $d = 0.1$. For **A**, $r_1 = 0.3$ and $r_2 = 0.303$. For **B**, $r_1 = 0.3$ and $r_2 = 0.33$.

Perhaps the key message from Figure 4.2 is that even a very small growth advantage (r_2 is 1% larger than r_1 in Figure 4.2A) will drive population R1 to extinction in a relatively few years. What we have just (re)discovered is Darwin's concept of natural selection mediated by the "struggle for existence." Two quotations from *The Origin of Species* illustrate this relationship. In the first, Darwin describes the concept of the exponential growth of populations we modeled in the preceding chapter. In particular, he emphasizes how such growth will always overwhelm any possible environment:

> A struggle for existence inevitably follows from the high rate at which all organic beings tend to increase. Every being, which during its natural lifetime produces several eggs or seeds, must suffer destruction during some period of its life, and during some season or occasional year, otherwise, on the principle of geometrical increase, its numbers would quickly become so inordinately great that no country could support the product.
>
> Darwin, *The Origin of Species*, Chapter III [2]

In other words, the exponential growth of a population, no matter how small the exponential parameter (r in Equation 4.1), will eventually overwhelm any possible environment. Darwin goes on to consider what effects this struggle would produce if one assumes that there are differences among individuals or populations in their ability to successfully reproduce:

> [Can] we doubt (remembering that many more individuals are born than can possibly survive) that individuals having any advantage, however slight, over others, would have the best chance of surviving and procreating their kind? On the other hand, we may feel sure that any variation in the least degree injurious would be rigidly destroyed.
>
> Darwin, *The Origin of Species*, Chapter IV [2]

Arguably, Figure 4.2 represents a mathematical synthesis of these two ideas. As $R1 + R2$ approaches K, the young individuals in $R1_y$ and $R2_y$ are increasingly less likely to survive to reproduce, corresponding to Darwin's struggle for existence. Meanwhile, even when r_2 is only very slightly larger than r_1, R1 will be driven to extinction.

I do not mean to suggest that Equation 4.9 is the only mathematical model consistent with Darwin's verbal model, or even that it is a particularly good model. Rather, I think it instructive that even our relatively limited exposure to models in the past three chapters has allowed us to pose a model capturing one of the most important ideas in biology.

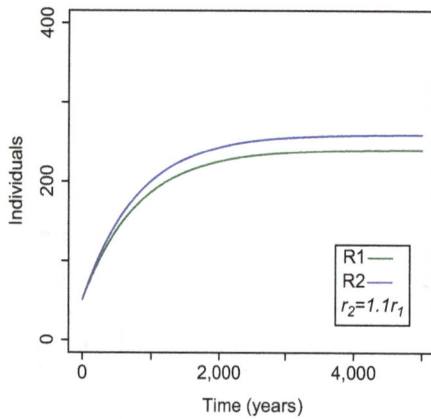

Figure 4.3 Competition between two populations with identical demographics save for a difference in reproductive rates and no developmental process (Equation 4.10). The initial conditions were $R1 = R2 = 50$; we set $r_1 = 0.3$ and $r_2 = 0.33$. Compare to Figure 4.2B.

Invalid Models

Before we discuss this model further, however, it is instructive to consider a *bad* potential model. That model is a simplified version of Equation 4.9 where we omit the youth stage:

$$\frac{dR1}{dt} = r_1 \cdot \left(1 - \frac{(R1 + R2)}{K}\right)$$

$$\frac{dR2}{dt} = r_2 \cdot \left(1 - \frac{(R1 + R2)}{K}\right)$$

(4.10)

When we solve this model using a 10% growth advantage for population $R2$ (**Figure 4.3**), we see only a slight advantage of $R2$ over $R1$ in the final conditions, quite different from what is seen with the 1% advantage in Figure 4.2A. Moreover, even this slight difference takes many more years to accumulate than that seen in Figure 4.2A. Why is this?

What we have (inadvertently) created in Equation 4.10 is a model of populations where the individuals are infinitely long-lived. In such a model, as the sum of the two populations approach K, reproduction effectively ceases. As a result, the differences in size between the two populations is only what could accumulate during the growth phase where $R1 + R2 \ll K$. It was this lack of dynamics in births and deaths that led us to include the $R1_y$ and $R2_y$ states in the original model.

I have shown the results from Equation 4.10 to caution us that mathematical models are not an invariable improvement on verbal ones. The two earlier quotations from Darwin provide a better understanding of how populations evolve in time than does Equation 4.10; equations are not a substitute for a proper understanding of your system.

A MODEL OF NATURAL SELECTION: ADVANTAGES AND PROBLEMS

Although the models we have just developed are not too far from the ideas first laid out by Darwin, we now are going to drop the topic of modeling evolution directly for several chapters. The reason, as you may have already guessed, is that there are two serious gaps in these models. These gaps are a lack of grounding in the molecular nature of heredity and the neglect of chance or random effects in the evolutionary process [3, 4]. While we do not yet have the tools to add random effects to our models, we can at least consider how we might think about the differences between the various possible models of the evolutionary process.

We have already said that the models in Equations 4.9 and 4.10 represent different possible mathematical representations of Darwin's essentially verbal model. As mentioned, however, they are not *only* models that might do so. As an obvious example, we represented the advantage in one of the two populations as being due to an increased growth rate $(r_2 > r_1)$. However, it would arguably be closer to Darwin's thinking to have r_2 equal to r_1 and instead model the advantage of the $R2$ population as having greater rate of survival into adulthood. In that view, we would write that $g_2 > g_1$.

This problem of determining whether a mathematical model does or does not exactly reflect a verbal model is a very complicated one. The presence or absence of chance effects or the use of a mathematical parameter with the same name but differing underlying meanings can make it difficult to tell if the model actually reflects, say, the original idea Darwin had in mind in his verbal explanations [5, 6].

In this book, fortunately, we are concerned with a slightly easier problem: comparing different models that are all expressed in some quantitative form. In the previous chapter we saw another such model of evolution:

$$Q = \frac{N_e s}{N} \cdot \frac{1}{1 - e^{-2N_e s}} \tag{4.11}$$

Here, Q is the probability that a new *mutation* that confers an advantage $(s > 0)$ or a disadvantage $(s < 0)$ will replace the original version of a gene in a population [7]. Adding to our discussion in the last chapter, we can mention a few more details about Equation 4.11. First, our population is not infinite in size $(N < \infty)$. Second, effects such as unequal sex ratios and inbreeding reduce the apparent, or *effective*, population size even further $(N_e \leq N)$ [4]. Finally, we express the selective advantage s in terms of the proportional excess of offspring that the carriers of the new mutation have relative to the original individuals. Since we assume that N is constant, this assumption places an upper bound on s [8].

Are the models in Equations 4.9 and 4.11 equivalent under any circumstances? The first obvious difference is that Equation 4.11 has terms for the population size: in fact, it predicts that in small populations some advantageous genetic changes $(s > 0)$ will not replace an original genetic variant while some disadvantageous ones $(s < 0)$ will. The model in Equation 4.9 does not include a term for population size. Therefore, we might assume at first that our model matches that of Equation 4.11 when the population size is infinitely large. However, there is a more important and subtle difference between the two. Our model compares two populations that differ in growth rate and do *not* interbreed with each other. Equation 4.11, on the other hand, is considering a new gene in an existing, interbreeding population. Hence, the models are actually considering different processes: the fate of populations versus the fate of genetic variants in a single population.

Should we then discard the model in Equation 4.9? Remembering a variant of Box's adage from the beginning of the last chapter ("All models are wrong ... [but some are useful]"), we should perhaps pause for a moment. The model in Equation 4.11 applies well to species like humans, where new genetic variations appear in an existing interbreeding population and where reproduction is necessarily coupled to sexual exchanges of genetic material. But what about a clonal bacterial population? In fact, bacterial species are generally not truly clonal [9], but for a large growth advantage, a local population might behave in a manner like that of Figure 4.2B. The reason is that the rate of spread of a new genetic variant by DNA exchanges between bacterial individuals is much slower than the rate at which the faster-growing population replaces the slower-growing one.

Now, perhaps you find this proposed application of Equation 4.9 to bacteria too glib. Is there a situation where it clearly does apply? In fact, we are about to find an application of this model that is both interesting in itself *and* has practical value.

TREATMENT OF A SLOWLY GROWING TUMOR CONSISTING OF MULTIPLE CELL POPULATIONS

One of the exciting things about studying biology is when you see unexpected connections between very different parts of the field. One such example is using models of evolution by natural selection to describe the progression of a cancerous growth in an individual.

One of the key evolutionary transitions in Earth's history was the appearance of multicellularity [10]. Until its appearance, the fortunes of a genome and a cell were tightly linked: anything that allowed the cell to divide faster or survive longer provided a similar benefit to the genes that organism carried. But in a complex multicellular organism, the genetic success of the whole organism in producing offspring is decoupled from the success of individual cells [10]. In fact, the success of the organism in its reproduction *requires* that individual cells forgo the control of their own reproduction. Instead, they must obey a genetic program that tells them when to divide, when not to divide, and even when to terminate themselves [10]. If this idea seems surprising to you, consider that in humans, like all animals, every cell in our body is fated to die, except for a handful of eggs or sperm. And yet those few surviving cells determine the evolutionary success or failure of the genetic program we carry in our genomes. In this view, the genome of the organism can be thought of as propagating itself by using most of the cells in the organism as powerful yet disposable vehicles. The organism's genetic survival program requires individual cells to sacrifice themselves for the program's benefit.

Multicellularity thus sets the stage for the possibility of conflict between individual cells and the organism as a whole. We can loosely think of the realization of this conflict as being the disease of cancer. In this disease, individual cells lose the genetic and physiological controls that have constrained their division to occur only in those instances when it is advantageous for the organism as a whole [11]. Figure 4.4 shows the *cell cycle*: the stages by which a cell decides to divide (known as *mitosis*) and performs the events needed to accomplish that division. For our purposes, the most important of those phases is G_0: when the cell has entered a resting phase with no immediate plan to divide. Although our discussions of this cycle tend to focus on its cyclic nature, a large majority of cells in an adult human body are not cycling at any given time [12]. Cells such as the ones comprising muscles and nerves are actually *terminally differentiated*, meaning they are no longer capable of cell division under normal conditions [12, 13]. On the other hand, other types of cells, including lymphocytes and fibroblasts, can remain in a nondividing, quiescent state until signaled to divide by some external stimulus [12]. In either case, we can think of cancer as a disease whereby a group of cells are proceeding through the cell cycle outside the genetic and environmental controls the organism's genome has established on that division for the benefit of the organism as a whole [14].

The genetics and molecular biology of how these controls are lost in cancer are very complicated [11]. However, they are not of interest to us right now. Instead, let us consider the world from the perspective of a group of cancer cells. Such a population is going to be quite small relative to the entire organism. As a result, at least early in the cancer's progression, the tumor can act as a parasite on organism's ability to obtain nutrients from the environment and deliver those nutrients to its cells. Because the tumor is small, the host will effectively be able to nourish it without a noticeable impact on that host's overall energy budget. In the language of a predator–prey model, we might think of the cancer cells as existing in an ecological niche that is almost empty [14].

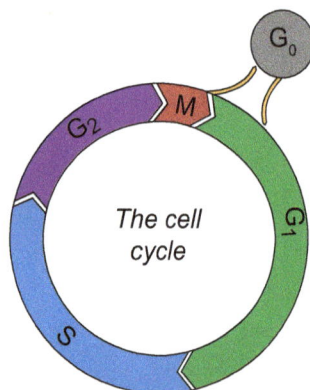

Figure 4.4 The cell cycle. Phases G_1 and G_2 are "gap" phases where the cell assesses its condition and decides whether or not to proceed with division. S phase is the DNA synthesis phase, and M is the mitosis event itself, where division occurs. Many cells are temporarily or permanently in G_0, meaning that they are neither dividing nor preparing to divide [12]. When a cell passes through this cycle despite external signals that it should not, that can be thought of as a cancerous behavior. (Adapted from Weinburg RA (2007) The Biology of Cancer. Garland Science, New York.)

Cancer Is an Evolutionary Disease

Now, it is probably the case that most of these cancerous cell populations derive from a single cell that has escaped control [15]. However, in such an open niche, that cell can start to reproduce. Some of the individuals in this growing population of cells will now experience mutations. And some of those mutations can alter the growth rate of their possessor. We thus have the situation where a tumor will start to behave like any other population whose individuals have variation in their growth or survival rates [15]. These arguments tell us that we need to think of cancer as an *evolutionary* disease. Once we have this realization, models like the one we created in Equation 4.9 begin to give us some important insights into the disease's progression [14].

Treating a Cancer with Susceptible and Resistant Cell Populations

To allow us to make simple models, we will consider a cancer that is reasonably slow growing. Certain forms of prostate cancer are examples of this type of disease [11]. We will further assume that we possess a drug D that can completely kill a set of *susceptible* tumor cells S. However, we will also assume the existence of a small population of *resistant* cells (R) for which the drug is totally ineffective [16]. While this resistance can take a variety of forms, it is very commonly seen after cancer treatment, especially when only a single drug is used [17].

A very important assumption about the model we are about to build is that the resistance we observe is costly. In this context, costly means that the cells possessing the resistance are at a disadvantage relative to the susceptible cells when no drug is present. This assumption of costly resistance is borne out in at least a few of the cases where the mechanism of cellular resistance to a drug is understood [18–21], and we will discuss one such mechanism in detail in Chapter 11. Here, we can just think of the costs as being due to factors such as a costly metabolic step that detoxifies the drug or the costly synthesis of a transporter to remove the drug from the cell.

Under the assumption of costly resistance, we can model the tumor as consisting of two competing populations of cells with dynamics similar to 4.9. In other words, in the absence of the drug, the resistant cells will have slower growth of $R1$ relative to $R2$ (Figure 4.2). However, in our tumor model we will not use the fixed carrying capacity of Equation 4.9. Instead, we will assume that the cancer is able to induce the host to feed it [11] in an amount proportional to the tumor size. In this framework, at any given moment, the carrying capacity for the tumor will be the sum of the susceptible and resistant cell populations, $S + R$.

The reason for structuring the model this way is that we will allow the tumor to grow indefinitely by inducing the host to provide nutrients proportionally. However, we also force the two populations into competition with each other for those nutrients. When the drug is not present, the S cells are more efficient than the R cells in collecting those nutrients and converting them to growth. We will represent this competitive advantage with a parameter a ($1 \leq a < \infty$). When $a = 1$, there is no advantage for S; the advantage will then be maximal as a approaches infinity. Mathematically, the parameter a will scale the proportion of the new cells born into the tumor that belong to S. We can write a system of equations as follows:

$$\frac{dS_y}{dt} = -g \cdot S_y + r \cdot (S+R) \cdot \left(\frac{S}{S+R} \right)^{1/a}$$

$$\frac{dR_y}{dt} = -g \cdot R_y + r \cdot (S+R) \cdot \left(1 - \left(\frac{S}{S+R} \right)^{1/a} \right)$$

$$\frac{dS}{dt} = g \cdot S_y - d \cdot S$$

$$\frac{dR}{dt} = g \cdot R_y - d \cdot R$$

$$(4.12)$$

These equations look a bit complex, but notice that the rightmost terms in dS_y / dt and dR_y / dt can be added as follows:

$$\left(\frac{S}{S+R}\right)^{1/a} + 1 - \left(\frac{S}{S+R}\right)^{1/a} = 1 \tag{4.13}$$

The interpretation of this equality is that we are allowing the tumor to grow at a rate r proportional to its total size $S + R$. We partition this growth between R and S via the parameter a, so that when $a = 1$, we have:

$$\frac{dS_y}{dt} = -g \cdot S_y + r \cdot (S+R) \cdot \left(\frac{S}{S+R}\right)$$
$$\frac{dR_y}{dt} = -g \cdot R_y + r \cdot (S+R) \cdot \left(1 - \left(\frac{S}{S+R}\right)\right) \tag{4.14}$$

which is just:

$$\frac{dS_y}{dt} = -g \cdot S_y + r \cdot (S+R) \cdot \left(\frac{S}{S+R}\right)$$
$$\frac{dR_y}{dt} = -g \cdot R_y + r \cdot (S+R) \cdot \left(\frac{R}{S+R}\right) \tag{4.15}$$

or:

$$\frac{dS_y}{dt} = -g \cdot S_y + r \cdot S$$
$$\frac{dR_y}{dt} = -g \cdot R_y + r \cdot R \tag{4.16}$$

In other words, when $a = 1$, we have two populations with identical reproduction rates. How would these two populations behave as the tumor grows? Assume that we have an initial population of susceptibles S_i and of resistants R_i. What we would find is the initial relative ratio of susceptibles to resistants (S_i / R_i) remains unchanged as the disease progresses. On the other hand, if we have $a > 1$, the R cells will be at a disadvantage relative to the S cells. This is because the total resources provided to the population $S + R$ will be preferentially used for growth by the S cells. In Exercise 4.3, you will solve Equation 4.12 for differing values of a: the results of this exercise are shown in **Figure 4.5**. As expected, increasing a increases the disadvantage to the R cells, causing them to be driven to extinction faster.

Exercise 4.3
Using R, solve Equation 4.12 for an initial condition of $S_y = R_y = 1$, $S = R = 50$, $a = 1.1$, $g = 1.0$, $d = 0.1$, and $r = 0.2$. Repeat your analysis with the same parameters except that $a = 1.5$. Plot your results for both cases. Example R commands:

- `library(deSolve)`
- `parameters<-c(a=1.1, r=0.2, g=1, d=0.1)`
- `state<-c(Sy=1, Ry=1, S=50, R=50)`
- `times <- seq(0, 200, by = 0.01)`
- `CancerModelND<-function(t,state,paramters)`
  `{with(as.list(c(state,parameters)), {`
- `dSy <- -g*Sy +r*(S+R)*(S/(S+R))^(1/a)`
- `dRy <- -g*Ry +r*(S+R)*(1.0-((S/(S+R))^(1/a)))`

```
• dS <- g*Sy -d*S
• dR <- g*Ry - d*R
• list(c(dSy, dRy,dS,dR))})}
• out <- ode(y=state, times=times, func=CancerModelND,
  parms = parameters)
• plot(out[,"time"], out[,"S"], type="l", xlab="time",
  ylab="CellMass", col="green", xlim=c(0,80),yli
  m=c(0,20000))
• points(out[,"time"], out[,"R"], type="l", xlab="-
  time", ylab="Individuals", col="red")
```

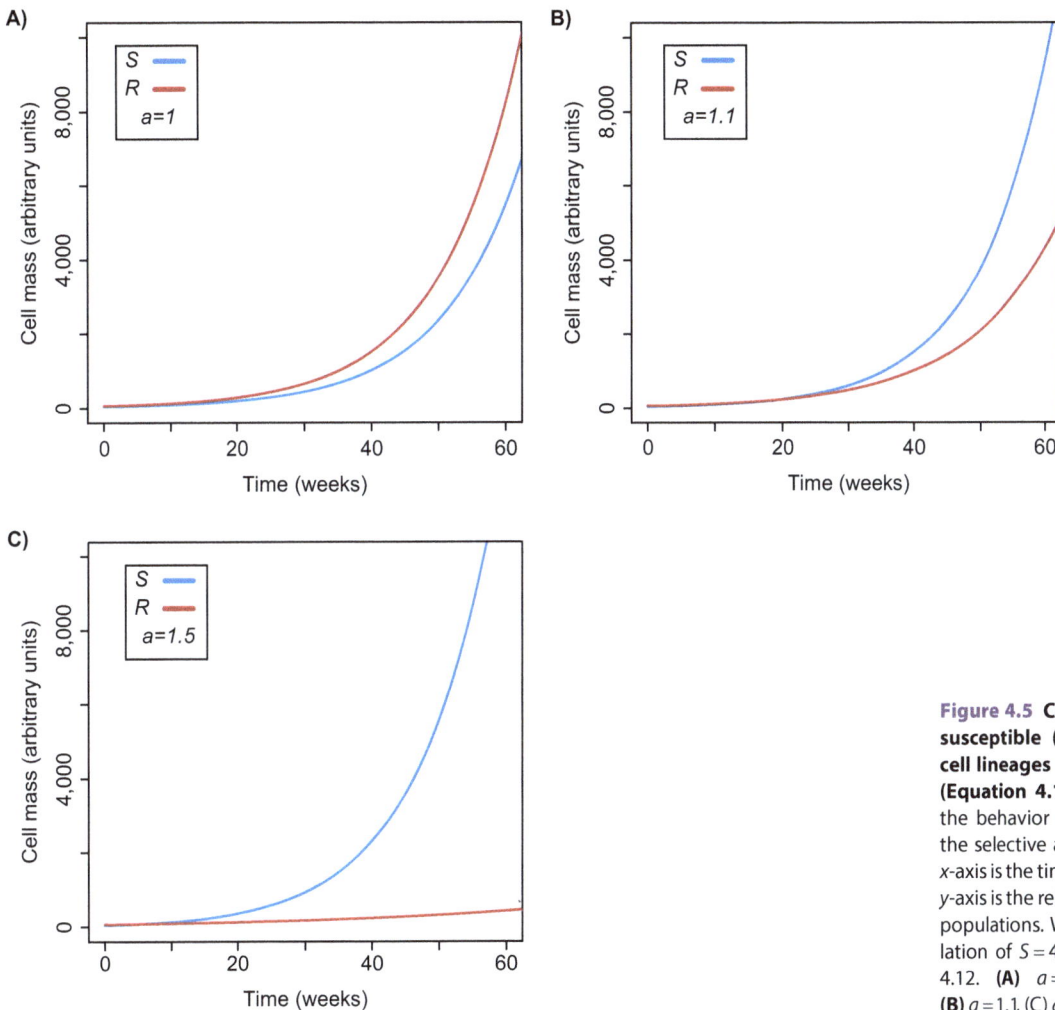

A)

B)

C)

Figure 4.5 **Competition between drug-susceptible (S) and drug-resistant (R) cell lineages in the absence of the drug (Equation 4.12).** The three panels show the behavior under different values of a: the selective advantage enjoyed by S. On x-axis is the time of tumor growth, while on y-axis is the relative mass of cells of the two populations. We assumed an initial population of $S = 40$ and $R = 60$. See Equation 4.12. **(A)** $a = 1.0$ (i.e., no advantage), **(B)** $a = 1.1$, (C) $a = 1.5$.

Treatment with a Drug That Acts Only on S

Now we will suppose that we have a drug D. When present, this drug will prevent any of the S_y cells from growing into adult S cells: in other words, it completely kills S cells but has no effect on R cells. We will assume that we will only start to give this drug when the tumor has grown by a factor of 10 relative to its initial size $(S + R = 1000)$. So, we have $D(0) = 0$ and $D(t) \to 1$ when $S + R = 1000$. We model the effect of the drug on S as a canceling term in dS / dt:

$$\frac{dS_y}{dt} = -g \cdot S_y + r \cdot (S+R) \cdot \left(\frac{S}{S+R}\right)^{1/a}$$

$$\frac{dR_y}{dt} = -g \cdot R_y + r \cdot (S+R) \cdot \left(1 - \left(\frac{S}{S+R}\right)^{1/a}\right)$$

(4.17)

$$\frac{dS}{dt} = g \cdot (1-D) \cdot S_y - d \cdot S$$

$$\frac{dR}{dt} = g \cdot R_y - d \cdot R$$

We can solve Equation 4.17 for an initial population of 90% S cells and 10% R cells, corresponding to a tumor with very low initial levels of resistance (Exercise 4.4). These solutions are shown in **Figure 4.6**, and they are somewhat dispiriting. While the drug successfully shrinks the tumor and keeps it very small for the better part of a year, eventually the resistant cells produce a tumor several times bigger than the original and for which there is no treatment with our drug.

Exercise 4.4

Solve Equation 4.17 for an initial condition of $S_y = 0.9$, $R_y = 0.1$, $S = 90$, $R = 10$, $a = 2.01$, $g = 1.0$, $d = 0.1$, and $r = 0.2$. You will need to use the root and event functions in R to apply the drug D once $S + R = 1000$:

```
• library(deSolve)
• kHigh=1000
• parameters<-c(a=2.01, r=0.2, g=1, d=0.1)
• state<-c(Sy=0.99, Ry=0.01, S=99, R=1, D=0)
• times <- seq(0, 200, by = 0.01)
• CancerModelD<-function(t,state,paramters)
  {with(as.list(c(state,parameters)), {
• dSy <- -g*Sy +r*(S+R)*(S/(S+R))^(1/a)
• dRy <- -g*Ry +r*(S+R)*(1.0-((S/(S+R))^(1/a)))
• dS <- g*(1-D)*Sy -d*S
• dR <- g*Ry - d*R
• dD <- 0
• list(c(dSy, dRy,dS,dR, dD))})}
• rootfunc <- function(t, state, parms) {
• yroot <- state[3] - kHigh
• return(yroot) }
• eventfunc <- function(t, state, params) {
• state[5] <- 1
• return(state) }
• out <- ode(y=state, times=times, func=CancerModelD,
  parms = parameters, rootfun = rootfunc,events =
  list(func = eventfunc, root = TRUE))
• plot(out)
```

Origins of the Resistant Cells

You might have wondered where these resistant cells came from and why they were waiting around for the drug treatment. In fact, the actual situation is more complicated, but cancer cells tend to be both quickly dividing (hence their cancerous nature) and subject to many mutations, both small and large [14, 22]. As a result, even were there *not* a preexisting population of R cells in our tumor, the drug treatment and continual appearance of new mutations would very likely drive the appearance of such cells before the drug killed all the tumor cells. Our model of a small initial population of R is hence a reasonable shortcut to modeling such mutations.

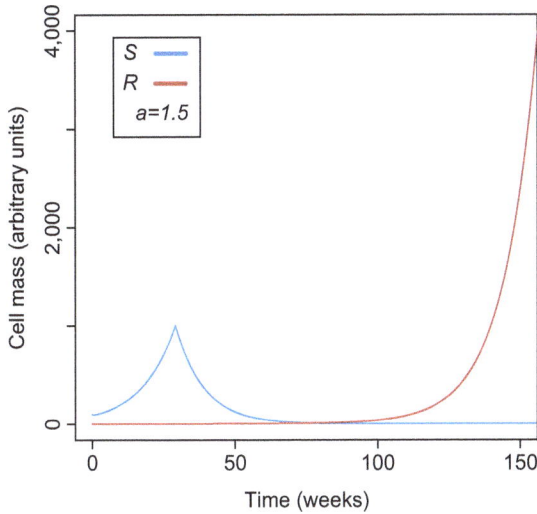

Figure 4.6 Treatment of a tumor with a highly effective drug against *S* that is ineffective for *R* (Equation 4.17). We initiate treatment when the total tumor mass $(S+R)=1000$. The *S* cells are quickly killed off, opening the environment to allow the *R* cells to proliferate and harm the patient. The *S* cells have an advantage $a=1.5$ over the *R* cells in the absence of the drug. See Exercise 4.4.

Exploiting Evolution for Cancer Treatment

Should we give up then? The drug we are considering here is, by any reasonable standard, excellent, killing almost all the tumor cells. The reason the treatment we just tried did not work is that we made a mistake, which is very natural for humans treating a disease. To correct it, we need to follow another path: *adaptive therapy*, where we exploit the evolutionary nature of the cancer rather than fighting it [16].

Our mistake was to try to achieve an absolute cure to the cancer, wiping it out completely with a single drug. The reason this choice was in error is the evolutionary nature of the disease: as soon as we subject the tumor to a chemical attack, any mutations that protect the cells from that attack will become evolutionarily beneficial and rise in frequency in the population, just as we see in Figure 4.6. In essence, we have tried to fight evolution, arguably the most powerful creative force on our planet.

What if we take a different approach? Specifically, we might ask whether our goal should be to destroy the tumor entirely or rather to control its size to limit its harmful effects. With control as our goal, we have two tools available to us. Rather than just the drug, we can add the competitive advantage of the susceptible cells to our drug toolkit. When we treat with the drug, those cells tend to die off, leaving *R* cells. But if we *stop* the drug treatment before all the *S* cells have died, those *S* cells should start to grow again. If their advantage over the *R* cells is then large enough, the tumor should again become dominated by *S* cells. When that tumor again grows too large, we can reintroduce the drug. To achieve this new treatment plan, we will use the same modeling equations from Equation 4.17 but with a more complex drug regime:

$$
\begin{array}{cc}
D=0 & D=1 \\
if\ (S(t)>k_{High}):D\rightarrow 1 & if\ (S(t)<k_{Low}):D\rightarrow 0 \\
else\ D\rightarrow 0 & else\ D\rightarrow 1
\end{array}
\tag{4.18}
$$

Thus, our new scheme is again to apply the drug treatment when the susceptible population reaches $k_{High}=1000$, but then to remove that treatment when *S* reaches k_{Low}. Exercise 4.5 allows you to analyze the system with these treatment rules. The results are shown in **Figure 4.7A, B**. With this new treatment, instead of a very rapid growth in *R*, we see that the periodic removal of the drug slows the growth in *R*, yielding the periodic waves of growth in Figure 4.7B. In this treatment regime, the resistant cells do increase in number, but rather slowly. If the cancer itself is also slow growing, such a damping of the resistant cell growth may be enough; we may prolong the time to the tumor becoming dangerous to such a degree that the patient will pass away from other causes before that happens.

Exercise 4.5

Solve Equation 4.17 for an initial condition of $S_y = 0.9$, $R_y = 0.1$, $S = 90$, $R = 10$, $a = 2.01$, $g = 1.0$, $d = 0.1$, and $r = 0.2$. See Exercise 4.4 for the base model commands. The root and event functions needed are given:

- ```kHigh=1000```
- ```kLow=200```
- ```parameters<-c(a=2.01, r=0.2, g=1, d=0.1)```
- ```state<-c(Sy=0.9, Ry=0.1, S=90, R=10, D=0)```
- ```times <- seq(0, 500, by = 0.01)```
- ```rootfunc <- function(t, state, parms) {```
- ```yroot <- c(state[3] - kLow, state[3] - kHigh)```
- ```return(yroot)```
- ```}```
- ```eventfunc <- function(t, state, params) {```
- ```state[5] <- if (abs(state[3]-kHigh)<1e-4) 1 else 0```
- ```return(state)}```
- ```out <- ode(y=state, times=times, func=CancerModelD,```
  ```parms = parameters, rootfun = rootfunc,events =```
  ```list(func = eventfunc, root = TRUE))```
- ```plot(out)```

But can we do even better than this? Releasing the S cells from the constraint of the drug allows them to outcompete the R cells and reduce the number of those R cells. It seems reasonable that one way to use this effect would be to apply the drug only when S cells are relatively abundant. In **Figure 4.7C, D** we increase our lower boundary for drug treatment to $k_{Low} = 400$. This higher boundary requires us to stop and start the drug treatment more often. When we do this, higher levels of the S cells that we start with allows them to outcompete the R cells. As a result, the overall number of cells of R declines with time. For at least this combination of model parameters, we can then keep the tumor in check indefinitely through periodic drug application. In fact, if we look more closely at Figure 4.7D, we might even have the chance to cure the disease entirely if we can cleverly apply the drug over intervals, keeping the R population in check while slowly reducing S until both are gone. That example is not worked out here but is an active area of research [23].

Implications of Adaptive Therapy

The examples in Figure 4.7 are of course slightly contrived in our choice of model parameters and assumptions, but the general message should not be lost. Cancer is an evolutionary disease: a conflict between the host and the cancer cells as well as among the cancer cells themselves. Treatments that we devise to fight this evolutionary paradigm, such as drugs for which resistance can evolve, may not work if the cancer has time to deploy evolution against them [23]. On the other hand, if we see our goal as the preservation of the host's health rather than the total destruction of the tumor, other treatment options may present themselves—treatments that complement, rather than oppose, the evolutionary forces at work.

REFLECTIONS AND PREVIEWS

In this chapter we saw again that one of the great powers of mathematical modeling is that models from one branch of biology can be used to give insights into others. With relatively minor modifications, our ecological Lotka–Volterra model could be converted into a model that reflects one of the most important discoveries in biology: evolution by natural selection. In what might be a further surprise, we could immediately use that evolutionary model to plan an improved form of cancer therapy. In the next chapter we will see another example of a common mathematical

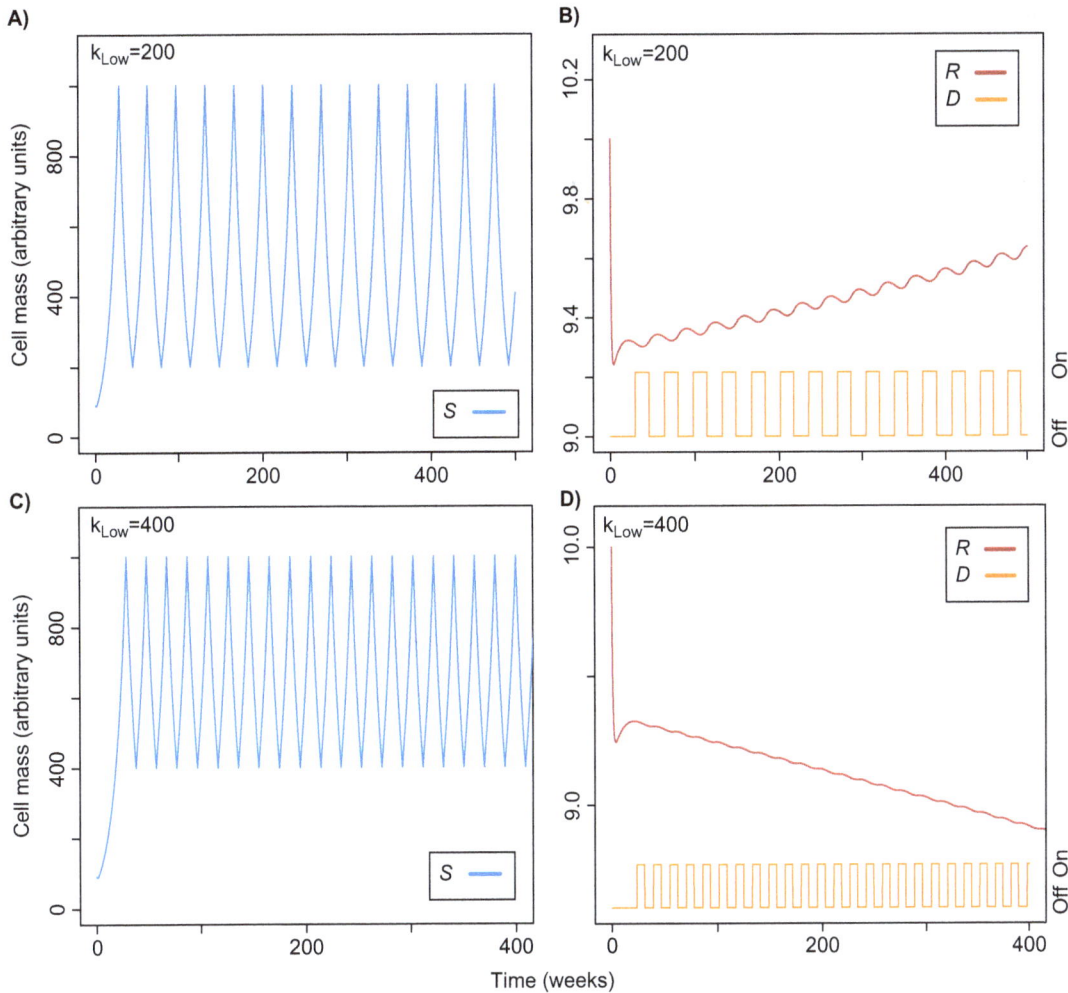

Figure 4.7 Treatment of a tumor through periodic applications of a drug against *S* that is ineffective for *R*. We initiate treatment when the total tumor mass $(S + R) = 1000$ and remove it when S drops below a threshold k_{Low} (200 in A/B and 400 in C/D). **(A)** The dynamics of S when $k_{Low} = 200$. On the x-axis is time in weeks and on the y-axis is tumor mass in arbitrary units. **(B)** The (slow) growth of the resistant cells in the regime of **A**. Shown in orange is the cyclic application of the drug (right axis). **(C)** As for **A** but with $k_{Low} = 400$. **(D)** As for **B** but with $k_{Low} = 400$. See Exercise 4.5.

framework being used in a completely new context. There we will start to delve into how life functions at a fundamental level, exploring the biochemistry of the cell. As we do so, one thing we will notice is the reappearance of some of the mathematical forms we used in Chapters 2–4. More interestingly, we will begin to address one of the key weaknesses of the evolutionary model from this chapter.

As we saw, our model did not consider the spread of a novel genetic variant in a single population, primarily because we had not yet considered that genes are inherited in a *particulate* fashion. In the next chapter we will find a deep and surprising link between the nature of biochemistry and one of the key discoveries that illustrated the particulate nature of inheritance.

REFERENCES

1. Murray JD (2001) *Mathematical biology, I: an introduction* 3rd Edition (Springer, New York).
2. Darwin C (1859) *The origin of species by means of natural selection* (John Murry, London).
3. Futuyma DJ (1998) *Evolutionary biology*: 3rd Edition (Sinauer Associates, Inc, Sunderland, MA).
4. Hartl DL & Clark AG (1997) *Principles of population genetics.* 3rd Edition (Sinauer associates, Sunderland MA).
5. Ariew A & Lewontin RC (2004) The confusions of fitness. *The British Journal for the Philosophy of Science* 55(2): 347–363.
6. Walsh DM, Ariew A, & Matthen M (2017) Four pillars of statisticalism.
7. Charlesworth B (2009) Effective population size and patterns of molecular evolution and variation. *Nature Reviews Genetics* 10(3):195–205.

8. Patwa Z & Wahl LM (2008) The fixation probability of beneficial mutations. *Journal of The Royal Society Interface* 5(28): 1279–1289.

9. Diop A, Torrance EL, Stott CM, & Bobay L-M (2022) Gene flow and introgression are pervasive forces shaping the evolution of bacterial species. *Genome Biology* 23(1):239.

10. Grosbreg RK & Strathmann RR (2007) The evolution of multicellularity: a minor major transition. *Annual Review of Ecology, Evolution and Systematics* 38:621–654.

11. Weinburg RA (2007) *The biology of cancer* (Garland Science, New York).

12. Yao G (2014) Modelling mammalian cellular quiescence. *Interface Focus* 4(3):20130074.

13. Marieb EN (1991) *Human anatomy and physiology* (The Benjamin/Cummings Publishing Company, Redwood City, CA), p. 1040.

14. Merlo LM, Pepper JW, Reid BJ, & Maley CC (2006) Cancer as an evolutionary and ecological process. *Nature Reviews Cancer* 6(12):924–935.

15. Nowell PC (1976) The clonal evolution of tumor cell populations: acquired genetic lability permits stepwise selection of variant sublines and underlies tumor progression. *Science* 194(4260):23–28.

16. Gatenby RA, Silva AS, Gillies RJ, & Frieden BR (2009) Adaptive therapy. *Cancer Research* 69(11):4894–4903.

17. Mansoori B, Mohammadi A, Davudian S, Shirjang S, & Baradaran B (2017) The different mechanisms of cancer drug resistance: a brief review. *Advanced Pharmaceutical Bulletin* 7(3):339.

18. Read AF & Taylor LH (2001) The ecology of genetically diverse infections. *Science* 292(5519):1099–1102.

19. Arion D, Kaushik N, McCormick S, Borkow G, & Parniak MA (1998) Phenotypic mechanism of HIV-1 resistance to 3′-azido-3′-deoxythymidine (AZT): Increased polymerization processivity and enhanced sensitivity to pyrophosphate of the mutant viral reverse transcriptase. *Biochemistry* 37:15908–15917.

20. Kerr SG & Anderson KS (1997) Pre-steady-state kinetic characterization of wild type and 3′-azido-3′-deoxythymidine (AZT) resistant human immunodeficiency virus type 1 reverse transcriptase: implication of RNA directed DNA polymerization in the mechanism of AZT resistance. *Biochemistry* 36(46): 14064–14070.

21. Chmielecki J, Foo J, Oxnard GR, Hutchinson K, Ohashi K, Somwar R, Wang L, Amato KR, Arcila M, & Sos ML (2011) Optimization of dosing for EGFR-mutant non–small cell lung cancer with evolutionary cancer modeling. *Science Translational Medicine* 3(90):90ra59.

22. Mitelman F (2000) Recurrent chromosome aberrations in cancer. *Mutation Research* 462(2–3):247–253.

23. Gatenby RA & Brown JS (2020) Integrating evolutionary dynamics into cancer therapy. *Nature reviews Clinical Oncology* 17(11):675–686.

Emergence
Genetic Dominance as an Emergent Property of Biochemical Models

5

[Colonel Ross:] "Is there any point to which you would wish to draw my attention?"
[Holmes:] "To the curious incident of the dog in the night-time."
[Colonel Ross:] "The dog did nothing in the night-time."
"That was the curious incident," remarked Sherlock Holmes.
—Arthur Conan Doyle, "Silver Blaze," The Memoirs of Sherlock Holmes

THE CENTRALITY OF BIOCHEMISTRY TO LIFE ON EARTH

Figure 5.1A is a microscopic image of *cyanobacteria*: cells without a nuclear membrane that are photosynthetic. Cyanobacteria are fascinating for several reasons. They can, using only sunlight, some dissolved carbon dioxide, and a bit of phosphate, nitrogen, sulfur, and a few other trace atoms, make complete copies of themselves [1]. Perhaps more startlingly, they and their relatives were responsible for terraforming the Earth itself. Some of the oldest evidence of life on Earth are fossils of *stromatolites* [2, 3]: aggregations of cyanobacteria that can still be found in a few unusual places on Earth today (**Figure 5.1B**) [4]. At the time stromatolites appeared, the Earth's atmosphere contained essentially no oxygen [5]. Working over hundreds of millions of years, the photosynthetic labor of trillions upon trillions of these tiny cells, slowly absorbing carbon dioxide and building new cells, produced molecular oxygen as a waste product and gave Earth its oxygen-rich

A)

B)

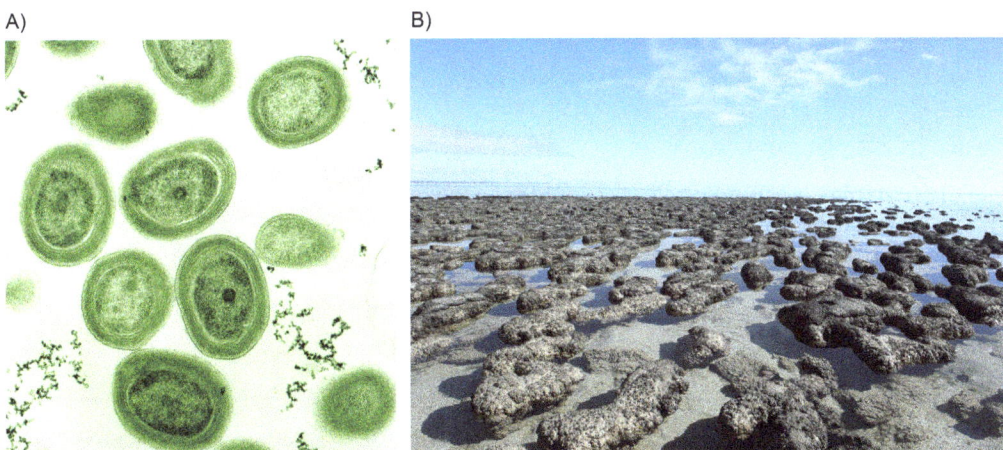

Figure 5.1 Cyanobacteria and the centrality of biochemistry. (A) The cyanobacteria *Prochlorococcus marinus*. These photosynthetic bacteria ingest sunlight, dissolve carbon dioxide, and synthesize from them complete new cells. Image taken from [37]. **(B)** Modern stromatolites in Australia, formed over time by aggregations of cyanobacteria. Those aggregations trap small particles, giving rise to structures of various shapes. A wide variety of microbial taxa are found in these structures [4]. (Image courtesy of Rob Bayer/Shutterstock, with permission).

DOI: 10.1201/9781003687504-5

atmosphere. This *great oxygenation event* occurred approximately 2.5 billion years ago [6, 7], and all of the evolution of life since then has operated under the biochemical rules first imposed on our planet by these tiny cells.

Cyanobacteria still play an enormous role in shaping our global ecosystem: roughly half of the carbon dioxide that is removed from the atmosphere each year is removed by ocean-dwelling photosynthetic life (cyanobacteria and eukaryotic algae) [8, 9]. Hence, understanding our planet at the largest scales of space and time requires an understanding of biochemistry: the controlled chemical reactions that maintain and energize all life. But of course, biochemistry touches our lives every day in a multitude of other ways as well. Here are just three examples:

- Of the human diseases caused by defective genes, in hundreds of cases the genes in question are those that encode enzymes for metabolic reactions [10].

- Plants in the mustard family have evolved to synthesize compounds that protect them from insect predators; certain insects have then subsequently evolved the ability to detoxify those protective compounds [11]. Ironically, these compounds have an attractive flavor for some humans and are the basis of products like mustard and horseradish [12].

- Baker's yeast has evolved to convert the sugar glucose into ethanol in an oxygen-rich atmosphere, making it the basis of fermenting beer and wine [13].

BIOCHEMICAL REACTIONS

The basis of all these phenomena is a complex set of chemical reactions contained in, and facilitated by, the cell. We will orient ourselves to these reactions by starting with a single, simple one, a schematic of which is given in **Figure 5.2**. The reaction shown in Figure 5.2 is *reversible*. This means that when there are quite a lot of the *reactants* (on the left) present but very little of the *products* (on the right), the reaction will proceed left to right and produce those products. However, if the situation is reversed, reactants can be produced. Surprisingly, a large fraction of the reactions in your cells are reversible in this way [14]. Reactions in biological systems are often *catalyzed* by a special type of protein called an *enzyme*, which allows the reaction to occur more quickly than it would on its own. We will consider these catalysts more carefully later in the chapter.

While Figure 5.2 on its own is not too difficult to wrap our heads around, organisms can carry out many, many such reactions. A species of cyanobacteria called *Synechocystis* sp. PCC 6803 has about 760 reactions known to occur in its cells [1], while the current estimate of the number of reactions that humans are capable of is approximately 7,000 [14]. In this and the next two chapters, we will explore the potential of mathematical and computational modeling to describe and explore the metabolic capacities of life.

The Energetics and Kinetics of Biochemistry

Before trying to model something as complex as a human cell, we should probably start with a refresher on the basic chemistry of a single reaction. We will therefore be reaching back to things we learned in our introductory chemistry classes. However, what we are about to find is that some of that chemistry is going to be an example of the importance of "leaving things out" of our models. Rather than

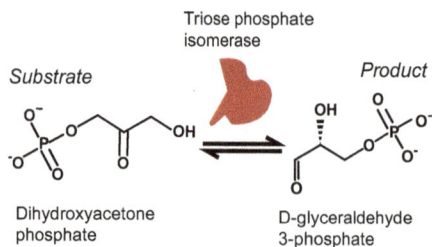

Figure 5.2 A simple chemical reaction that converts between two 3-carbon sugars. The reaction is reversible and is catalyzed by the enzyme triose phosphate isomerase.

simply and silently omitting that part of the story, we are going to walk through it to understand why we do not need it for the rest of the chapter.

We will start by returning to the question of reaction reversibility and being a bit more formal with the reaction direction than we just were. First, we need to consider the chemical energy of the reactants and that of the products: we refer to the energy stored in the chemical bounds of a compound (measured at constant pressure) as its *enthalpy*, symbolized by H [15]. If we compute the difference between the enthalpy of the products $\left(H_p\right)$ and the reactants $\left(H_r\right)$, we have:

$$\Delta H = H_p - H_r \tag{5.1}$$

If ΔH is negative, the products have the lower energy state. They will then be favored because chemical systems prefer to converge to low energy states [16]. How then can the reaction be reversible?

The reversibility comes from a second factor, *entropy* (represented with S), which measures the degree of disorder in the chemical system. As those of us with small children know, our universe prefers disordered states to ordered ones. For the reaction of Figure 5.2, we can represent the total number of molecules of reactants $\left(n_r\right)$ and products $\left(n_p\right)$ as $n_r + n_p = n$. We can then imagine our system as consisting of n boxes, each of which holds either a molecule of reactant $\left(r\right)$ or product $\left(p\right)$. In the limiting cases where n_p (or n_r) $= 0$, all the boxes are filled with the letter r (or p). Since we assume that every r is identical to every other one, there is only $N = 1$ way to fill the boxes, and entropy is minimal. On the other hand, when $n_r = n_p = n/2$, there are:

$$N = \frac{n!}{\left(\left(\frac{n}{2}\right)!\right)^2} \tag{5.2}$$

ways to arrange the molecules: at that point entropy $\left(S\right)$ is maximal. We will return to this *binomial* equation in Chapter 8 and explain its origins. We will then consider entropy more fully in Chapter 16. For the moment, we should just notice that entropy will drive the system toward a state of equal numbers of reactant and product molecules.

Hence, in predicting the direction of a reaction, we must consider both enthalpy and entropy—forces that may be working in opposition to each other. How do we compute the net direction? Let us compare two states of the system, s_1 and s_2. For the reaction in Figure 5.2, these states would correspond to two different combinations of the concentration of the reactant and the product. To determine if the system will spontaneously transition between s_1 and s_2, we can compute the change in *free energy* between them with the equation:

$$\Delta G = \Delta H - T\Delta S \tag{5.3}$$

Here, we have T as the temperature of the system. As with ΔH above, ΔS measures the change in entropy between the two states [15]. ΔG therefore combines the effects of entropy and enthalpy to allow us to predict reaction direction. Equation 5.3 is a core result of basic chemistry, one we would spend considerable time on in a chemistry class [16]. And yet, as we are about to see, it will be of rather little use to us in our quest to model the cell.

Chemical Equilibrium

One issue with Equation 5.3 is that measuring ΔH and ΔS is rather difficult. We can get around that problem to a degree by realizing that the "push–pull" between entropy and enthalpy will generally eventually reach an equilibrium at a macroscopic level, where the energetic favorability of the products will be balanced by the lack of disorder inherent in having almost the whole system as products. At equilibrium, the microscopic conversions of products to reactants will balance the conversion of reactants to products, and the system will be in macroscopic equilibrium with the concentration of $R = [R_e]$ and $P = [P_e]$. Notice that at a microscopic level the reaction has not stopped: reactants are still being converted to products and products to reactants: it is just that these two processes are occurring at the same rate. We can represent these equilibrium concentrations with the equilibrium constant K_e:

$$K_e = \frac{[R_e]}{[P_e]} \qquad (5.4)$$

This equilibrium point therefore will be where the ΔH and ΔS terms in Equation 5.3 reach a balance: conveniently, we can measure K_e with a simpler set of experiments than H and S. However, for the models we intend to build, K_e is not particularly more useful than is the ΔG of Equation 5.3.

Reaction Kinetics and Catalysts

The problem with both Equations 5.3 and 5.4 is that they only describe the final state of the system. As such, they cannot tell us how *long* it will take for the system to equilibrate to that final state. A simple example of this problem is a log of wood. We are all aware that a log can be burned in a fireplace to warm a room. The heat we get from burning the log makes it clear that the final energy of the ash and gases that result from burning the log have a lower free energy (G) than did the original log. And yet, the log does not (appreciably) spontaneously convert itself to ash and gases at room temperature. What is missing from the picture is *kinetics*: how fast the process of conversion is.

In biological systems, kinetics are critical—more important, in fact, than equilibrium conditions. Why is this? Well, notice first that all cells are constantly importing nutrients, converting them to cellular building blocks and excreting waste products. It is very rare for *any* of the reactions to come to equilibrium under these circumstances. In fact, we might say only slightly flippantly that the only cell that is in equilibrium is a dead cell. Equally importantly, because of this continuous flux of material inside the cell, what is limiting the cell is not the free energy of Equation 5.3, but rather how fast the various reactions are occurring. This leads us to the major factor that is missing from Equation 5.3: *activation energy* (**Figure 5.3**).

To know how fast a reaction proceeds, we need to know the activation energy, which is loosely the energy we need to apply to the reactants to allow them to convert to products. That energy can be contributed in a number of ways, but conceptually it can be thought of as coming from molecular collisions. The kinetic energy of these collisions is absorbed and converted to the energy of rearranging chemical bonds. In the log example, if we heat the log enough, the molecular motion resulting from the increased temperature allows more of the log's molecules to overcome their activation energy and be converted to the lower-energy water, carbon dioxide, and ash. Hence, the *speed* of a reaction is less a function of ΔG than of the required activation energy: a log does not appreciably burn at room temperature despite its large negative ΔG because of the high activation energy of converting wood to CO_2, ash, and water vapor.

Perhaps the key biochemical innovation that life has made through evolution is the use of biological polymers as *catalysts* [15]. These catalysts are most commonly proteins but also occasionally ribonucleic acid (RNA). A catalyst is a compound that is not consumed in a reaction but that lowers the activation energy of that reaction (see Figure 5.3). One of the most important points about the role of biological catalysts is that they cannot change ΔG: they increase the *rate* of a biochemical reaction but do not change its direction. When we think of an enzyme in our cells, we should recognize that in many cases that enzyme can catalyze a reaction in *either* direction, depending on the relative concentrations of the substrates and products [14]. However, the addition of enzymes lets processes very like the burning of wood happen at appreciable rates even at room temperature, making life itself possible.

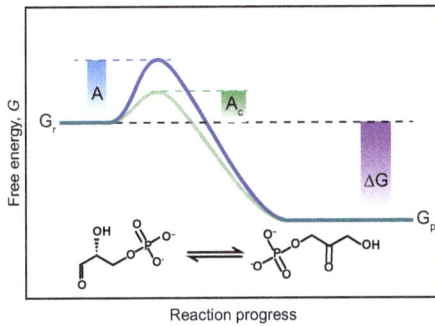

Figure 5.3 Activation energy of a chemical reaction with and without a catalyst. The ΔG of the reaction (the difference between G_r and G_p on the y-axis) is an intrinsic property of the reactants and products; it is not altered by the catalyst. The activation energy (A), on the other hand, determines the reaction velocity. The addition of a catalyst decreases the activation energy (A_c) and hence accelerates the reaction. Notice that I have reversed the chemistry of Figure 5.2 because the compound that is physiologically the product in that figure has a higher energy state in test-tube conditions.

BUILDING BIOCHEMICAL MODELS

Having built this background, we can now start to think about potential models of biochemical kinetics. We can start by noting what we plan to exclude from our models. Parameters like ΔG and K_e do not directly drive the reaction kinetics and thus do not need to be explicitly included in the model. We should be clear that the models we are about to build do not *violate* the rules of Equations 5.3 and 5.4: rather, the kinetic parameters we will fit into our models implicitly depend on ΔG, such that our models will converge to Equations 5.3 and 5.4 if we let them come to equilibrium over very long time periods.

We should also stop to consider *why* we might want to construct such biochemical models—or, to phrase it differently, the applications of these models. Potentially they can be used to:

- Predict the response of cells to changes in nutrition

- Predict the nutritional requirements and growth rates of single-celled organisms

- Model the evolutionary costs of biochemical pathways

- Predict the effects of mutations in enzyme-coding genes

- Understand the general role of catalysts in biological systems.

There are, of course, many, many more applications. In fact, a fully realized biochemical model would, in theory, predict and describe almost all aspects of life, up to and including the chemical and electrical functions of the brain.

As we will see, there are enormous technical, experimental, and computational hurdles to developing a biochemical model that is even approximately complete in this sense. And yet even the very limited models we have in hand tell us fascinating and unexpected things about the nature of life, genetics, and evolution.

The Model of Michaelis and Menten

Models of enzyme-catalyzed reactions can become very complicated, but much of the conceptual machinery and insight can be built with the simple reaction framework involving the reversible binding of a substrate (S) to an enzyme (E) to form an enzyme and substrate complex $(E + S = C)$. Unfortunately, I am now switching notation on you: we will use *substrate* rather than *reactant* from this point to be consistent with most of the literature in this area. (I had avoided using S for substrate earlier to avoid confusion with entropy.) Then some fraction of the complexes undergoes the irreversible production of a product (P):

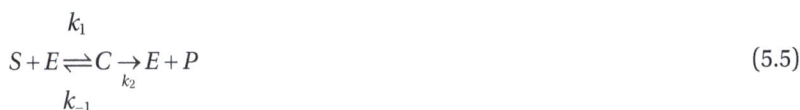

$$S + E \underset{k_{-1}}{\overset{k_1}{\rightleftharpoons}} C \overset{k_2}{\rightarrow} E + P \qquad (5.5)$$

This equation is not strictly accurate for the reaction of Figure 5.2 because that reaction is reversible, but we can conceptually assume an irreversible version for the next few pages and return to the reversible case later. In fact, the original work of Michealis and Menten in 1913 that we are about to consider analyzed the (reversible) conversion of sucrose to glucose and fructose. However, these authors only considered the earliest phase of the reaction, where that reverse reaction could be neglected [17, as translated by 18]. Again, I emphasize that the three rate parameters k_1, k_{-1}, and k_2 in Equation 5.5 are functions of both ΔG and the activation energy: none of what we discussed above is lost in this model.

We can immediately express Equation 5.5 as a set of four differential equations in S, E, C, and P:

$$\frac{dS}{dt} = -k_1 \cdot S \cdot E + k_{-1} \cdot C$$

$$\frac{dE}{dt} = -k_1 \cdot S \cdot E + \left(k_{-1} + k_2\right) \cdot C$$

$$\frac{dC}{dt} = k_1 \cdot S \cdot E - \left(k_{-1} + k_2\right) \cdot C \qquad (5.6)$$

$$\frac{dP}{dt} = k_2 \cdot C$$

We notice again how the basic form of these equations mirrors those of the SIR and Lotka–Volterra models. In all three cases the rate of events is proportional to the *product* of the two entities involved. For the SIR model those entities are susceptible and infected individuals, while in the Lotka–Volterra model they are predator and prey. Here they are reactants and enzymes. Finally, in all three cases we can think of the event as being initiated by a conceptual "collision" of the two entities.

In this case, the $S \cdot E$ term in fact represents actual collisions of molecules of S and E. To understand this fact better, we can stop and note the units of all the quantities in Equation 5.6. The quantities of S, E, C, and P are *concentrations*: for simplicity, we will write them as moles/liter (Mol/L). If S, E, C, and P themselves are measured in Mol/L, the units of their derivatives will be changes in concentration per unit time, or Mol/(L·sec) (moles per liter per second). As a result, the two constants k_{-1} and k_2 are the rates of the complex C breaking down either into free enzyme and substrate or free enzyme and product. Their units are hence 1/sec, which, when multiplied by the concentration of C, gives the Mol/(L·sec) units of the left side of Equation 5.6. The units of k_1 are a bit more complex: they effectively scale the affinity of the substrate and enzyme for each other by their actual concentrations. The units of k_1 are hence L/(Mol·sec). We could express this unit as being the number of liters we would need for the system's kinetics yield a mole of the complex forming every second. With k_1 measured in these units, the product $S \cdot E$ in Equation 5.6 is the frequency of collisions in the solvent between E and S. We then see that k_1 tells us how often such a collision results in complex formation.

Rather than immediately exploring the equations comprising Equation 5.6, our work will be easier if we consider the nature of the system a bit more carefully. In particular, look closely at the equations for dE/dt and dC/dt. They differ only by a negative sign. The reason is that E is a catalyst for this reaction: it is not consumed over the reaction's course. If we denote the initial or total enzyme concentration as E_0, we know that it must be the case that $E + C = E_0$ at all times. We can use this fact to eliminate the expressions for E from Equation 5.6, replacing them with $E_0 - C$. At the same time we can remove the equation for dE/dt, since it is simply the inverse of dC/dt:

$$\frac{dS}{dt} = -k_1 \cdot S \cdot \left(E_0 - C\right) + k_{-1} \cdot C$$

$$\frac{dC}{dt} = k_1 \cdot S \cdot \left(E_0 - C\right) - \left(k_{-1} + k_2\right) \cdot C \qquad (5.7)$$

$$\frac{dP}{dt} = k_2 \cdot C$$

We can then rewrite dS/dt and dC/dt as:

$$\frac{dS}{dt} = -k_1 \cdot S \cdot E_0 + \left(k_1 \cdot S + k_{-1}\right) \cdot C$$

$$\frac{dC}{dt} = k_1 \cdot S \cdot E_0 - \left(k_1 \cdot S + k_{-1} + k_2\right) \cdot C \qquad (5.8)$$

$$\frac{dP}{dt} = k_2 \cdot C$$

We can simplify these equations even more, but to understand why, it is helpful to explore their behavior a bit first. **Table 5.1** gives some hypothetical but plausible values for the system, which we will assume starts out with $C(0) = P(0) = 0$. Conceptually, we can view these conditions as starting with a test tube of enzyme at concentration E_0 into which we drop $S(0)$ Mol/L of substrate at $t = 0$.

TABLE 5.1 PARAMETERS OF EQUATION 5.6 FOR ANALYSIS

Parameter	Value
k_1	$10^6 \cdot$L/(Mol·sec)
k_{-1}	$1000 \cdot$s^{-1}
k_2	$10 \cdot$s^{-1}
E_0	$0.0001 \cdot$Mol/L
$S(0)$	$0.01 \cdot$Mol/L

The Steady-State Approximation

Figure 5.4 shows the results of modeling the equations comprising Equation 5.8. At first glance it looks as expected, with a decline in the concentration of the substrate $\left(S(t)\right)$ that is matched by an increase in the concentration of the product $\left(P(t)\right)$. Because we assumed that the concentration of enzyme was small relative to S, the concentration of the enzyme/substrate complex $C(t)$ never rises to a large

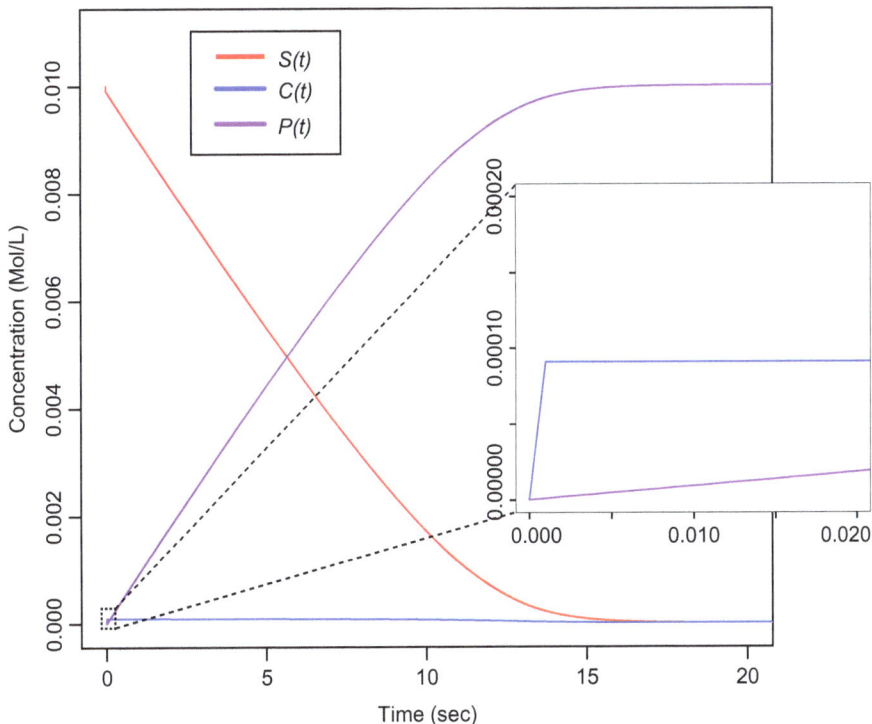

Figure 5.4 Dynamics of a Michaelis–Menten reaction model with the parameters from Table 5.1. Shown are the three variables from Equation 5.8, with S corresponding to the substrate, C the enzyme/substrate complex, and P the product. Over the long time scale of the main panel, the substrate is consumed and the product is generated. However, there is an initial transient period where the concentration of C changes in less than 100th of a second from the initial state of $C(0) = 0$ up to what visually appears to be a steady state. You can regenerate this figure in Exercise 5.1.

value. However, if we look more closely at the period immediately after $t = 0$ (inset of Figure 5.4), we see something less expected: the concentration of the enzyme/substrate complex rapidly jumps from $C(0) = 0$ (our initial conditions) up to a level that is close to the total enzyme concentration $(E_0 = 0.0001 \, \text{Mol}/\text{L})$ and then appears to remain reasonably constant thereafter. There are two related reasons for this behavior. The first reason lies in the relationship of k_{-1} and k_2. If we map our scheme of a catalyst driving the reduction in the activation energy of a reaction from Figure 5.3 onto the reaction in Equation 5.5, we can see that the constant k_2 actually expresses the rate at which the enzyme allows the substrate to be converted to product. On the other hand, k_{-1} describes the rate at which the enzyme/substrate complex dissociates back into free enzyme and substrate. Since that dissociation involves no large input of activation energy, we expect k_{-1} to be much bigger than k_2 for almost any biochemical reaction. Under such circumstances, the initial formation of C from S and E will occur much faster than the formation of P. The second reason for the spike in Figure 5.4 lies in the relative concentrations of E and S. In normal biological systems, the enzyme is present at much lower concentrations than the substrate is [15, 18]. This difference means that, if we neglect this initial jump in $C(t)$, changes in concentration in entities involving the enzyme will be much smaller than changes that involve the substrate or the product alone.

In 1925, Briggs and Haldane [19] argued that it would be reasonable to neglect this initial transient phase and treat the first part of Equation 5.5 as being in a steady state. This approximation is reasonable for two reasons: enzyme molecules are rare relative to substrates and products, and product is created much more slowly than the enzyme/substrate complexes form and dissociate. As a result, the fusion of $E + S$ into C and back again remains at an effective equilibrium over the course of the reaction. We can state this claim mathematically as $dC/dt \cong 0$. If we make this substitution into the dC/dt equation of Equation 5.8 and move the term in C to the left side of the equation, we get:

$$(k_1 \cdot S + k_{-1} + k_2) \cdot C = k_1 \cdot S \cdot E_0 \tag{5.9}$$

We can now solve for the steady-state value of C, which is:

$$C = \frac{k_1 \cdot S \cdot E_0}{k_1 \cdot S + k_{-1} + k_2} \tag{5.10}$$

If we multiple the numerator and denominator of Equation 5.10 by $1/k_1$, we have:

$$C = \frac{S \cdot E_0}{S + \dfrac{k_{-1} + k_2}{k_1}} \tag{5.11}$$

We can now introduce a new symbol K_m that relates our three rate constants k_1, k_{-1}, and k_2:

$$K_m = \frac{k_{-1} + k_2}{k_1} \tag{5.12}$$

This new quantity allows us to rewrite Equation 5.11 as:

$$C = \frac{S \cdot E_0}{S + K_m} \tag{5.13}$$

Putting dP/dt from Equation 5.8 aside for a moment, we can now replace C in the equation for dS/dt from Equation 5.8 with its new steady-state value from Equation 5.13, giving:

$$\frac{dS}{dt} = -k_1 \cdot S \cdot E_0 + (k_1 \cdot S + k_{-1}) \cdot \frac{S \cdot E_0}{S + K_m} \tag{5.14}$$

Multiplying the first term on the right side by $S + K_m$, we get:

$$\frac{dS}{dt} = \frac{\left(-k_1 \cdot S \cdot E_0\right) \cdot \left(S + K_m\right) + \left(k_1 \cdot S + k_{-1}\right) \cdot S \cdot E_0}{S + K_m} \tag{5.15}$$

We can multiply through the $S + K_m$ on the upper left and $k_1 \cdot S + k_{-1}$ on the upper right to obtain:

$$\frac{dS}{dt} = \frac{\left(-k_1 \cdot S^2 \cdot E_0\right) + \left(-k_1 \cdot S \cdot E_0\right) \cdot K_m + k_1 \cdot S^2 \cdot E_0 + k_{-1} \cdot S \cdot E_0}{S + K_m} \tag{5.16}$$

Now the $-k_1 \cdot S^2 \cdot E_0$ and the $k_1 \cdot S^2 \cdot E_0$ terms cancel out, and if we expand the expression for K_m in the numerator, we have:

$$\frac{dS}{dt} = \frac{\left(-k_1 \cdot S \cdot E_0\right) \cdot \left(\dfrac{\left(k_{-1} + k_2\right)}{k_1}\right) + k_{-1} \cdot S \cdot E_0}{S + K_m} \tag{5.17}$$

We can then cancel the k_1 in the upper left of the right-hand side:

$$\frac{dS}{dt} = \frac{-\left(S \cdot E_0\right) \cdot \left(k_{-1} + k_2\right) + k_{-1} \cdot S \cdot E_0}{S + K_m} \tag{5.18}$$

Which, when we cancel the positive and negative pair of terms in $k_{-1} \cdot S \cdot E_0$ and move k_2 and a negative sign, gives:

$$\frac{dS}{dt} = -k_2 \cdot \frac{S \cdot E_0}{S + K_m} \tag{5.19}$$

The new Equation 5.19 represents our expression for the derivative of S under the assumption that C is in a steady state. What happens when we make the same substitution into dP / dt from Equation 5.8?

$$\frac{dP}{dt} = k_2 \cdot \frac{S \cdot E_0}{S + K_m} \tag{5.20}$$

In other words, when we assume that the concentration of C is in a steady state, any S that disappears must, because our system is mass-balanced, become P. And indeed, we see that Equation 5.20 is simply the inverse of Equation 5.19, fulfilling this requirement.

What has all of this algebra accomplished? Well, first we have taken a system that was apparently described by four differential equations and shown that it can be understood using only dP / dt. But we can also understand the behavior more easily now. In particular, as $S \to \infty$, the K_m term in the denominator of Equation 5.20 becomes negligible, and the value of dP / dt reaches its maximum value. Intuitively, this result makes sense: for a fixed enzyme concentration, the reaction will have its highest rate when the concentration of substrate is unlimited. We can define a new model parameter $V_{max} = E_0 \cdot k_2$ to be this maximal reaction velocity.

Figure 5.5 shows the reaction velocity as a function of the substrate concentration (in other words, a plot of Equation 5.20). As expected, as the substrate concentration increases, the reaction velocity approaches the plateau of $dP / dt = E_0 \cdot k_2$. We will refer to this maximal rate as V_{max}. Less expected is the concentration of the substrate at which the reaction achieves ½ of its V_{max}. We can find this value by solving Equation 5.20 for the substrate concentration at which $dP / dt = \frac{1}{2} E_0 \cdot k_2$ (i.e., ½ V_{max}):

$$\frac{1}{2} k_2 \cdot E_0 = k_2 \cdot \frac{S \cdot E_0}{S + K_m} \tag{5.21}$$

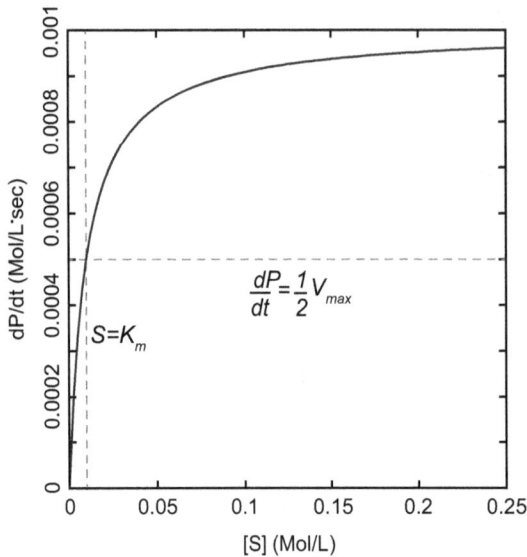

Figure 5.5 **Reaction velocity (*dP/dt*) as a function of substrate concentration *S* under the Michaelis–Menten model (Equation 5.20).** In this plot $V_{max} = 0.001$ Mol/(L·sec) and $K_m = 0.0101$ Mol/L. The velocity approaches V_{max} as the substrate concentration approaches infinity. At $S = K_m$ the reaction achieves half-maximal velocity (intersection of the dashed lines).

Canceling the $E_0 \cdot k_2$ terms on both sides and moving the $(S + K_m)$ to the left-hand side gives us:

$$\frac{1}{2}(S + K_m) = S \tag{5.22}$$

Which, if we move ½S to the right side and multiply by 2, gives:

$$K_m = S \tag{5.23}$$

In other words, K_m gives the substrate concentration at which velocity (dP/dt) is half-maximal. It is, in fact, this equality that accounts for our introduction of the quantity K_m in Equation 5.12 in the first place.

There is also a more subtle implication of Figure 5.5. The quantities in Table 5.1 include the three reaction rate constants k_1, k_{-1}, and k_2. If we wanted to apply this model to actual biochemical reactions, we need somehow to measure or infer the values of these model parameters. Unfortunately, measuring such instantaneous rate constants can be very difficult and is prone to significant error [15]. On the other hand, V_{max} can be reasonably approximated by a straightforward experiment. In that experiment, the substrate concentration is serially increased, and the rate of product formation is measured at the very beginning of the reaction (allowing of course for the concentration of C to reach its quasi-steady state). Once one finds that an increase of substrate concentration no longer increases the initial reaction velocity, V_{max} (for that enzyme concentration) has been found. K_m can then be estimated as the substrate concentration giving a velocity of $1/2\ V_{max}$. Hence, unlike the Lotka–Volterra model, we have (in principle) a reasonably clear way to apply our model to many actual biochemical reactions. (As an aside, it is mostly straightforward to modify the differential equations in Equation 5.6 for more complex reactions with multiple substrates and products [20] while still measuring V_{max} and K_m for each.)

MANY ASSUMPTIONS, ONE MODEL: DIFFERING PATHS TO THE MICHAELIS–MENTEN MODEL

There is another surprising point regarding Equation 5.20 worth noting. Following Briggs and Haldane [19], we derived that equation under the assumption that the concentration of the enzyme-substrate complex is small relative to the sum of the concentrations of substrates and products, and likewise that the rate of change of the concentration of that complex is also small relative to the other differential values in Equation 5.6. However, prior to their work, the same equation had been independently derived by Michaelis and Menten [17] and by Van Slyke and Cullen [21]. Yet each of these three research teams made a different set of assumptions to reach Equation 5.20. Michaelis and Menten assumed that the formation and breakdown of C from E and S were very rapid relative to the formation of P from C, treating the first part of the reaction in Equation 5.5 as being nearly in equilibrium at all times. Since another way to describe that equilibrium is to have $dC/dt \cong 0$, it is not surprising that their derivation produced the same result of that of Briggs and Haldane. Yet Van Slyke and Cullen made a very different assumption, namely that even the formation of C from E and S was an irreversible process, just as the formation of P from C was assumed to be. This assumption corresponds to setting k_{-1} in Equation 5.5 to zero. How is it then that their expression for dP/dt was so similar to that of the other two teams? The answer is one that is not always recognized when thinking about scientific models: there can be multiple ways to reach the same mathematical model, just as there can be many possible ways to verbally describe that model. In this case, the three different derivations all resulted in a common (or nearly common) equation, allowing us to notice the mathematical equivalency of the three models. However, for more complex models it is possible to create models with apparently differing forms that give rise to identical predictions. We will see an example of this situation in Chapter 13. We should also learn from this example to be wary of comparing models only through their verbal descriptions. The fact that these three teams used differing terms to describe their models in some sense hid the fact that the underlying models were the same.

EMERGENT PROPERTIES: AN APPLICATION OF THE MICHAELIS–MENTEN MODEL

It is naturally satisfying to have a compact mathematical model of enzyme-catalyzed reactions. But the implications of these types of models are more general and extend further beyond biochemistry than might be expected. To consider why, we will analyze the deceptively simple system shown in **Figure 5.6**. This system is simply three chained copies of the reaction in Equation 5.5, with the slight modification that all three reactions are now reversible.

Figure 5.6 is an example of a *pathway*: a sequence of reactions a biological system uses to break down or synthesize a compound for energy extraction or cellular construction. As we will see in later chapters, the notion of a pathway is an oversimplification of biochemical reality. However, it is appropriate in many circumstances and will be a useful framework for this example.

So far, we have made "test-tube-like" models of biochemistry, in the sense that we start with a fixed quantity of substrate that is allowed to convert into product until it is exhausted. However, a more common biological system is one in which some nutrient is provided at a relatively constant rate, say by the blood supply, which also removes any final waste products. So, we will propose a version of Figure 5.6 where we assume that the concentration of A is fixed as an input nutrient and the concentration of D is also fixed because the organism at large drains off D as it accumulates. We could write differential equations modeling this situation in R. However, because biochemistry is such an important part of biology, there are pre-built software tools that allow us to enter biochemical reactions in an intuitive chemical layout and have the software handle the conversion of the model to a computation. The tool we will use is called COPASI [22], which is freely available and will also be useful in Chapter 7 when we start to consider the possibility of random effects in biochemistry.

If we model Figure 5.6 with some plausible enzyme concentrations and rate constants, we can measure the *flux* (moles per second per liter) through the pathway. We are particularly interested in how the concentration of the middle enzyme, E_2, might change this flux (**Figure 5.7**). What we find is quite a surprise: if, keeping the concentrations of E_1 and E_3 constant, we change the concentration of E_2 from 10-fold *less* than that of E_1 and E_3 to 10-fold *more* that these two enzymes, the overall flux through the pathway changes less than 2%!

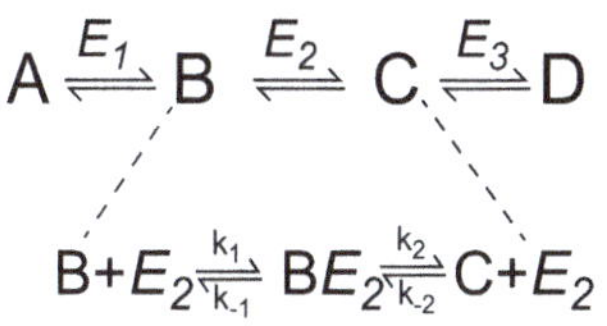

Figure 5.6 A simple chain of three biochemical reactions. We will assume that the larger cell is able to hold the concentrations of A and D constant and explore the effects of enzyme concentration on the *flux* through the middle reaction.

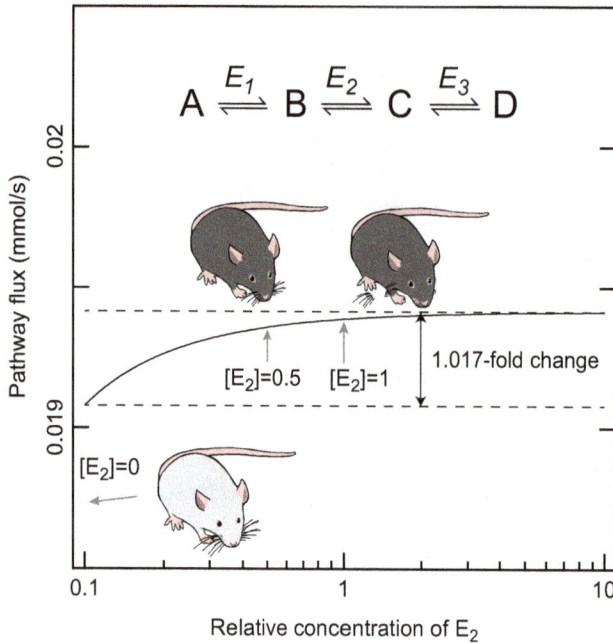

Figure 5.7 Distributed robustness in a biochemical pathway. We assume a fixed initial concentration of A and D of 1 mmol/ml. For the reaction scheme in Figure 5.6, $k_1 = 0.2\,\text{ml}/(\text{mmol}\cdot\text{s})$, $k_{-1} = 0.11/\text{s}$, $k_2 = 10\,1/\text{s}$, and $k_{-2} = 0.012\,\text{ml}/(\text{mmol}\cdot\text{s})$. The concentrations of E_1 and E_3 are fixed at 0.1 mmol/L, while the concentration of E_2 is allowed to vary along the x-axis from 0.01 mmol/ml to 1 mmol/ml. Plotted on the y-axis is the flux through the pathway across this change. The underlying data for this plot were generated in COPASI [22] and the figure itself adapted from [38]. If we assume the pathway produces a brown fur pigment, we can see that wild-type individuals having $\left[E_2\right] = 0.1\,\text{mmol}/\text{L}$ and heterozygote ones for which one copy is nonfunctional $\left(\left[E_2\right] = 0.05\,\text{mmol}/\text{L}\right)$ would probably be phenotypically indistinguishable. On the other hand, a complete absence of E_2 might produce a white mouse, with pathway flux off this chart because of the logarithmic scale of the x-axis [39, 40]. Lab mouse art courtesy of NIAID, used with permission. You can generate data similar to this figure in Exercise 5.2.

In a more mathematically thorough and rigorous way, Henrik Kacser and James Burns [23] showed in 1980 that this insensitivity to the concentrations of intermediate enzymes in biochemical pathways is actually the expected condition: as more enzymes are added to a pathway, the degree to which the concentration of any one of those enzymes is determining the overall pathway output (flux) becomes smaller and smaller. The conceptual reason behind this fact is essentially the many points of flexibility in the system. If we imagine partly blocking the conversion of B to C in Figure 5.6, the result will be a build-up of B. But, as we see in Figure 5.5, as substrate concentration increases, the reaction will produce its product C more rapidly. Moreover, if we assume the reactions are reversible, the equilibrium will shift to favor C, since B is now in relative excess. Since, in a complex pathway, there is not just a single reaction that can increase in this way, but many, the result is that partially blocking one reaction will tend to drive other reactions to change their behavior in a compensating manner. Hence, many biochemical systems are self-correcting, or, to use another word, *robust* [24]. We will now take a brief detour to an apparently unrelated area of biology, namely genetics, to see just what the implications of that robustness can be.

DIGRESSION: BIOCHEMICAL ROBUSTNESS AND GENETIC DOMINANCE

Gregor Mendel's work [25] is fascinating because of the degree to which it presciently outlines many core ideas of modern genetics [26]. One aspect of this prescience is particularly striking because it is in conflict with our intuitive notions of the nature of inheritance. That concept is that of genetic *dominance* (**Figure 5.8**).

Mendel mated, or *crossed*, pairs of lineages of pea plants where each lineage had shown constancy in a particular characteristic of the plants (such as seed shape) over several generations but where the two lineages differed in that characteristic (see Figure 5.8). As an example, he mated a pea lineage that always produced smooth seeds to one that always produced wrinkled seeds [25]. To understand his results, we should first remember that pea plants, like humans, inherit genetic material from two parents and hence have two copies of every gene. We call this condition *diploid*; in humans, one copy of the gene comes from our fathers and one from our mothers [27].

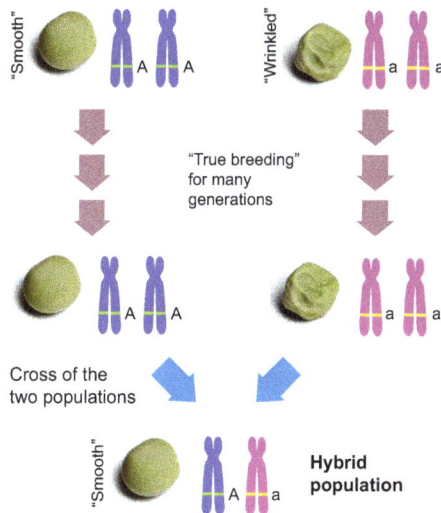

Figure 5.8 A cartoon of the first part of Mendel's experiment illustrating genetic dominance. Two populations of peas differ in their seed shape, and these differences persist unchanged over many generations in each population. However, when the two populations are mated (or *crossed*) with each other, the wrinkled form is lost and all the offspring in the resulting *hybrid population* are smooth. To the right of each lineage is a proposed genetic explanation: the two populations differ in the form of one gene: the **A** form gives rise to the smooth seeds and the **a** form gives rise to wrinkled ones. However, when the two populations are mated with each other, the hybrids (**Aa**) show the smooth form. (Image courtesy of Martin Shields/ Alamy, used with permission).

Curiously, the plants produced in Mendel's crosses did not show any mixing of the two parental forms. Instead, in this example, all the offspring had smooth seeds. He wrote:

> In the case of the seven [pea] crosses the hybrid-character resembles that of one of the parental forms so closely that the other either escapes observation completely or cannot be detected with certainty.
>
> (25)

Mendel referred to this form of the trait as being (in Bateson and Blumberg's English translation, www.mendelweb.org) *dominant* over the other, which he referred to as being *recessive*.

The critical factor to keep in mind about these crosses, however, is that the recessive form was not permanently lost. If members of the resulting *hybrid* population were mated with each other, the recessive form reappeared in their offspring at a ratio of 1:3 with the dominant form (**Figure 5.9**).

We can understand this result better if we make a diagram of this second kind of mating. We have named the version of the gene (or *allele*) that produces smooth seeds **A** and the version producing wrinkled ones **a** (see Figure 5.8). The smooth-seeded parent had two copies of the **A** allele: we know this because that lineage only ever produced smooth seeds. By the same reasoning, the wrinkled parent had two **a** alleles (see Figure 5.8). Hence, when we mated these two lineages together, *all*

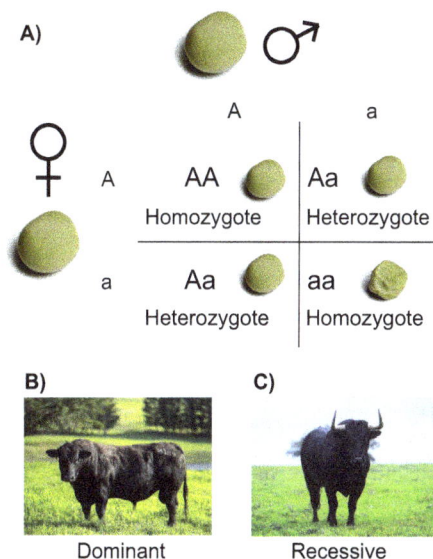

Figure 5.9 Dominant and recessive forms in a genetically variable population. (**A**) An illustration of the effects of crossing two individuals from the hybrid population of Figure 5.8. Each parent has one **A** and one **a** allele, meaning that the offspring can have any of the four gene combinations shown. However, due to dominance, roughly 75% of the offspring will show the dominant form and 25% the recessive. (**B** and **C**) In cattle, the *polled* condition (lacking horns) is dominant over the horned condition [41]. (Images courtesy of Alamy and William Edge/Shutterstock, used with permission).

the offspring in the resulting hybrid population must have had one **A** and one **a** allele, since each parent could only contribute either an **A** or an **a**.

What happens if one individual with the **Aa** gene combination mates with another with the same combination? Well, when a diploid organism makes an egg or a sperm, it effectively flips a coin to decide which of its two copies of each gene to pass on [27]. As we can see in Figure 5.9A, there are then four possible outcomes for such a mating: **AA, Aa, aA**, and **aa**. Each of these four outcomes occurs at effectively the same frequency of one-quarter. (We will return to this problem in Chapter 11.) However, we assume that **Aa** and **aA** are indistinguishable, so we could at most observe three types of offspring.

Mendel's observation was that very often we only observe *two* offspring types: the dominant form and the recessive form we saw in the parental lineages. Moreover, the fact that we see three times as many dominant individuals as recessive ones strongly suggests that the **Aa** gene combination effectively behaves the same way the **AA** combination does [27]. This hypothesis also explains the original matings: since all the offspring were **Aa**, they were all indistinguishable from the form of the dominant parent.

Explanations for Dominance

The phenomenon of dominance is rather common in genetic systems and yet runs counter to our intuitive expectation that offspring will be intermediate in form relative to their parents. One of the first geneticists to propose an explanation for the prevalence of dominance was R. A. Fisher. He, along with Sewall Wright and J. B. S. Haldane (whom we met earlier refining the Michealis–Menten model of enzyme kinetics), built the core mathematical and statistical framework used by geneticists today [28]. Fisher proposed that dominance represented an *evolved* response of organisms to protect themselves from the effects of mutation [29, 30]. In this view, mutations that disable gene function would be detrimental to the organism carrying them. However, since mutations are rare, most of the individuals carrying them would do so in the *heterozygous* state: that is, the **Aa** case above. Hence, here heterozygous is shorthand for having one mutant copy that we assume to be malfunctioning in some way and a *wild-type* copy that is much more common in the population as a whole and that performs the corresponding function well (see Figure 5.9A). Fisher argued that because such mutations had been occurring continuously, albeit rarely, throughout evolutionary history, natural selection would have altered other genes to protect the organism from the deleterious effects a heterozygous loss-of-function mutation. Hence, in his view, dominance is observed in modern organisms because they have evolved to protect themselves from mutations in the heterozygous state.

Both Wright and Haldane, among others, immediately criticized this idea. They pointed out that if mutations are rare, the evolutionary value of having genetic variants that protect against those mutations is so small that natural selection will not be able to create or maintain them [31, 32]. We will explore how the size of the benefit of having a gene variant affects its evolutionary history in more detail in Chapter 11. For the moment, we will simply note that modern models and tests have effectively confirmed their criticism: selection to protect against mutation does not explain dominance [33, 34].

DOMINANCE AS AN EMERGENT PROPERTY

We can now see the real power of the work of Kacser and Burns [23]. Although I expressed their results earlier in the chapter in the dry terms of biochemistry, the title of their paper is actually "The Molecular Basis of Dominance." Their insight was that most traits of an organism arise from biochemical pathways. If those pathways are robust to reductions in the levels of intermediate enzymes, they will display dominance. Looking back at our example, think of E_2 in Figure 5.7 as existing in both a wild-type (say, **A**) and a nonfunctional form (say, **a**). The reaction that E_2 catalyzes might, for instance, act as an intermediate step in the production of a pigment. Because of this nature of biochemical pathways, an individual with one working copy of E_2 will produce effectively as much pigment as those with two copies. This idea is illustrated in Figure 5.7 with the two gray arrows, where the heterozygote with $[E_2] = 0.5$ having effectively the same pathway flux as the homozygous wild-type individual ($[E_2] = 1.0$). As a result, those heterozygous individuals will *look* identical to individuals with two wild-type versions of E_2. We would therefore declare the wild-type enzyme allele to be dominant over the broken mutant copy. What Kacser and Burns have demonstrated, therefore, is that dominance is an emergent property of biochemistry. Dominance is not a feature of organisms that needs an evolutionary or genetic explanation: it *emerges* because of the nature of biochemistry itself. Depending on your point of view, many other features of biology are emergent. Even the fact that life can exist might be seen as an emergent property of the physical and chemical rules of our universe.

REFLECTIONS AND PREVIEWS

In this chapter we found that differential equation models similar in form to those of the SIR and Lotka–Volterra models could be applied in a wholly different domain: subcellular biochemistry. We also took our first steps toward developing more practical models that can be applied to actual biological data. In so doing, we saw that it is often necessary to adjust such models to state them in a form where the parameters can be measured in the laboratory. We also found a new emergent property: genetic dominance results from the inherent flexibility of biochemical pathways.

Genetics, Biochemistry, and Emergence

This connection between the apparently very different fields of genetics and biochemistry also speaks to important ideas in evolutionary theory. In fact, Mendel's discovery of dominance was just one piece of his key insight: the *particulate* nature of inheritance. Earlier in this chapter and in Chapter 4, we contrasted that view with another view that was common in his time: *blended* inheritance. Blended inheritance essentially reflects our common experience that offspring generally have characteristics, such as height, that are intermediate between their parents. We will provide a theoretical explanation for this observation in Chapter 9. However, were all organismal traits inherited in a blended manner, evolution by natural selection would function very poorly. Such inheritance would cause any new beneficial trait to be averaged back to the original state over the generations. Darwin was aware of this difficulty and even edited later editions of *The Origins of Species* to address this problem [35]. Ironically, Mendel had offered a clear solution to Darwin's difficulty, but it was not noticed because Mendel's work languished unnoticed until after Darwin's own death [36]. Hence, with the conclusion of this chapter, we have two of the three pieces needed for a true quantitative theory of natural selection: the competing populations from Chapter 4 and the particulate inheritance of Mendel, as explained by Kacser and Burns. It remains to add the random effects, a task we will take up in Chapter 11.

Bigger Models

Meanwhile, in the next chapter we will consider the practical limitations of the models we have developed here for understanding biochemistry at the full cell level. In so doing, we will start to explore a new type of mathematical modeling: the tools of *linear algebra*. We will then use them to discover an altogether different type of biological robustness.

Exercise 5.1

Using the parameter values in Table 5.1, solve the equations in Equation 5.8 using R over a time scale of 20 s, with a timestep size of 0.01 s. Plot the results both for the full time sequence and for the interval t = 0 s to t = 0.02 s. Example R commands:

- `library(deSolve)`
- `parameters<-c(k1=1e6,km1=1000, k2=10, E0=0.0001)`
- `state<-c(S=0.01,C=0,P=0)`
- `MicMen_model<-function(t,state, parameters) {with(as.list(c(state,parameters)), {`
- `dS <- -k1*S*E0+(k1*S+km1)*C`
- `dC <- k1*S*E0-(k1*S+km1+k2)*C`
- `dP <- k2*C`
- `list(c(dS,dC,dP))})}`
- `times <- seq(0, 25, by = 0.001)`
- `out <- ode(y=state, times=times, func=MicMen_model, parms = parameters)`
- `plot(out[,"time"], out[,"S"], type="l", xlab="-time", ylab="Concentration", col="red", xlim=c(0,20),ylim=c(0,0.011))`
- `points(out[,"time"], out[,"P"], type="l", xlab="-time", ylab="Concentration", col="purple")`
- `plot(out[,"time"], out[,"S"], type="l", xlab="-time", ylab="Concentration", col="red", xlim=c(0,0.02), ylim=c(0,0.0002))`
- `points(out[,"time"], out[,"P"], type="l", xlab="-time", ylab="Concentration", col="purple")`
- `points(out[,"time"], out[,"C"], type="l", xlab="-time", ylab="Concentration", col="blue")`

Exercise 5.2

Using COPASI, model the biochemical pathway in Figure 5.6 using the kinetic constants and enzyme concentrations given in Figure 5.7. Find the pathway flux for E_2 = 0.01 mmol/ml, 0.05 mmol/ml, 0.1 mmol/ml, 0.5 mmol/ml, and 1 mmol/ml. Plot the results.

REFERENCES

1. Knoop H, Gründel M, Zilliges Y, Lehmann R, Hoffmann S, Lockau W, & Steuer R (2013) Flux balance analysis of cyanobacterial metabolism: the metabolic network of Synechocystis sp. PCC 6803. *PLoS Computational Biology* 9(6):e1003081.

2. Awramik SM (2006) Palaeontology: respect for stromatolites. *Nature* 441(7094):700–701.

3. Allwood AC, Walter MR, Kamber BS, Marshall CP, & Burch IW (2006) Stromatolite reef from the Early Archaean era of Australia. *Nature* 441(7094):714–718.

4. Papineau D, Walker JJ, Mojzsis SJ, & Pace NR (2005) Composition and structure of microbial communities from stromatolites of Hamelin Pool in Shark Bay, Western Australia. *Applied and Environmental Microbiology* 71(8):4822–4832.

5. Ligrone R & Ligrone R (2019) The great oxygenation event. *Biological innovations that built the world: a four-billion-year journey through life and earth history*:129–154.

6. Holland HD (2006) The oxygenation of the atmosphere and oceans. *Philosophical Transactions of the Royal Society B: Biological Sciences* 361(1470):903–915.

7. Cavalier-Smith T (2006) Cell evolution and Earth history: stasis and revolution. *Philosophical Transactions of the Royal Society B: Biological Sciences* 361(1470):969–1006.

8. Field CB, Behrenfeld MJ, Randerson JT, & Falkowski P (1998) Primary production of the biosphere: integrating terrestrial and oceanic components. *Science* 281(5374):237–240.

9. Behrenfeld MJ, O'Malley RT, Siegel DA, McClain CR, Sarmiento JL, Feldman GC, Milligan AJ, Falkowski PG, Letelier RM, & Boss ES (2006) Climate-driven trends in contemporary ocean productivity. *Nature* 444(7120):752–755.

10. Hoffmann GF, Nyhan WL, Zschocke J, Kahler SG, & Mayatepek E (2002) *Inherited metabolic diseases* (Lippincott Williams & Wilkins, Philadelphia).

11. Erlich HA, Gelfand D, & Sninsky JJ (1991) Recent advances in the polymerase chain-reaction. *Science* 252:1643–1651.

12. Edger PP, Heidel-Fischer HM, Bekaert M, Rota J, Glöckner G et al., (2015) The butterfly plant arms-race escalated by gene and genome duplications. *Proceedings of the National Academy of Sciences, USA* 112(27):8362–8366.

13. Piškur J, Rozpedowska E, Polakova S, Merico A, & Compagno C (2006) How did *Saccharomyces* evolve to become a good brewer? *Trends in Genetics* 22:183–186.

14. Thiele I, Swainston N, Fleming RM, Hoppe A, Sahoo S, Aurich MK, Haraldsdottir H, Mo ML, Rolfsson O, & Stobbe MD (2013) A community-driven global reconstruction of human metabolism. *Nature Biotechnology* 31(5):419–425.

15. Mathews CK & Van Holde KE (1996) *Biochemistry* 2nd Edition (The Benjamin/Cummings Publishing Company Inc., Menlo Park).

16. Levine IN (1995) *Physical chemistry*. 4th Edition (McGraw-Hill, Inc, New York), p. 901.

17. Michaelis L & Menten ML (1913) Die Kinetik der Invertinwirkung. *Biochemische Zeitschrift* 49:333–369.

18. Johnson KA & Goody RS (2011) The original Michaelis constant: translation of the 1913 Michaelis–Menten paper. *Biochemistry* 50(39):8264–8269.

19. Briggs GE & Haldane JBS (1925) A note on the kinetics of enzyme action. *Biochemical Journal* 19(2):338.

20. Teusink B Passarge J, Reijenga CA, Esgalhado E, van der Weijden CC et al., (2000) Can yeast glycolysis be understood in terms of *in vitro* kinetics of the constituent enzymes? Testing biochemistry. *European Journal of Biochemistry* 267:5313–5329.

21. Van Slyke DD & Cullen GE (1914) The mode of action of urease and of enzymes in general. *Journal of Biological Chemistry* 19(2):141–180.

22. Hoops S, Sahle S, Gauges R, Lee C, Pahle J, Simus N, Singhal M, Xu L, Mendes P, & Kummer U (2006) COPASI--a complex pathway simulator. *Bioinformatics* 22(24):3067–3074.

23. Kacser H & Burns JA (1981) The molecular basis of dominance. *Genetics* 97(3–4):639–666.

24. Wagner A (2005) *Robustness and evolvability in living systems* (Princeton University Press, Princeton, NJ).

25. Mendel G (1866) Versuche über Pflanzen hybriden. *Verhandlungen des naturforschenden Vereins in Brünn*, pp 3–47.

26. Hartl DL & Orel V (1992) What did Gregor Mendel think he discovered? *Genetics* 131(2):245.

27. Pierce BA (2008) *Genetics: A conceptual approach* (Macmillan).

28. Crow JF (1987) Population genetics history: a personal view. *Annual Review of Genetics* 21(1):1–22.

29. Fisher RA (1928) The possible modification of the response of the wild type to recurrent mutations. *American Naturalist* 62(679):115–126.

30. Fisher RA (1928) Two further notes on the origin of dominance. *American Naturalist* 62(683):571–574.

31. Wright S (1929) Fisher's theory of dominance. *American Naturalist* 63(686):274–279.

32. Haldane J (1930) A note on Fisher's theory of the origin of dominance, and on a correlation between dominance and linkage. *American Naturalist*:87–90.

33. Orr HA (1991) A test of fisher's theory of dominance. *Proceedings of the National Academy of Sciences* 88(24):11413–11415.

34. Charlesworth B (1979) Evidence against Fisher's theory of dominance. *Nature* 278(5707):848–849.

35. Gould SJ (1991) Fleeming Jenkin Revisited. *Bully for brontosaurus* (W. W. Norton and Company, New York), pp 340–353.

36. Simunek M, Hoßfeld U, & Wissemann V (2011) 'Rediscovery' revised–the cooperation of erich and armin von tschermak-seysenegg in the context of the 'rediscovery'of Mendel's laws in 1899–1901 1. *Plant Biology* 13(6):835–841.

37. Thompson L & Watson N (2017) TEM image of "prochlorococcus marinus", a globally significant marine cyanobacterium. https://commons.wikimedia.org/wiki/File:Prochlorococcus_marinus.jpg

38. Pires JC & Conant GC (2016) Robust yet fragile: expression noise, protein misfolding and gene dosage in the evolution of genomes. *Annual Review of Genetics* 50(1):113–131.

39. NIAID Visual and Medical Arts (9/26/2024) Lab Mouse. in *NIAID BIOART Source*. https://bioart.niaid.nih.gov/bioart/283

40. NIAID Visual and Medical Arts (9/26/2024) Lab Mouse (Brown). in *NIAID BIOART Source*. https://bioart.niaid.nih.gov/bioart/281

41. Schafberg R & Swalve H (2015) The history of breeding for polled cattle. *Livestock Science* 179:54–70.

Growing Too Big
Full-Cell Metabolic Models

6

Tho' much is taken, much abides; and tho'
We are not now that strength which in old days
Moved earth and heaven, that which we are, we are;
One equal temper of heroic hearts,
Made weak by time and fate, but strong in will
To strive, to seek, to find, and not to yield.
—Alfred, Lord Tennyson, Ulysses

WHEN SHOULD I SWITCH MODELS?

In the preceding chapters we have discussed the value of quantitative models, how to develop them, and how to analyze them. But one of the important scientific skills we have yet to discuss is when it is time to abandon a model. Recall one of Chapter 3's epigraphs, by George Box: if all models are wrong, the fact that a model makes the error of omitting some details is not necessarily a reason to stop using it. So now we will discuss two reasons for abandoning a model. First, the model may become too complex, such that either the computational burden of analyzing it or the experimental burden of trying to match the model to the real biological system becomes prohibitive. Second, the model assumptions may preclude understanding a behavior we are interested in. We will explore the first problem now, trying to scale *up* the biochemical models we built in the last chapter to model the whole cell. In Chapter 7 we will explore the second problem: what happens when our biochemical systems become so *small* that we need to track the behavior of individual molecules, something we cannot do with the models of the previous chapter.

Too Big: Modeling Metabolism at the Cell or Organismal Scale

Figure 6.1 is an (incomplete) illustration of the complexity of the biochemical *map* or *network* of the cyanobacteria *Synechocystis* spp. PCC 6803 [1]. This illustration is called a *network* diagram, because we have simplified the cell's metabolism to a set of compounds (dots) and the reactions that connect them (lines). We will explore this idea of networks in considerably more detail in Chapter 14: for the moment, the main message of Figure 6.1 is precisely how *difficult* this image is to understand. As mentioned, there are hundreds of different compounds found in this type of cell, with a similarly large number of different biochemical reactions that convert among them.

DOI: 10.1201/9781003687504-6

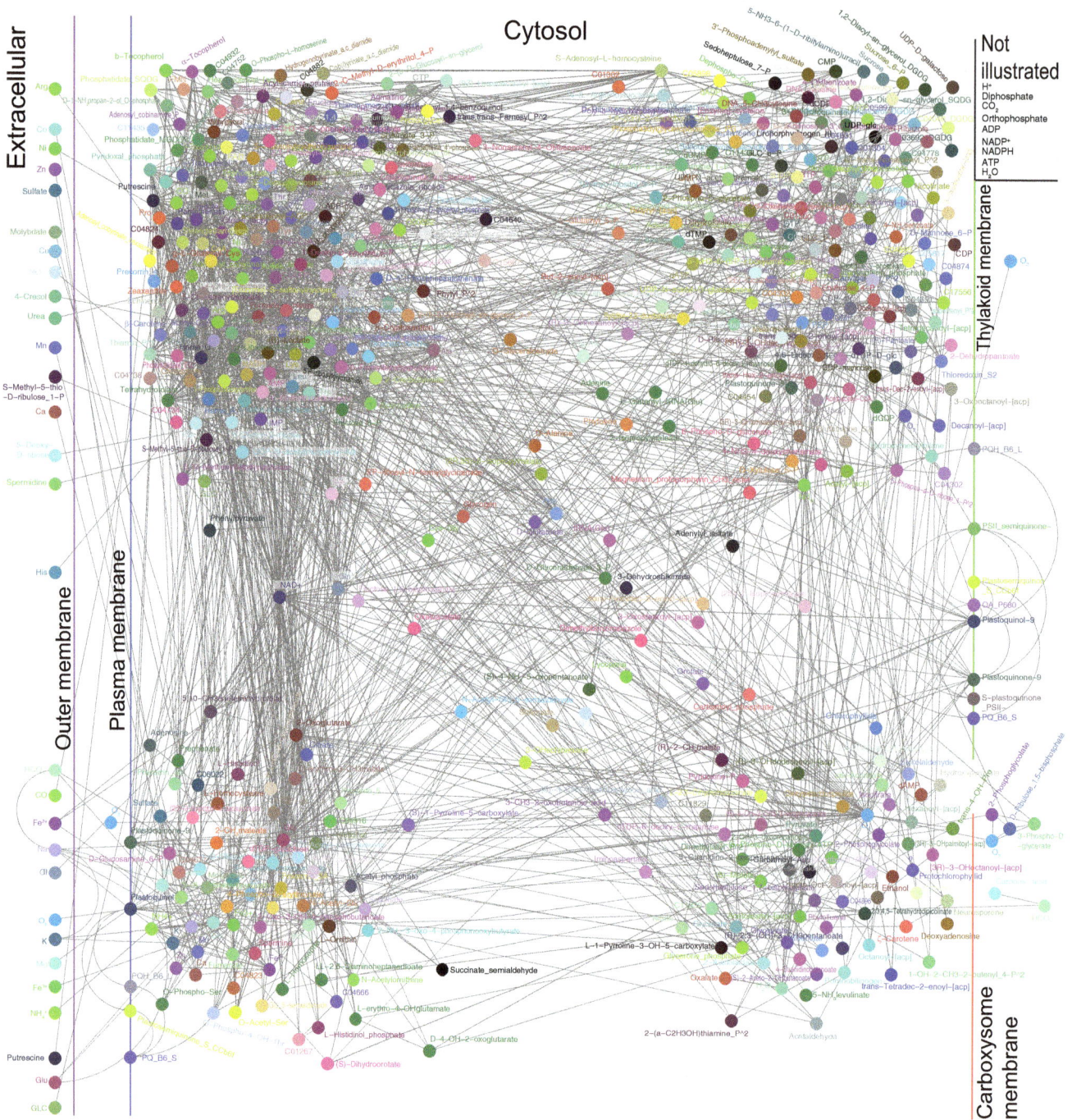

Figure 6.1 **A diagram of the metabolic map of *Synechocystis* sp. PCC 6803 [1].** Each of the small molecules this bacterium is known to use or synthesize is represented as a colored dot. We then join a pair of compounds by a gray line if they are both present in the same biochemical reaction. (We will discuss these *network diagrams* more in Chapter 14.) The map is compartmentalized, meaning that we distinguish between metabolites that are imported or exported by the cell, which are found outside the bacteria's outer membrane. Most metabolites are found in the cytosol, but there are also a few reactions found in the thylakoid membrane, the thylakoid lumen, and the carboxysome [2]. A few compounds like water and ATP are so prevalent in the map that their depiction would be confusing: these omitted metabolites are listed on the upper right. Likewise, compounds with very long names are listed by their compound ID from the original dataset [2].

CAN WE SCALE UP THE MODELS WE MADE IN CHAPTER 5 TO ANALYZE A FULL BACTERIAL (OR HUMAN) CELL?

In principle, the answer to this question is yes. Solving the system of a few thousand differential equations that describe the cell's thousands of compounds and reactions is not computationally infeasible. In fact, a model of the metabolism of human red blood cells was published in the late 1980s [3]. In practice, however, there is a significant problem. To understand that problem, we need to consider a simpler model that was published about 25 years ago by Teusink et al. [4]. The system of equations in Equation 6.1 describe this model, which is one of the largest metabolic systems yet analyzed—that of the glycolysis pathway from bakers' yeast:

$$\frac{dGlc}{dt} = v_{HXT} - v_{HXK}$$

$$\frac{dG6P}{dt} = v_{HXK} - v_{PGI} - 2v_{trehalose} - v_{glycogen}$$

$$\frac{dF6P}{dt} = -v_{PGI} - v_{PFK}$$

$$\frac{dF1,6bP_2}{dt} = v_{PFK} - v_{ALD}$$

$$\frac{dTrio}{dt} = 2v_{ALD} - v_{GraPDH} - v_{glycerol}$$

$$\frac{dBPG}{dt} = v_{GraPDH} - v_{PGK}$$

$$\frac{d3GriP}{dt} = v_{PGK} - v_{PGM}$$

$$\frac{d2GriP}{dt} = v_{PGM} - v_{ENO}$$

$$\frac{dPEP}{dt} = v_{ENO} - v_{PYK}$$

$$\frac{dPYR}{dt} = v_{PYK} - v_{PDC}$$

$$\frac{dAcAld}{dt} = v_{PDC} - v_{ADH} - 2v_{succinate}$$

$$\frac{dP}{dt} = -v_{HK} - v_{PFK} + v_{PGK} + v_{PYK} - v_{ATP_{ase}}$$
$$- v_{trehalose} - v_{glycogen} - 4v_{succinate}$$

$$\frac{dNADH}{dt} = v_{GraPDH} - v_{ADH} + v_{glycerol} + 3v_{succinate}$$

$$\frac{dNAD}{dt} = -\frac{dNADH}{dt}$$

$$(6.1)$$

Kinetic Parameters Are Difficult to Measure

We will see in Chapter 15 that analyzing Equation 6.1 computationally takes only seconds. Hence, it seems clear that computation is not the limiting factor for such models. Instead, the difficulty with expanding models like the one in Equation 6.1 to the cellular scale is in obtaining the values of the v parameters. As we saw in Chapter 5, each of these reaction velocities depend on a V_{max}-like parameter, as well as several binding affinities related to the k_1 and k_{-1} parameters. (You might also notice that, unlike our example in that chapter, these reactions involve multiple substrates and products, requiring potentially more parameters per reaction.) The model also requires an estimate of the initial concentration for each metabolite. Teusink et al. [4] obtained all of these values for Equation 6.1 by compiling results from other researchers who had performed many complex and difficult experiments to measure them. Likewise, the red blood cell model incorporated the results of experiments over the course of nearly 20 years [3]. Yet, as Figure 6.1 and **Figure 6.2** illustrate, the biochemistry described by the equations comprising Equation 6.1 is vastly less complicated than that of a full animal, plant, or fungal cell.

Hence, while it is possible to develop large kinetic models of cells, those models cannot predict real biology without correct values for the model parameters. And those parameters must be measured experimentally. Unfortunately, the effort required to do so would be enormous [5], likely months of laboratory work for each parameter.

Complicating things further are the results presented by Kacser and Burns [6]. We explored their work in the context of changing the concentration of E_2 in our

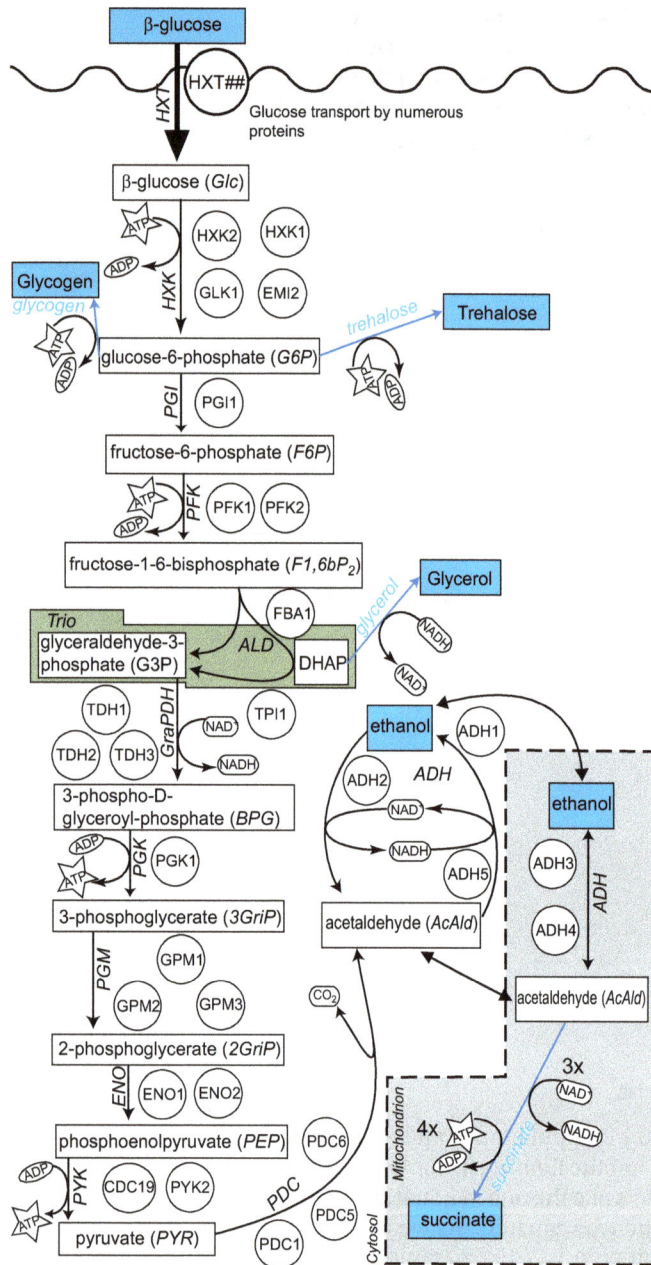

Figure 6.2 A diagrammatic view of the set of reactions described by the equations comprising Equation 6.1. Boxes represent the metabolites modeled: the italicized abbreviations correspond to the differential equations in Equation 6.1. Blue boxes are metabolites assumed to be in a steady state and hence not included in the equations. Reaction names from Equation 6.1 are shown next to reaction arrows while names in circles correspond to the gene or genes in baker's yeast catalyzing that reaction [2]. In the cases of glycogen, glycerol, trehalose, and succinate, several enzymatic steps are combined, and hence no corresponding genes are indicated. The aldolase reaction produces two products, but the model assumes that these can freely interconvert, hence they are treated as a single *Trio* metabolite (green box). (Adapted from Conant GC & Wolfe KH (2007) *Molecular Systems Biology* 3:129).

toy model in Chapter 5. But recall that in that model, V_{max} and enzyme concentration change in direct proportion to each other. As a result, the insensitivity of the model to the value of $[E_2]$ holds equally well for our estimates of V_{max} of that reaction. In other words, the model may tolerate errors *some* of our parameter estimates [7]. Unfortunately, it is not easy to work out in advance which parameters we do need precise values for, making models such as that represented by Equation 6.1 easier to write down than to apply to real biological problems.

Kinetic Parameters Are Also Species Specific

Another serious problem is that most of the metabolic models researchers have created are derived from a handful of well-studied species. Our two examples illustrate this fact, being drawn from humans and from bakers' yeast. Why bakers' yeast is such a well-studied organism is a fascinating topic in its own right: to over-simplify slightly, this species both grows well in a laboratory and has the economic importance of being humanity's key domesticated microorganism [8]. Now, many of the biochemical reactions found in humans are also found in bakers' yeast; even more reactions are common to both humans and other mammals. Unfortunately, while those enzymes are shared, their kinetics are not. As a result, every time we wanted to build a metabolic model of a new species, we would have to repeat all of the complex kinetic measurements we just described.

Data in Quantitative Biology

The preceding paragraphs remind us of a topic we have neglected so far in this book: quantitative approaches in biology are most powerful when we can couple them with biological data we have collected experimentally. For the SIR and Lotka–Volterra models, the necessary data are simply a few parameter values, although, as we saw in those chapters, those values could easily be specific to a particular disease outbreak or forest. As we move deeper into this book, however, the datasets involved will be getting progressively larger. For even just the model of yeast glycolysis, we have just seen that dozens of parameters, measured with dozens of experiments, are needed. What if we wanted to consider *all* the reactions that an organism could carry out? The natural place to start to address that question is the organism's *genome*: the full contents of its genetic inheritance from its parent(s).

BIOCHEMISTRY IN THE ERA OF UBIQUITOUS GENOMES

The problem of needing to make hundreds of kinetic measures to model the metabolism of each new species would not be quite so galling were it not for the fact that it has become enormously easier to identify all the different enzymes in almost any species. Both the technical difficulties and the costs associated with DNA sequencing have dropped in an extraordinary fashion in the past two decades or so [9] (see Chapter 12). As a result, there are now complete genome sequences available for many thousands of organisms [10, 11]. Through techniques similar to those we will discuss in Chapter 12, we can compare genes among genomes and identify genes in a new genome that match the known enzyme-coding genes in another. Such comparisons are relatively robust even for distantly related species because metabolic functions change very slowly over evolutionary time [12, 13]. The reason for this conservation is probably that metabolic reactions evolved early in the history of life [14] and then were passed down to modern organisms relatively unchanged. Hence, if we have a gene in a new organism with a similar sequence to that for a known enzyme-coding gene in another species, we can be fairly certain that the new organism also possesses the reaction in question.

Thus, we find ourselves in the irritating position of being able to computationally reconstruct the metabolic potential of an organism directly from its genome [15, 16] but unable to model it further because of the difficulties with the differential equation models just mentioned. Currently, there are approximately 100 complete, genome-scale models of metabolism constructed from genome sequences [17], but none of them can be modeled with the approaches of Chapter 5 due to the lack of kinetic parameters.

Is there a way around this problem? As we will see in the following sections, paradoxically, we can work through this problem by *simplifying* our metabolic model.

Constraint-Based Metabolic Models: Dispensing with Dynamics

The key simplification enabling us to study metabolism at the whole-cell scale across many organisms is to assume that the metabolite concentrations are unchanging in time. This assumption may appear to be silly if not altogether absurd: surely cells are not metabolically static! Yet the assumption is less ridiculous that it appears. The analogy I use is that of a well-run factory. In a car factory, component parts of the automobiles are delivered at some daily rate. Those parts are then assembled into cars that are driven away. If the rate of arrival of any part is too low, the factory will be idled waiting for that part: the car cannot be driven without all its parts. But having parts arrive too fast is no better: sooner or later, the factory will run out of space to store them. In the same way, a cell does not want either to run out of any given metabolite or to be so full of that metabolite that its membrane breaks open. So at least at a high level, cells do exist in some sort of metabolic steady state with controlled metabolite levels.

The effects that constraining metabolite concentrations to be fixed has on our modeling approach are best understood with an example. **Figure 6.3** shows a simple example where a cell "eats" two compounds, A and B, and uses them to synthesize compounds C and D. From C and D, new copies of the cell are made. We can write this system as a set of six chemical equations:

$$A + B \xrightarrow{r_1} C$$
$$B + C \xrightarrow{r_2} D$$
$$C + D \xrightarrow{r_3} X$$
$$A_o \xrightarrow{r_4} A$$
$$B_o \xrightarrow{r_5} B$$
$$X \xrightarrow{r_6} X_o$$

(6.2)

We have represented the external supply of A and B by defining two *external* metabolites A_o and B_o. We then further define two transport reactions r_4 and r_5 that bring those nutrients into the cell. We have also replaced the new cells with a pseudo-metabolite X and added an exporting reaction for it: r_6. The necessity of these transport reactions is probably not obvious at the moment but should become clear as we proceed.

We can easily write a system of differential equations for the five metabolites indicated in Equation 6.2:

$$\frac{dA}{dt} = \upsilon_{r_4} - \upsilon_{r_1}$$
$$\frac{dB}{dt} = \upsilon_{r_5} - \upsilon_{r_1} - \upsilon_{r_2}$$
$$\frac{dC}{dt} = \upsilon_{r_1} - \upsilon_{r_2} - \upsilon_{r_3}$$
$$\frac{dD}{dt} = \upsilon_{r_2} - \upsilon_{r_3}$$
$$\frac{dX}{dt} = \upsilon_{r_3} - \upsilon_{r_6}$$

(6.3)

Here υ_{r_x} represents how often reaction r_x turns over per unit time (perhaps measured in moles per liter per second). In a full model, this reaction's velocity (or *flux*) would depend on the concentrations of all the metabolites. But in the framework we are about to build, that complexity is not needed. To see why, recall what it means for the concentrations of these metabolites to be unchanging. We can write this constraint simply as $dA/dt.dX/dt = 0$.

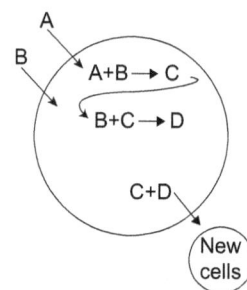

Figure 6.3 An example system for constraint-based modeling. Two metabolites A and B are imported from the environment and used to synthesize C and D, which are the components of new cells.

$$0 = v_{r_4} - v_{r_1}$$
$$0 = v_{r_5} - v_{r_1} - v_{r_2}$$
$$0 = v_{r_1} - v_{r_2} - v_{r_3} \qquad (6.4)$$
$$0 = v_{r_2} - v_{r_3}$$
$$0 = v_{r_3} - v_{r_6}$$

Before we start analyzing the equations in Equation 6.4, we should pause to orient ourselves. A rate of change of zero does not mean that the underlying concentration is zero, just that it is constant. In other words, constant metabolite concentrations do not mean that there is no *flow* through the system. In our factory analogy, the number of windshields at the factory on any given day is the same. Every day some number of windshields $+x$ are delivered and exactly the same number $-x$ are installed into new cars and leave the factory.

Figure 6.4 gives another analogy for this problem: we have a set of buckets connected by a series of pipes. The questions are whether there is a set of flow rates through the pipes that leaves the level of the water in every bucket unchanged and, if so, what those flow rates are. In this example, the top pipe corresponds to the input of metabolites into the system (A_o and B_o in Equation 6.2), and the bottom pipes indicate the formation of new cells (X_o in Equation 6.2). How can we find these flow rates? To answer that question might require a trip down memory lane.

Digression: High School Mathematics and Linear Equations in the Real World

To work with the equations comprising Equation 6.4, we should first look at them more closely. The values we are interested in are the reaction velocities ($v_{r_x}s$), which is another way to describe the flow rates through the pipes. Notice that these $v_{r_x}s$ are, in some sense, "nice": they do not appear in complex functions

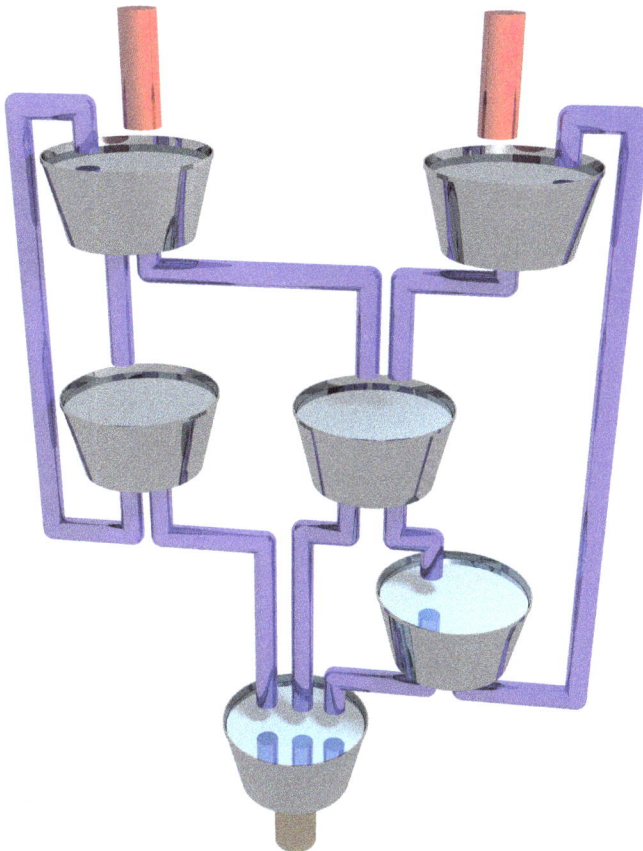

Figure 6.4 A metaphor for a metabolic system in a steady state. Water enters through the upper pipes and leaves through the lower ones: we are seeking a combination of flow rates through the pipes that leaves the level of water in each bucket unchanged.

(sin, cos, e^x), nor are they multiplied together or raised to powers of themselves. Instead, they are what we refer to as *linear equations*. Depending on your age and educational history, that phrase may bring up memories: some good, some bad.

Many of us studied linear equations in secondary school, for the most part concentrating on examples such as:

$$y = -4x + 3$$
$$y = 2x - 4$$
(6.5)

The goal then was to see if there was some combination of values of x and y that satisfied both equations. Geometrically, we can understand the problem by first plotting the set of points that satisfies the first equation in Equation 6.5 and then plotting those that satisfy the second equation on that same plot. There are an infinite number of points in each set, and they each form a line in the x–y plane, as shown in **Figure 6.5**.

In the two dimensions of the x–y plane, there are only three ways these two lines can behave: they may be identical (the two equations are different forms of each other), the lines may be parallel and never cross (if the slope of x is the same in both cases but the points of intersection on the x-axis differ), or they may cross at exactly one point, as shown in Figure 6.5.

Such *systems* of linear equations involving only two variables can be useful but are of limited relevance to biological systems with hundreds of potential variables. However, there is an analogous set of rules and procedures for dealing with systems having arbitrary numbers of variables. This area of mathematics is called *linear algebra*. An example with only one more variable (and hence dimension) is sufficient for a conceptual understanding of this generalization. We will start with a pair of equations in x, y, and z:

$$z = -2x + y - 1$$
$$z = 2x - y + 1$$
(6.6)

If we make the required three-dimensional plot of Equation 6.6 (**Figure 6.6B**), we realize that these equations each describe the three-dimensional analog of a line: a plane. Moreover, in this case, the two planes in question intersect each other along a line. On the other hand, we can also create a situation where the two planes do not intersect:

$$z = 2x + y - 10$$
$$z = 2x + y + 10$$
(6.7)

These equations are illustrated in **Figure 6.6A**. The two planes are parallel to each other, and hence there are no values of $x, y,$ and z that satisfy both equations in Equation 6.7.

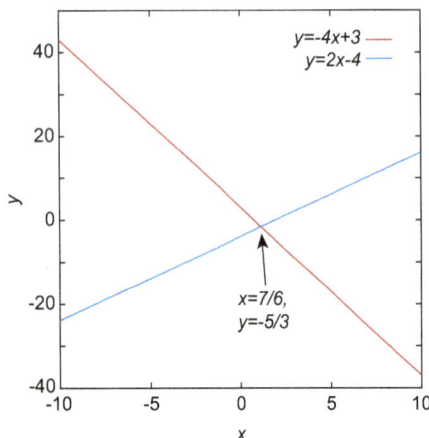

Figure 6.5 Two equations with two unknowns. The set of equations in Equation 6.5 generate two lines, which cross at exactly one point in the x–y plane.

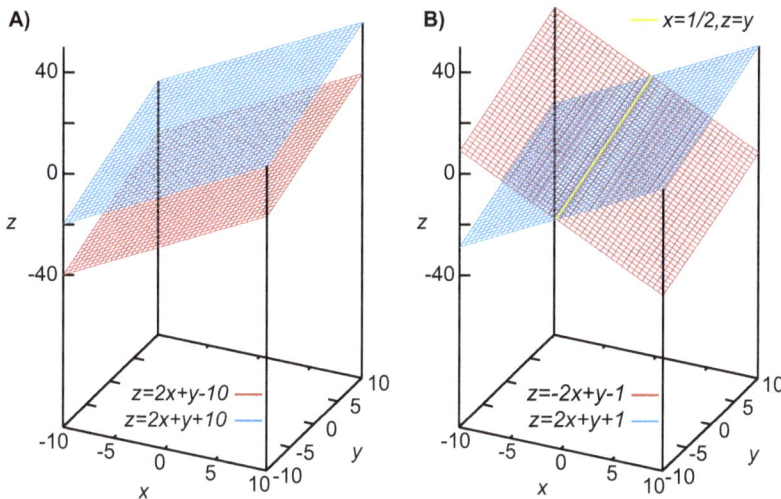

Figure 6.6 **Systems of linear equations in three variables.** In **A**, the equations describe parallel planes: there are no common points between the two. In **B**, the two planes intersect at a line (*yellow*). In Exercise 6.1, you can solve for the equation of this line. To do so, try first solving for the value of x when y=z=0. Then solve for y and z. (Note the figure mistakes the solution as x=1/2 instead of x=-1/2).

From these examples we can enumerate some general rules about the relationships between the number of equations we have, the number of variables involved, and the form of the solutions. If the number of variables is *greater* than the number of equations, there may be *no* solution (as is the case with the parallel planes of Figure 6.6A) or there will be infinitely many solutions. Such is the case in Figure 6.6B, where we have three variables ($x, y,$ and z) but only two equations. If there are as many equations are variables, which is the case for the two equations in Equation 6.5, we can actually have no solutions, one solution, or infinitely many solutions. If the number of equations is *greater* than the number of variables, there will generally be no solution, but in some cases there will be one or more solutions (see Chapter 15). The details of determining, for any given set of equations, how many solutions can be found are beyond the scope of our work here [18]. Instead, we will explore the most likely characteristics of metabolic models within this framework.

Steady-State Metabolic Models and Linear Algebra

In the system of Equation 6.4, notice that we have five equations (corresponding to the $m = 5$ metabolites in the model) but *six* unknown reaction velocities. Almost all metabolic models will behave this way. This is because the number of equations will be given by the number of different metabolites in the system, but the number of variables ($v_{r_x}s$) will be number of different reactions. Since nearly every metabolite will participate in more than one reaction, we will always have more variables (i.e., unknown reaction *fluxes*) than equations (i.e., metabolites). This fact alone does not guarantee that there will be multiple solutions to the metabolic problem, but it does mean that it is possible there will be.

Before we go further, we need to give a few definitions. I have already used the term *flux*: in fact, our goal is to find a set of reaction fluxes that give a metabolic steady state. Flux is best thought of as reaction turnovers per unit time. For a simple reaction like the import of a compound ($A_o \rightarrow A$), the units of flux can therefore be moles per second (or moles per liter per second if we prefer).

We are now able to describe our problem of interest in terms of a set of equations of the form:

$$b_m = m_{1,m}x_1 + m_{2,m}x_2 .. m_{n,m}x_n \qquad (6.8)$$

Each metabolite m has such an equation (for a total of five in our example). For each metabolite, the value of b_m is that metabolite's change in concentration in time. We are assuming steady state, such that $b_m = 0$ for every equation in Equation 6.4. The variables we are solving for are the $v_{r_1} .. v_{r_n}$ reaction fluxes, which we will now rename to $x_1 .. x_n$. Meanwhile the $m_{1,m} .. m_{n,m}$ slope values are simply the stoichiometric coefficients on the reactions from Equation 6.2. Looking back at

Equation 6.4, we see that it thus follows this form, keeping in mind that most of the $m_{i,j}$ values are 0 because that reaction does not involve the metabolite of interest.

SOLUTIONS TO THE STEADY-STATE PROBLEM

Unlike the examples in Figure 6.6, the equations comprising Equation 6.4 possess an important feature: the left side of each equation is 0. This condition means that the hyperplane (n-dimensional plane) that describes each equation passes through the origin. This fact has a very important corollary: we *know* that there is at least one solution for Equation 6.4. That solution is obvious when one notices that setting all the x_is to 0 always satisfies the equations in Equation 6.4 or any other metabolic steady-state model. Reassuringly, this observation accords with our intuition: a trivial way to make all the reaction concentrations unchanging in time is to have the flux through every reaction be equal to 0.

So, a properly constituted steady-state metabolic model is certain to have at least one solution; the question is whether there are more solutions. To find out, we need to apply some tools of linear algebra. The first step is to rewrite the equations in Equation 6.4 in a *matrix* format:

$$\begin{vmatrix} -1 & 0 & 0 & 1 & 0 & 0 \\ -1 & -1 & 0 & 0 & 1 & 0 \\ 1 & -1 & -1 & 0 & 0 & 0 \\ 0 & 1 & -1 & 0 & 0 & 0 \\ 0 & 0 & 1 & 0 & 0 & -1 \end{vmatrix} \cdot \begin{vmatrix} x_1 \\ x_2 \\ x_3 \\ x_4 \\ x_5 \\ x_6 \end{vmatrix} = \begin{vmatrix} 0 \\ 0 \\ 0 \\ 0 \\ 0 \end{vmatrix} \tag{6.9}$$

Which we could rewrite as:

$$M \cdot \vec{x} = \vec{0} \tag{6.10}$$

Here, our matrix **M** is just all the $m_{i,j}$ values from Equation 6.8, with $\vec{x}$ and $\vec{0}$ being vectors of x_is and 0s, respectively.

While the matrix format of Equation 6.9 may look foreboding at first, it is actually simply a compact way of writing down linear algebra problems. **Figure 6.7** illustrates how to convert from the matrix format back to the standard format of Equation 6.4. In essence, we take the column of the n unknown $x_1..x_n$ values, rotate

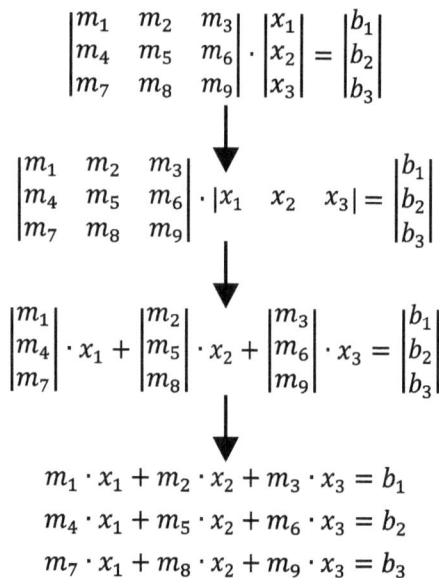

$$\begin{vmatrix} m_1 & m_2 & m_3 \\ m_4 & m_5 & m_6 \\ m_7 & m_8 & m_9 \end{vmatrix} \cdot \begin{vmatrix} x_1 \\ x_2 \\ x_3 \end{vmatrix} = \begin{vmatrix} b_1 \\ b_2 \\ b_3 \end{vmatrix}$$

$$\downarrow$$

$$\begin{vmatrix} m_1 & m_2 & m_3 \\ m_4 & m_5 & m_6 \\ m_7 & m_8 & m_9 \end{vmatrix} \cdot \begin{vmatrix} x_1 & x_2 & x_3 \end{vmatrix} = \begin{vmatrix} b_1 \\ b_2 \\ b_3 \end{vmatrix}$$

$$\downarrow$$

$$\begin{vmatrix} m_1 \\ m_4 \\ m_7 \end{vmatrix} \cdot x_1 + \begin{vmatrix} m_2 \\ m_5 \\ m_8 \end{vmatrix} \cdot x_2 + \begin{vmatrix} m_3 \\ m_6 \\ m_9 \end{vmatrix} \cdot x_3 = \begin{vmatrix} b_1 \\ b_2 \\ b_3 \end{vmatrix}$$

$$\downarrow$$

$$m_1 \cdot x_1 + m_2 \cdot x_2 + m_3 \cdot x_3 = b_1$$

$$m_4 \cdot x_1 + m_5 \cdot x_2 + m_6 \cdot x_3 = b_2$$

$$m_7 \cdot x_1 + m_8 \cdot x_2 + m_9 \cdot x_3 = b_3$$

Figure 6.7 The procedure for multiplying an *mxn* matrix by a vector of length *n*. Each element of the vector is multiplied by all the elements of its corresponding matrix column: then the per-column results of the multiplication are added to produce a final vector of length *m*.

them $90°$, and multiply each column of the matrix by the corresponding x_i. Hence, there will be n columns in the matrix (one per x_i). This column of x_is is referred to as a *vector*, and the operation shown in Figure 6.7 is *matrix-vector multiplication*. The result of that multiplication is a solution vector, which in our case is the column of 0s in Equation 6.9. The solution vector has the same m rows as does the matrix, which again is the number of metabolites in the model.

We can now see that this matrix notation allows us to see the same trivial solution of the problem we had already identified, namely $x_1..x_n = 0$:

$$\begin{vmatrix} -1 & 0 & 0 & 1 & 0 & 0 \\ -1 & -1 & 0 & 0 & 1 & 0 \\ 1 & -1 & -1 & 0 & 0 & 0 \\ 0 & 1 & -1 & 0 & 0 & 0 \\ 0 & 0 & 1 & 0 & 0 & -1 \end{vmatrix} \cdot \begin{vmatrix} 0 \\ 0 \\ 0 \\ 0 \\ 0 \\ 0 \end{vmatrix} = \begin{vmatrix} 0 \\ 0 \\ 0 \\ 0 \\ 0 \end{vmatrix} \tag{6.11}$$

Are there other solutions? This is a difficult question for the arbitrary problems we are about to study. There can in fact be a very large number of possible solutions, and the mathematics of how to compute all those solutions and represent them as vectors is beyond the scope of this book. However, the advantage of this simple problem is that it is rather easy to find another potential solution (Exercise 6.2).

$$\begin{vmatrix} -1 & 0 & 0 & 1 & 0 & 0 \\ -1 & -1 & 0 & 0 & 1 & 0 \\ 1 & -1 & -1 & 0 & 0 & 0 \\ 0 & 1 & -1 & 0 & 0 & 0 \\ 0 & 0 & 1 & 0 & 0 & -1 \end{vmatrix} \cdot \begin{vmatrix} 2 \\ 1 \\ 1 \\ 2 \\ 3 \\ 1 \end{vmatrix} = \begin{vmatrix} 0 \\ 0 \\ 0 \\ 0 \\ 0 \end{vmatrix} \tag{6.12}$$

In Exercise 6.2, you can also confirm that the flux vector in Equation 6.12 does give a zero of 0s for the metabolite concentrations. On reflection, we can also see that the solution of Equation 6.12 is only one among many: any constant multiplied by the solution vector in Equation 6.12 is also a solution.

$$\begin{vmatrix} -1 & 0 & 0 & 1 & 0 & 0 \\ -1 & -1 & 0 & 0 & 1 & 0 \\ 1 & -1 & -1 & 0 & 0 & 0 \\ 0 & 1 & -1 & 0 & 0 & 0 \\ 0 & 0 & 1 & 0 & 0 & -1 \end{vmatrix} \cdot c \cdot \begin{vmatrix} 2 \\ 1 \\ 1 \\ 2 \\ 3 \\ 1 \end{vmatrix} = \begin{vmatrix} 0 \\ 0 \\ 0 \\ 0 \\ 0 \end{vmatrix} \tag{6.13}$$

In essence, this result is just an illustration of the idea that in most cases these metabolic models have an infinite number of potential solutions. In this simple example, those solutions all lie on a single line, but for real systems they will comprise what is known as a *hyperplane*: a subspace of the full n-dimensional space containing many possible combinations of fluxes that satisfy the steady-state requirement.

Exercise 6.1

Try to find the set of points in Figure 6.6B that satisfy both the equations in Equation 6.6.

Exercise 6.2

Using R, find a solution to Equation 6.9, where not all of the x_i values are 0. You can represent the matrix and constant concentrations in R with the command:

- ```
  A<-matrix(c(-1,-1,1,0,0,0,-1,-1,1,0,0,
  0,-1,-1,1,1,0,0,0,0,1,0,0,0,0,0,0,0,-1),5,6)
  ```
- ```
  b<-c(0,0,0,0,0,0)
  ```

Using the library MASS, we can compute the *null* space of *A*. The meaning of a null space is slightly beyond our scope here: it just represents all the *x* vectors we could multiply *A* by to get 0 (see Chapter 15).

- ```
  library(MASS)
  ```
- ```
  x<-Null(t(A))
  ```

We can see that this *x* value solves our problem in Equation 6.9:

- ```
  A%*%x
  ```

If we multiply *x* by 9/2, we get our solution in Equation 6.12, which also satisfies Equation 6.9:

- ```
  x2 = x*9/2
  ```
- ```
  A%*%x2
  ```

Modeling Real Cells with Steady-State Models: Flux Balance Analysis

The example we just completed gives us the underlying mathematical framework to model metabolism at a whole-cell scale, if we are willing to accept the assumption of the various metabolites being in steady state. Before we elaborate on the model, however, we should consider some of its important biological properties.

First, the model needs to obey the rules of chemical systems, particularly the conservation of matter: no atoms can be created or destroyed in our metabolic models. Interestingly, we do not enforce this rule in the way we solve the equations, but rather in the equations themselves. In a properly formulated steady-state metabolic model, we can trace the origin and fate of every atom and show that no reactions involve either the creation or destruction of any of those atoms. If our formulation of the model fails to enforce this conservation rule, its predictions will not be meaningful.

A second issue is that the general solutions for Equation 6.13 are just that: general. Thus, negative values of *c* are acceptable, as are arbitrarily large values of *c*. Negative values of *c* would imply a cell is effectively consuming itself. In those circumstances the cell would be converting the complex polymers and lipids it is made up of *back* into the nutrients it used to construct those molecules. While some self-catabolism in cells does occur, a full-cell degradation process is not compatible with life. Moreover, as we saw in the previous chapter, while many or most biochemical reactions are reversible, not all are. As a result, such backward metabolism is not possible chemically. Similarly, cells are not arbitrarily large, so there must be some upper limit on the amount of biochemical flux that can pass through them.

Both these problems suggest that our steady-state model needs to incorporate *boundaries*: limits on the direction and magnitude of the reaction fluxes. These boundaries change the problem from one of standard linear algebra to the more complex case of finding solutions subject to constraints on the values of some x_is [19]. The mathematics of doing so is again beyond the scope of this book, but there are severally freely available software tools for solving this problem [19]. For concreteness, we can call this modeling approach *flux balance analysis*, or *FBA* [19].
  ```

To run FBA, we will need to add three features to our model:

1. We will bound any irreversible reactions i to have flux $x_i \geq 0$.

2. We will define one or more critical nutrient import reactions to have a finite value to prevent arbitrarily large solutions.

3. We will require that the flux through the *biomass* reaction be strictly positive $(x_{biomass} > 0)$.

An Example FBA Analysis

These three added features of FBA modeling are best understood with an example. The one we will use is the metabolic model of Figure 6.1 from *Synechocystis* sp. PCC 6803. The QBio website offers a version of this metabolic map containing 613 reactions and 575 metabolites for which you can run FBA using some code from my laboratory [20] and the free numerical solver GLPK [21]. On the website, all irreversible reactions are correctly bounded to have positive flux. At the same time, nutrients to the model are limited. Nutrients for this photosynthetic organism consist in large part of light, so our boundaries are set to allow only a fixed number of photons to feed the cell. For a simple condition (Exercise 6.3) where the cells get their energy from photons from the sun and their carbon from CO_2, 322 of the 613 reactions are used in the production of new cells (i.e., biomass; Figure 6.8).

Exercise 6.3

Find a set of fluxes that maximize the production of new cells of *Synechocystis* from light, dissolved carbon dioxide (HCO_3^-), sulfur, phosphorous, oxygen, potassium, magnesium, and nitrogen either using the `opt_biomass_rxn` (See http://qbio.statgen.ncsu.edu/data/FBA/) program or the QBio website (see below). Using the program, type

- `opt_biomass_rxn SynechocystisPCC6803_reactions_epa.`
  `txt SynechocystisPCC6803_bounds.txt R90`

Here, R90 is the reaction describing the compounds needed to construct new *Synechocystis* biomass, so the program will seek a steady-state solution that maximizes the production of biomass from the fixed inputs above (and where light is assumed to be the limiting reagent).

Now explore the effects of "knocking out," or removing different reactions from the metabolic model. Pick a few reactions from those shown in the first run (e.g., reactions with nonzero flux in the initial model) and remove them from the model:

- `opt_biomass_rxn SynechocystisPCC6803_reactions_epa.`
  `txt SynechocystisPCC6803_bounds.txt R90 -k:R722`

Using the QBio website, select the *FBA* app (http://qbio.statgen.ncsu.edu/FBA). Then select the *Synechocystis* metabolic network. Running without selecting a knockout reaction will give a histogram of reaction fluxes. Knockout reactions to try:

- R12
- R130
- R206
- R478 (see Figure 6.9A)

You can also sequentially knock out every reaction with flux in the original model ("ALL") to generate Figure 6.9B. In each case, try to interpret your results.

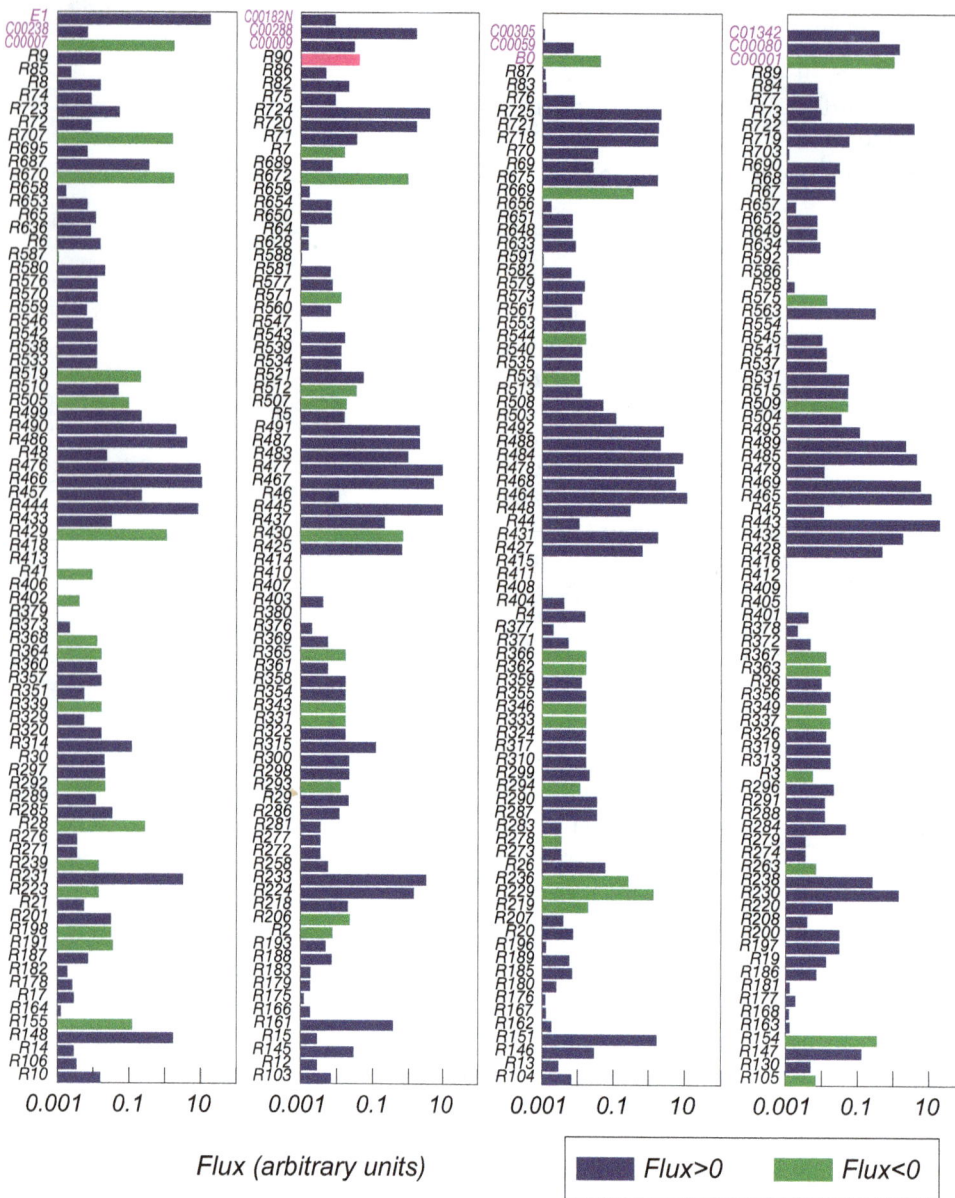

Figure 6.8 Modeling the _Synechocystis_ metabolic network with FBA. Of the 613 reactions in the model, 322 are used when 18.7 units of photons are supplied as an energy source. Under those conditions, a total of 0.04 units of biomass are produced (_magenta_). Reactions that run in the forward direction in the model are shown in blue, and those running in reverse are shown in green. The compounds the cell is able to absorb or excrete are _exchange_ reactions and shown in purple text. You can generate this plot at http://qbio.statgen.ncsu.edu/FBA.

MULTIPLE METABOLIC SOLUTIONS AND THE IMPORTANCE OF BIOMASS PRODUCTION

Simply knowing that a metabolic model has steady-state solutions (see Figure 6.8) is important: it implies that the set of described reactions is sufficient to allow the organism to use the given source of nutrition to build complete cells. However, because there are so many possible solutions to real metabolic models, it is relatively unlikely that an arbitrary solution found mathematically will match the set of reaction fluxes occurring in the cell. To address this problem, we need to revisit that odd reaction #3 in Equation 6.2, namely the _biomass_ component.

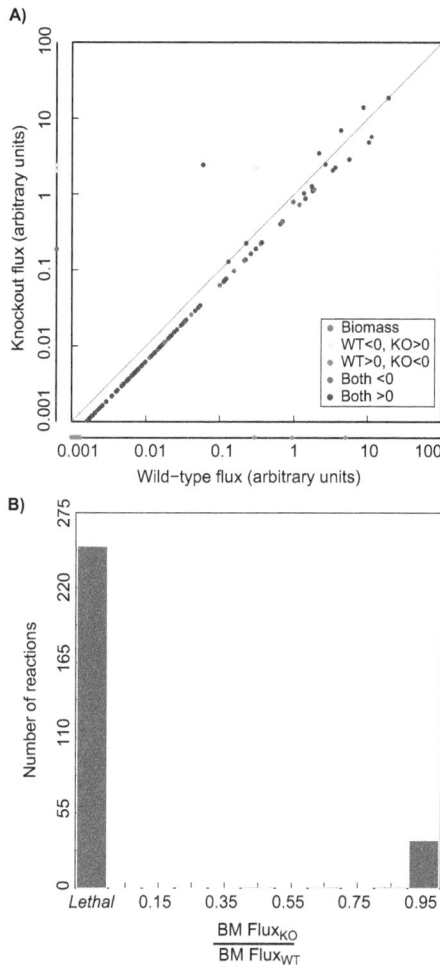

Figure 6.9 Flux-balance analysis and reaction knockouts. (**A**) The result of knocking out a single reaction (R478) from the *Synechocystis* metabolic network using FBA, showing how flux through the various other reactions changes in response. Here the *y*-axis gives the knockout flux and the *x*-axis the wild-type flux. Hence, points above $y = x$ line have higher knockout flux; below it, lower. Under these conditions, less biomass is produced: most of the fluxes, including that for biomass (*magenta dot*), fall below $x = y$. Reactions with nonzero flux for *both* the wild type and the knockout are shown in the main plot: reactions with zero flux in one or the other are shown below the respective axis. (**B**) Categorizing the results of individually knocking out all reactions in the *Synechocystis* metabolic model that showed flux in the wild-type model. On the *x*-axis is the ratio of the knockout biomass flux to the wild-type flux, while on the *y*-axis is the number of reactions with that flux ratio when knocked out. Notice that many reactions do not change the wild-type biomass production when knocked out. You can generate these plots at http://qbio.statgen.ncsu.edu/FBA.

BIOMASS REACTIONS AND EVOLUTIONARY ASSUMPTIONS

For the *Synechocystis* model, the biomass reaction consists of several parts. It includes all 20 amino acids and 8 DNA and RNA nucleotides, as well as a number of lipids and carbohydrates, cell wall components, and a resting steady-state requirement for cellular energy in the form of adenosine triphosphate (ATP) to maintain homeostasis. In other words, it lists the cellular components that must be synthesized to allow cellular growth and division. Our third requirement above means that the existence of a steady-state solution implies that the nutrients we have given to the *Synechocystis* model are sufficient for the synthesis all the biomass components needed.

But the biomass equation is more powerful that just this. To find a steady-state solution that more closely matches the *in vivo* reaction fluxes, we will make a key assumption: *these cells have evolved to maximize biomass production from fixed nutrient inputs* [22, 23]. As we will discuss shortly, this assumption is an evolutionary one rather than a biochemical or mathematical one. To employ it, we will ask our numerical solver to find a set of fluxes that, given the boundaries provided, route the maximum possible flux through the biomass reaction. Another way to describe this idea would be to refer back to Figure 6.4: we are looking for a set of flow rates that not only keep the bucket levels even but, among all such solutions, have the maximum flow rate through the bottom-most pipe, corresponding to the production of new cells.

Why is maximizing biomass production evolutionarily desirable? If we were to presume that nutrients are the limiting factor for microbes' growth, we would expect that those cells should want (evolutionarily speaking) to produce as many new cells as possible from a given input of nutrients. In such organisms, then, the ability to use assimilated nutrients to build new cells is perhaps *the* key aspect of the evolutionary competition between individuals.

The validity of this assumption of the maximization of biomass production was demonstrated in an elegant experiment by Ibarra, Edwards, and Palsson [24]. These authors started with cells of the bacterium *Escherichia coli* and measured their fitness and experimental metabolic fluxes on two different nutrient sources. The first nutrient, or *carbon source*, was malate, a sugar that *E. coli* has evolved to consume. The second carbon source, on the other hand, was glycerol, which is not a nutrient that *E. coli* has commonly encountered in its evolutionary history. For the first carbon source, asking FBA for a set of reaction fluxes that maximized biomass production for a given set of inputs gave predicted fluxes that matched the experimentally measured fluxes well. This result showed that these cells were indeed using the carbon source with maximal efficiency. However, for glycerol, the initial metabolic behavior the researchers measured did not match the flux balance predictions. Instead, the cells were using the glycerol inefficiently, making fewer new cells than the model predicted they could have made.

However, these authors now did something quite clever: they allowed evolution to occur in their laboratory. They let different populations of *E. coli* cells to compete with each other in an environment containing only glycerol for 40 days (recall our work on the *E. coli* generation time from Chapter 3). After this period of laboratory evolution, the researchers found that the metabolic measurements from these evolved cells had converged to the predicted optimum flux. In other words, evolution by natural selection had, in those 40 days, discovered the metabolically optimal way for an *E. coli* cell to grow on glycerol. (They had also incidentally disproven the frequently made claim that "no one has ever observed evolution occur in a laboratory".) Their results suggest that FBA and the maximization of biomass production are appropriate tools for modeling genome-scale metabolism at least under some assumed conditions.

Modeling Gene/Reaction Knockouts with Flux Balance Analysis

FBA thus provides us with a computational method for estimating the metabolic behavior of a cell. However, its true power lies in the ability to *explore* that metabolic potential under different conditions using computation to replace labor-intensive and time-consuming experimental work. We will consider one particular example of this power: the ability to computationally block different reactions and assess the resulting change in biomass production. Conceptually, making such a block can be thought of as computationally "knocking out" the gene encoding the enzyme for one of the system's reactions. However, to be precise, we do need to mention the caveat that some reactions have enzymes encoded by more than one gene (see **Figure 6.10**), meaning that multiple gene knockouts would be required to block that reaction completely.

Figure 6.9A shows the results of knocking out a reaction involving the production of NAHPH from reduced ferredoxin proteins and NAHP+. What is notable about removing this reaction from the model is that biomass production is reduced but does not fall to zero. As can be seen, flux through most of the reactions in the model is reduced in these circumstances, corresponding to their points falling below the $x = y$ line. But a few reactions *increase* in flux in the knockout (falling above the line), and some others are lost (seen on the line below the x axis: i.e., these reactions only show flux in the wild type). From Figure 6.9A we can infer that the model (and hence the cell) has found an alternate series of reactions that imperfectly replace

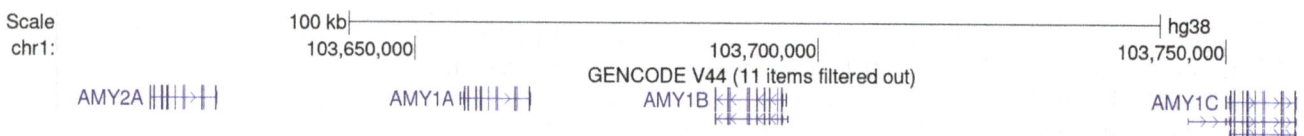

Figure 6.10 **A small region of human chromosome 1, showing multiple copies of genes for amylase (*AMY...*).** The thicker and thinner bars in each gene represent the *exons* and *introns*, respectively [25]. (Generated with the UCSC Genome Browser [26]).

the missing reaction, allowing growth, albeit at a lower rate. We can generalize this result by trying to sequentially remove *all* 321 reactions that show flux in Figure 6.8 (except the biomass reaction). Figure 6.9B shows the result of this analysis. On the *x*-axis of this figure is the ratio of the biomass produced by the knockout model relative to that of the wild type. Perhaps the most striking thing about Figure 6.9B is that it is *not* a single bar at $x = 0$. Remember that we only considered knockouts of reactions that showed flux in the wild-type model. We might expect that all those reactions would be *essential*, each one required for the cells to grow normally and divide. And indeed, more than 80% of the reactions completely block biomass production when knocked out ("*Lethal*" in Figure 6.9B); we can think of these knockouts as effectively killing the cells. But there are several reactions that do not block biomass production completely. What is happening in these cases? How might we describe this behavior in comparison with systems humans have built?

Biological Robustness

One of the most noticeable differences between the technology we engineer and living systems is the degree of *robustness* we find in life [27]. For instance, only about 1,100 of the 5,600 genes in baker's yeast are required for growth in rich media [28]. A similar degree of robustness is seen at the level of the protein sequence, with at least 65% of amino acid–changing substitutions not resulting in a loss of function in various proteins [29, 30]. In contrast, if you think of a human-engineered car or computer, they will generally not function at all if even a single part from the engine or motherboard is disabled.

We have already discussed one form of biological robustness in the previous chapter: the robustness to heterozygous enzyme losses described by Kacser and Burns [6]. A second type of robustness is that due to *duplicated* genes. Many enzymes in our genome are encoded by more than one gene [31], meaning that we would need to remove all the copies of that gene for a complete blockage of the reaction in question. A very interesting example of this phenomenon is shown in Figure 6.10. The amylase gene encodes an enzyme used by humans and other mammals to break down starch into its component sugars [32]. Humans possess a number of duplicated copies of this gene arranged one after the other on chromosome 1. Strikingly, human populations with longer histories of agriculture show more copies of amylase in this region than do populations that have a history of hunting and gathering [33]. The implication of this observation is that the farming populations have seen evolutionary pressure to improve starch digestion. That pressure was then resolved by increasing the level of the amylase protein through its production from multiple copies of its encoding gene.

Circling back to duplicate genes and robustness, we might expect that knocking out a gene with another duplicate copy would show lesser effects than would knocking out a gene without such a copy. Several studies suggest that this expectation is indeed fulfilled, with duplicate genes in a number of organisms being less likely to be essential when removed than other genes are [34–36].

A Third Type of Biological Robustness

What is striking is that we know that *neither* of these sources of robustness can be the source of the observed robustness in Figure 6.9B. Why is that? Well, recall that the main justification for the linear models of metabolism we have been developing in this chapter is that the more complex differential equation–based models that Kacser and Burns used cannot be easily scaled up. Hence, the dynamic sources of robustness these two scientists identified have been implicitly removed from our models here. Similarly, in our analysis of the *Synechocystis* model, we eliminated *reactions*, not *enzymes* coded for by particular genes. Hence, the robustness seen cannot be a result of duplicated genes. Instead, it is yet another type of *distributed robustness*. It is, in fact, robustness created by the existence of several different biochemical routes to the same end products.

The neat lists of biochemical pathways in books and posters and on websites belie the reality of biochemistry in the cell: there is an elaborate network of reactions that interact with each other in complex ways. Figure 6.1 is one (imperfect) attempt to visualize this complexity: modeling with flux balance analysis provides another (see Figures 6.8 and 6.9). And our results from Figure 6.9B are not atypical. In the many organisms where such metabolic models have been made, a large fraction of genes or reactions show the potential for metabolic flux to be rerouted around missing enzymes [22, 37]. We can therefore attribute at least part of the high degree of robustness found in living systems to these three different levels of biochemical robustness that we have seen in this and the preceding chapter.

REFLECTIONS AND PREVIEWS

In this chapter we again saw that a major challenge of modeling in biology is in creating models that can be effectively compared to observation. We found that the differential equation models of Chapter 5 are in some sense too detailed: the parameters needed to match them to experiment are not always accessible from those experiments at the scale required.

Interestingly, we saw that one route around this problem was to simplify our model, leaving out most of the dynamical complexity. The resulting models used linear rather than differential equations and allowed at least a degree of prediction from types of biological data that are relatively straightforward to collect. After creating these models, we found that they predicted yet another kind of biological robustness.

Robustness as a Prerequisite for Evolution

Given that we have now seen three kinds of biological robustness, we can consider questions such as "How does the robustness of the network of biological reactions shape the trajectory of evolution?" We will speak of the relationship of the organism's genes and its eventual makeup, or *phenotype*. Evolution would be much more difficult in the absence of robustness: if every mutation dramatically changed the organism's phenotype, we would expect that most of those changes would be damaging, slowing the rate of change in both genes and the phenotype. If, however, many mutations have little or no effect on the phenotype because organisms are robust, many more different genetic layouts, or *genotypes*, can exist. More subtly, those different genotypes, while equivalent in one environment, may not be equivalent if conditions change. We can then imagine that these genetically diverse but phenotypically similar populations could become phenotypically distinct in a changed environment [38].

This idea has some large consequences, some of which have been illustrated by a clever study by João Matias Rodrigues and my doctoral advisor, Andreas Wagner [39]. They explored both the ability of a single metabolic genotype to exploit many different environments and the ability of very different genotypes to exploit the same environment. In this case, by genotype we simply mean the list of biochemical reactions encoded by an organism's genome. Their first finding was that the set of genotypes (i.e., reaction sets) that could exist in the same environment was very large and genetically diverse. In other words, genomes encoding a very dissimilar set of reactions could nonetheless yield organisms that could live in the same environment. On the other hand, they also found that if they started with all the genotypes that could exist in a single environment, when they went to other environments, even very dissimilar ones, they could usually find at least one genotype from the original environment that was able to survive in that new environment. In other words, the phenomenon of phenotypic robustness allows genetically diverse populations to exist in a single environment. In those diverse populations there is a good deal of "preadaptation" to other environments. As a results, if the environment were to change, at least some of the individuals would be able to survive in it. A less genetically diverse population would not have such preadaptation. Hence, we can think of metabolic robustness as acting to prime evolution's pump by giving it a large space of potential genotypes and phenotypes to work with when evolving to exploit a novel environment.

Non-Robust Life

In the next chapter we will consider a case where organisms are not particularly robust and explore how their genes respond to an environment. The reason for this lack of robustness is the inherent randomness of molecular processes, and this molecular stochasticity will be our introduction to biological models that incorporate probability and statistics.

REFERENCES

1. Knoop H, Gründel M, Zilliges Y, Lehmann R, Hoffmann S, Lockau W, & Steuer R (2013) Flux balance analysis of cyanobacterial metabolism: the metabolic network of Synechocystis sp. PCC 6803. *PLoS Computational Biology* 9(6):e1003081.
2. Conant GC & Wolfe KH (2007) Increased glycolytic flux as an outcome of whole-genome duplication in yeast. *Molecular Systems Biology* 3:129.
3. Joshi A & Palsson BO (1989) Metabolic dynamics in the human red cell: Part I—A comprehensive kinetic model. *Journal of Theoretical Biology* 141(4):515–528.
4. Teusink B, Passarge J, Reijenga CA, Esgalhado E, van der Weijden CC, Schepper M, Walsh MC, Bakker BM, van Dam K, Westerhoff HV, & Snoep JL (2000) Can yeast glycolysis be understood in terms of *in vitro* kinetics of the constituent enzymes? Testing biochemistry. *European Journal of Biochemistry* 267:5313–5329.
5. Price ND, Papin JA, Schilling CH, & Palsson BO (2003) Genome-scale microbial in silico models: the constraints-based approach. *Trends in Biotechnology* 21(4):162–169.
6. Kacser H & Burns JA (1981) The molecular basis of dominance. *Genetics* 97(3–4):639–666.
7. Gutenkunst RN, Waterfall JJ, Casey FP, Brown KS, Myers CR, & Sethna JP (2007) Universally sloppy parameter sensitivities in systems biology models. *PLoS Computational Biology* 3(10):e189.
8. Duina AA, Miller ME, & Keeney JB (2014) Budding yeast for budding geneticists: a primer on the Saccharomyces cerevisiae model system. *Genetics* 197(1):33–48.
9. Goodwin S, McPherson JD, & McCombie WR (2016) Coming of age: ten years of next-generation sequencing technologies. *Nature Reviews Genetics* 17(6):333.
10. Aken BL Achuthan P, Akanni W, Amode MR, Bernsdorff F et al., (2017) Ensembl 2017. *Nucleic Acids Research* 45(D1):D635–D642.
11. Coordinators NR (2017) Database resources of the national center for biotechnology information. *Nucleic Acids Research* 45:D12.
12. Jeong H, Tombor B, Albert R, Oltvai ZN, & Barabási A-L (2000) The large-scale organization of metabolic networks. *Nature* 407:651–654.
13. Ma H & Zeng A-P (2003) Reconstruction of metabolic networks from genome data and analysis of their global structure for various organisms. *Bioinformatics* 19(2):270–277.
14. Caetano-Anollés G, Kim HS, & Mittenthal JE (2007) The origin of modern metabolic networks inferred from phylogenomic analysis of protein architecture. *Proceedings of the National Academy of Sciences* 104(22):9358–9363.
15. Francke C, Siezen RJ, & Teusink B (2005) Reconstructing the metabolic network of a bacterium from its genome. *Trends in Microbiology* 13(11):550–558.
16. Dias O, Rocha M, Ferreira EC, & Rocha I (2015) Reconstructing genome-scale metabolic models with merlin. *Nucleic Acids Research* 43(8):3899–3910.
17. King ZA, Lu J, Dräger A, Miller P, Federowicz S, Lerman JA, Ebrahim A, Palsson BO, & Lewis NE (2015) BiGG Models: a platform for integrating, standardizing and sharing genome-scale models. *Nucleic Acids Research* 44(D1):D515–D522.
18. Lay DC (1994) *Linear algebra and its applications* (Addison-Wesley, Reading, MA), p. 445.
19. Orth JD, Thiele I, & Palsson BØ (2010) What is flux balance analysis? *Nature Biotechnology* 28(3):245–248.
20. Bekaert M, Edger PP, Hudson CM, Pires JC, & Conant GC (2012) Metabolic and evolutionary costs of herbivory defense: systems biology of glucosinolate synthesis. *New Phytologist* 196(2):596–605.
21. Foundation G (2024) GLPK (GNU linear programming kit). https://www.gnu.org/software/glpk/
22. Edwards JS & Palsson BO (2000) The Escherichia coli MG1655 in silico metabolic genotype: its definition, characteristics, and capabilities. *Proceedings of the National Academy of Sciences, USA* 97(10):5528–5533.
23. Famili I, Forster J, Nielsen J, & Palsson BO (2003) *Saccharomyces cerevisiae* phenotypes can be predicted by using constraint-based analysis of a genome-scale reconstructed metabolic network. *Proceedings of the National Academy of Sciences, USA* 100(23):13134–13139.
24. Ibarra RU, Edwards JS, & Palsson BO (2002) Escherichia coli K-12 undergoes adaptive evolution to achieve in silico predicted optimal growth. *Nature* 420(6912):186.
25. Alberts B, Johnson A, Lewis J, Raff M, Roberts K, & Walter P (2002) *Molecular biology of the cell*. 4th Edition (Garland Science, New York)
26. Kent WJ, Sugnet CW, Furey TS, Roskin KM, Pringle TH, Zahler AM, & Haussler D (2002) The human genome browser at UCSC. *Genome Research* 12(6):996–1006.
27. Wagner A (2005) *Robustness and evolvability in living systems* (Princeton University Press, Princeton, NJ).
28. Giaever G, Chu AM, Ni L, Connelly C, Riles L et al., (2002) Functional profiling of the *Saccharomyces cerevisiae* genome. *Nature* 418(6896):387–391.
29. Guo HH, Choe J, & Loeb LA (2004) Protein tolerance to random amino acid change. *Proceedings of the National Academy of Sciences* 101(25):9205–9210.
30. Huang W, Petrosino J, Hirsch M, Shenkin PS, & Palzkill T (1996) Amino acid sequence determinants of β-lactamase structure and activity. *Journal of Molecular Biology* 258(4):688–703.
31. Bekaert M & Conant GC (2011) Copy number alterations among mammalian enyzmes cluster in the metabolic network. *Molecular Biology and Evolution* 28:1111–1121.
32. Mathews CK & Van Holde KE (1996) *Biochemistry*. 2nd Edition (The Benjamin/Cummings Publishing Company Inc., Menlo Park)
33. Perry GH, Dominy NJ, Claw KG, Lee AS, Fiegler H, Redon R, Werner J, Villanea FA, Mountain JL, Misra R, Carter NP, Lee C, & Stone AC (2007) Diet and the evolution of human amylase gene copy number variation. *Nature Genetics* 39(10):1256–1260.
34. Gu Z, Steinmetz LM, Gu X, Scharfe C, Davis RW, & Li W-H (2003) Role of duplicate genes in genetic robustness against null mutations. *Nature* 421:63–66.

35. Makino T, Hokamp K, & McLysaght A (2009) The complex relationship of gene duplication and essentiality. *Trends in Genetics* 25(4):152–155.

36. Conant GC & Wagner A (2004) Duplicate genes and robustness to transient gene knockouts in *Caenorhabditis elegans*. *Proceedings of the Royal Society, Biological Sciences* 271(1534): 89–96.

37. Duarte NC, Herrgård MJ, & Palsson BØ (2004) Reconstruction and validation of *Saccharomyces cerevisiae* iND750, a fully compartmentalized genome-scale metabolic model. *Genome Research* 14:1298–1309.

38. Wagner A (2008) Neutralism and selectionism: a network-based reconciliation. *Nature Reviews Genetics* 9(12): 965–974.

39. Matias Rodrigues JF & Wagner A (2009) Evolutionary plasticity and innovations in complex metabolic reaction networks. *PLoS Comput Biol* 5(12):e1000613.

Shrinking Too Small
Noise in Biochemical Systems

7

The disintegration of a single radioactive atom is observable...But if you are given a single radioactive atom, its probable lifetime is much less certain than that of a healthy sparrow. Indeed, nothing more can be said about it than this: as long as it lives (and that may be for thousands of years) the chance of its blowing up within the next second ... remains the same. This patent lack of individual determination nevertheless results in the exact exponential law of decay of a large number of radioactive atoms of the same kind.

In biology we are faced with an entirely different situation. A single group of atoms existing in only one copy produces orderly events, marvellously tuned with each other and with the environment according to the most subtle laws ... (This) situation is unprecedented, it is unknown anywhere else except living matter ...
—Erwin Schrödinger, What Is Life [1].

TIME AND CHANCE: INTRODUCTION TO RANDOMNESS

While numbers are an extraordinarily powerful tool for analyzing nature, the ease with which we can manipulate very small and very large quantities can give a false sense of confidence and blind us to how inadequate our intuition is in these unfamiliar domains. In science, the two places where our ordinary sense of numbers is most easily overwhelmed are the astronomical and molecular scales. At the astronomical scale, there are more stars in the universe than grains of sand on all the beaches of Earth [2]; at the molecular scale, the number of atoms in an ordinary glass of water is comfortably greater than the number of stars in the universe (**Figure 7.1**).

The Scale of Molecular Systems

Leaving (regretfully) the universe to its own devices, we ought to consider how the scale of molecular phenomena gives rise to patterns and behaviors that violate our intuition, built as it is from observations at the macroscopic scale. A good description of the problem is given by Schrödinger in the chapter epigraph. As one of the developers of quantum theory, he was intensely aware of how processes that are random at the level of individual molecules, like radioactive decay, give rise to very predictable behaviors at the macroscopic level. The reason for this predictability is simply the large number of molecules involved. We will explore *why* large numbers behave in this way in Chapter 9. For the moment, however, we should notice that all the analyses in Chapters 5 and 6 implicitly assumed that we did not need to worry about individual molecules. In those chapters, our units of analyses were always *concentrations*: moles per liter. We were hence treating systems made up of individual molecules as if they were continuous, just as with the SIR model

DOI: 10.1201/9781003687504-7

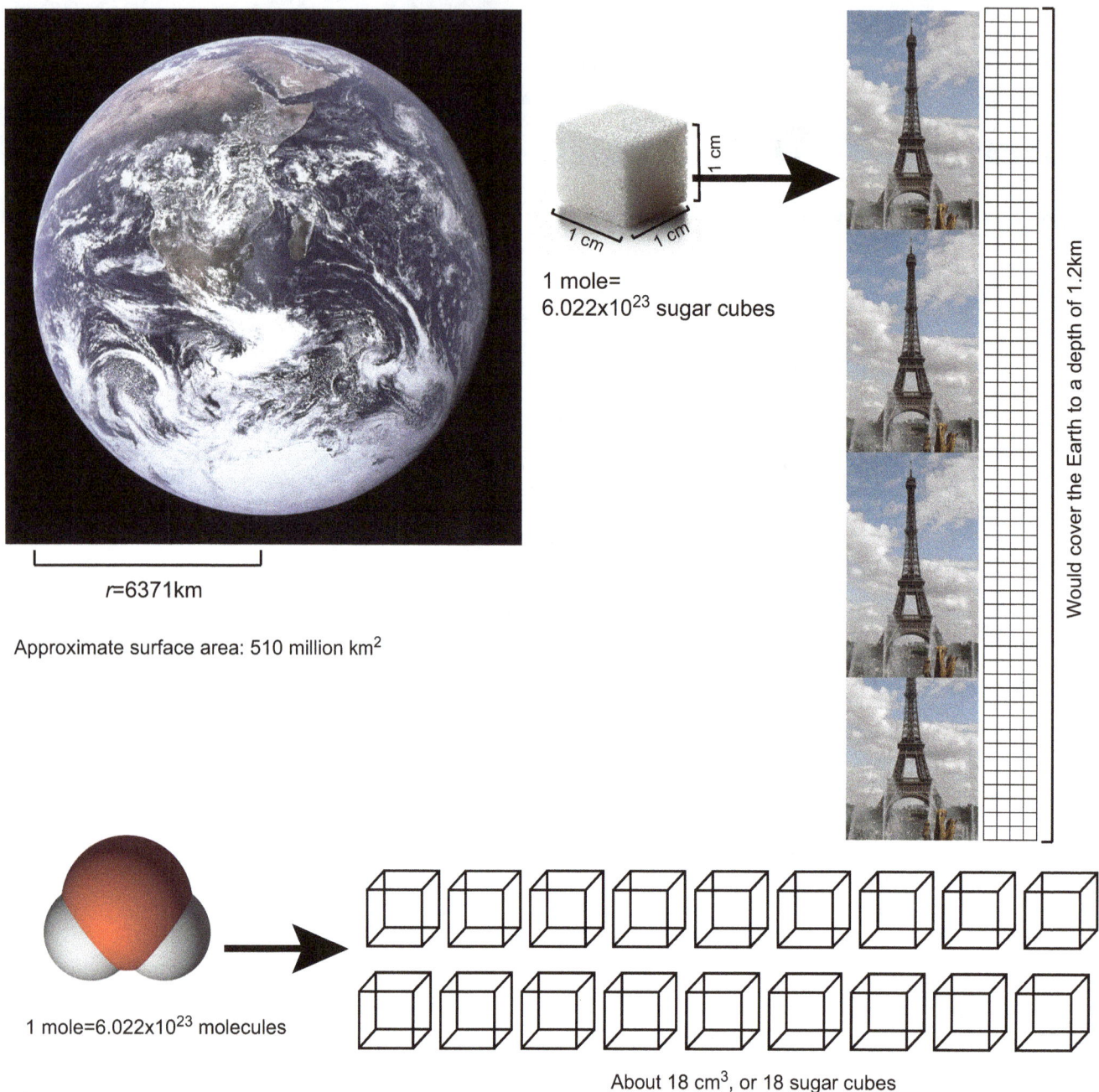

Figure 7.1 The scale of molecular phenomena. While we remember the molar conversion factors for our chemistry classes, the objects' sizes and the scale of the numbers we are considering can be easy to overlook. A mole of even a small common object like a sugar cube (assume for simplicity one that is 1 cm on a side) is an enormous amount of matter. If we approximate the Earth as a true sphere of radius 6,378 km [4], we find a surface area of $4 \cdot \pi \cdot (6{,}371 \text{ km})^2 \approx 511{,}186{,}000 \text{ km}^2$. One mole of sugar cubes would occupy 602,000,000 km³ (10^{15} cm³/km³ $\cdot$ $6.022 \cdot 10^{23}$ cubes of 1 cm³ volume). This number of cubes would cover the Earth to a depth of about 1.2 km, or about three and two-thirds Eiffel Towers. For comparison, 1 mole of water molecules has a volume of slightly more than 18 sugar cubes. ("Blue marble" image of the Earth courtesy of NASA.)

we treated populations of humans as continuous rather than being made up of discrete people.

There are several implications to assuming that a set of molecules can be treated as a concentration. For instance, we are claiming that if we were to divide our unit of study (think of a test tube, but also of a cell) in half, the behavior of each half would be identical to the other half and to the intact whole. Now, at some point this symmetry must break: if we divide a volume in half over and over, we will finally reach a point where we have only one molecule of our compound left, and it will either go into the right-hand bucket or the left-hand one. Just as our SIR models included the slightly nonsensical notion of half-people, our biochemical models have the possibility of considering half-molecules.

Is the Continuous Approximation Invalid?

As Schrödinger noted, in almost every case, this concern is entirely irrelevant, and one can treat the quantity of molecules in a sample as continuous. To understand why, let us consider the actual number of molecules in several small things. Figure 7.1 gives the number of water molecules in the volume of a sugar cube: if we move down to a drop of water, which has a volume of about 1/20th of a cubic centimeter, how many molecules will we find?

To answer this question (and indeed to compute the volume of a mole of water in Figure 7.1), we need to start by asking what the mass of a single water molecule actually is. As we may vaguely remember from high school chemistry, this value can be obtained from the atomic weights on a periodic table: water consists of two hydrogen molecules (1.0079 *daltons* each) and one oxygen atom (15.999 daltons), for a total of 18.015 daltons per molecule [3]. What is a dalton? It is a unit of mass, but to relate it to units with which we are familiar (like a gram), we need first to remind ourselves of what a mole is. By definition, it is N_A: the number of ^{12}C atoms in 12 grams of ^{12}C (here ^{12}C is simply one of the isotopes of carbon, namely the one with equal numbers of protons and neutrons). By comparing the charge of a mole of electrons to the charge of a single electron, an estimate of N_A can be made (which, as we also may vaguely remember, comes out to 6.022×10^{23}). Once we have this number is in hand [3], we find that a dalton is:

$$1 \, \text{dalton} = \frac{1}{12} \cdot \frac{12 \, \text{grams}}{N_A} \tag{7.1}$$

or 1.7×10^{-24} grams. Hence, one molecule of water has a mass of 3.0×10^{-23} grams. Similarly, and unsurprisingly, 1 mole of water molecules has a mass of 18.015 grams (3.0×10^{-23} grams/molecules $\cdot 6.022 \times 10^{23}$ molecules/mole).

Conveniently, 1 gram of water has effectively a volume of 1 cm^3 [4]. Hence, the volume required to contain 1 mole of (liquid) water is roughly 18 cm^3 (see Figure 7.1). From this value we can compute the number of water molecules in 1/20 cm^3 as:

$$\frac{6.022 \times 10^{23} \, \text{molecules} / \text{mole}}{18.015 \, \text{cm}^3 / \text{mole}} \cdot \frac{1}{20} \, \text{cm}^3 = 1.7 \times 10^{21} \, \text{molecules} \tag{7.2}$$

How many molecules is 1.7×10^{21}? One way to gain a feel for this kind of number is to use time as a reference. If we could look at one molecule per second, Table 7.1 tells us how long it would take to look at different numbers of molecules. For comparison, the current estimate of the age of the universe is 13.8 billion years [5], meaning that it would take almost 4,000 times the known lifetime of the universe to look at all the molecules in a drop of water.

TABLE 7.1 THE TIME IT WOULD TAKE TO CHECK DIFFERENT NUMBER OF MOLECULES AT A RATE OF 1 MOLECULE PER SECOND	
Number of Molecules	*Time (s)*
1,000	16 minutes
10^6 = 1 million	11.6 days
10^9 = 1 billion	31.7 years
10^{12} = 1 trillion	32,000 years
10^{15}	32 million years
10^{21}	32 trillion years

THE MOLECULAR SCALE OF CELLS

The above argument is certainly not a formal proof, but it strongly suggests that the use of the continuum approximation for chemical species dissolved in even as small a volume as one drop is highly unlikely to provoke problems. But cells are very much smaller than a drop of water. How certain are we that the continuous, concentration-based models we used in Chapters 5 and 6 are appropriate for a cell?

Again, without making a formal proof, we can make some rough calculations and see where they lead us. Mammalian cells come in a very wide variety of sizes [6]: some of the cranial nerves of giraffes can be over 4 meters long [7]. For the sake of argument, we can select a size at the smaller end of estimates of their volumes: 2,000 cubic micrometers (μm^3) [6]. A cubic micrometer is 10^{-18} cubic meters ($10^{-6} m / \mu m \cdot 10^{-6} m / \mu m \cdot 10^{-6} m / \mu m$), so this volume works out to 2×10^{-15} m^3. Since $1 mL = 1 cm^3$ and $1 cm^3 = 10^{-6} m^3$ ($10^{-2} m / cm \cdot 10^{-2} m / cm \cdot 10^{-2} m / cm$), we have a volume of 2×10^{-9} mL.

Exactly what might fit in this space is a difficult question to answer because cells are extraordinarily complex and crowded places [8–10]. But we can use our previous example of a water molecule as a rough gauge. Hence, if 18 mL contains a mole of water molecules, then:

$$\frac{6.022 \times 10^{23} \, molecules / mole}{18.015 \, mL / mole} \cdot 2 \times 10^{-9} \, mL = 6.7 \times 10^{13} \, molecules \qquad (7.3)$$

That is, 67 trillion water molecules could fit in the volume of such a cell. By our previous measure, it would take about 2 million years to look at all of them (see Table 7.1). Instinctively, the continuous approximation seems reasonable. However, because of the work we did in Chapter 5, we need not be satisfied with this intuition: we can at least partly check how well the continuous models of biochemistry work at this scale.

Biochemistry at the Scale of a Bacterial Cell

Let's make things a bit more challenging by going down in scale another couple of factors of 10 and considering biochemistry inside a bacterial cell. These cells are roughly 2,000 times smaller than our own cells, or about $1 \mu m^3$ in total. This volume works out to 10^{-12} mL. For simplicity, we will model a reversible, single-step biochemical reaction involving only a single reactant and product:

$$\begin{array}{c} k_1 \\ S \rightleftharpoons P \\ k_{-1} \end{array} \qquad (7.4)$$

Again, to make the problem a challenge for the continuous approximations, we will start with no product ([P] = 0) and very little substrate: [S] = 10^{-3} mmol/ml.

Before proceeding any further, we should calculate the number of molecules per (bacterial) cell these concentrations imply. First, notice that, given that $1 mmol = 10^{-3}$ mol, 10^{-3} mmol/ml = 10^{-6} mol/ml. Thus,

$$6.022 \times 10^{23} \, molecules / mole \cdot 10^{-6} \, mole / ml \cdot 10^{-12} \, mL \approx 602{,}000 \, molecules \quad (7.5)$$

This number of molecules is a reasonable place to start assessing whether continuous approximations in biochemical models are likely to be misleading in biology. It is a small enough number that, as we shall see, we can tractably model each of those molecules in software; it is also on the lower end of size for the biochemical systems we might realistically encounter.

In Exercise 7.1, you will use COPASI [11] to compute a time course of Equation 7.4 using reasonable values of $k_1 = k_{-1} = 0.1$ for the parameters. In **Figure 7.2A**, we see the results of the concentration-based model. The time course is rather uneventful because $k_1 = k_{-1}$, meaning that S and P simply converge to chemical equilibrium.

A)

B)

Time (s)

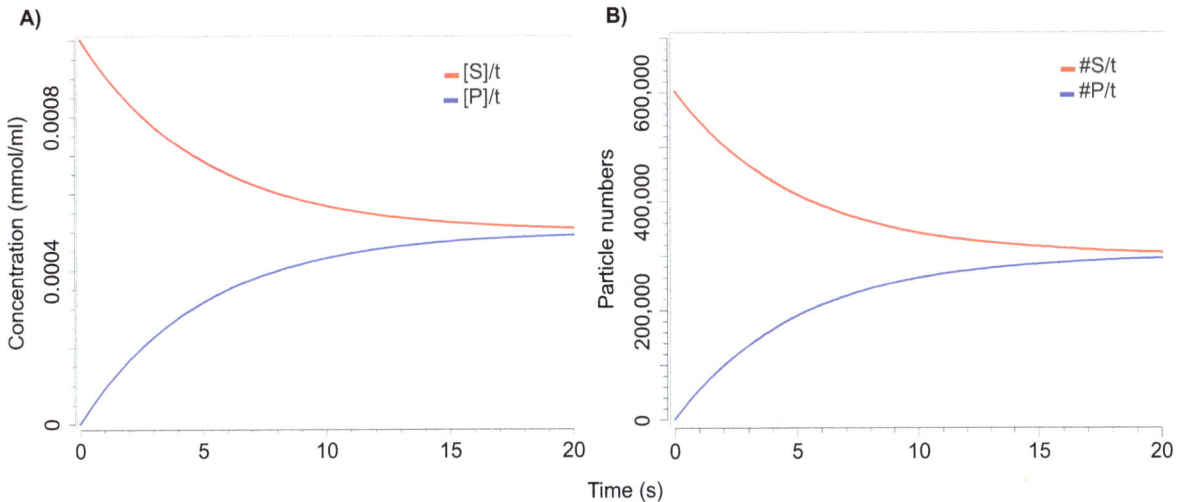

Figure 7.2 Modeling biochemistry with continuous (A) and discrete (B) models. Each model is performed in a volume approximating a bacterial cell: 10^{-12} ml. The reaction considered is the simple conversion of S to P (i.e., Equation 7.4). **(A)** In this panel a continuum approximation is used, with an initial concentration of $S = 0.001$ mmol/ml and $P = 0$ mmol/ml. The simulation is carried out for 20 s (x-axis); the resulting concentrations of S and P are shown on y. **(B)** In this panel we convert to COPASI's τ-Leap discrete-particle model that incorporates stochastic effects [11]. To do so, the initial concentration values from **A** are converted to particle counts, yielding the values on y.

Exercise 7.1

Using COPASI, model the reaction in Equation 7.4 using both a continuous approximation and a stochastic simulation. First, create a compartment with a volume of 10^{-12} ml. Create species S and P, giving S an initial concentration of 0.001 mmol/ml and P an initial concentration of 0. You will need to describe Equation 7.4 with a pair of irreversible reactions for the stochastic simulation: set k_1 for each such reaction to 0.1. Now create a plot showing the concentration of S and P on y and time on x. Run a time course using a deterministic model. Now change COPASI's settings to "Particle Numbers" at the top. Make sure that S and P are now shown as molecule numbers in the "Species" tab. Create a new plot where x is again time and y now gives particle numbers rather than concentrations. Run a "Stochastic τ-Leap" simulation for your time course and compare to your deterministic results.

Is it possible to study this system not in terms of the concentrations of S and P but rather in terms of the absolute number of molecules of each? As we computed above, the total number of molecules in the system is not extraordinarily large and therefore should be within the capacity of a computer simulation.

Simulating Randomness in Chemical Reactions

It might appear surprising that 600,000 molecules can be easily simulated, and in fact, the difficulty depends on the scope of your simulation. Were we to explicitly model the entire volume of the cell and simulate the motions of each individual molecule in that space, then indeed the problem would be very hard. However, if we are willing to assume that the volume is well mixed, it is possible to develop a simulation based on the expected time we would have to wait for the different chemical events. Hence, the software will create a *probability distribution* for the time you would have to wait before an S molecule changed to a P, and vice versa. It will do so given k_1 and k_{-1} and the number of each type of molecule currently present. This approach, formalized by Gillespie [12], is termed the *stochastic simulation algorithm*. It allows us to model our system from Equation 7.4 in COPASI quite efficiently (see Figure 7.2B) and to compare that result to the continuous approximation (see Figure 7.2A). We can see that visually the two curves are effectively identical, again suggesting that, even for a tiny bacterial cell with a relatively low-flux reaction, our continuum approximation is a reasonable model of the biochemistry.

RANDOM BIOCHEMISTRY, SCHRÖDINGER, AND THE CENTRAL DOGMA

Schrödinger's Paradox: The Biochemistry of Single Molecules

Now we can return to the paradox described by Schrödinger in the opening quotation. In so doing, we will find an instance where the continuum approximation does *not* hold. The reason is that there are molecules in a cell that are present not in hundreds of thousands of copies but in only one or two copies. These are, of course, the genes themselves. Somehow, despite their scarcity relative to all the other compounds in a cell, it is these molecules that orchestrate the intricate molecular dance that is life itself.

Schrödinger argues that other systems with such small numbers of molecules do not display intrinsic order. We will therefore use the simulation approach we just employed to explore that proposition and to start to understand where the biochemical organization comes from. For this purpose, we will simulate one of the key discoveries of 20th-century biology: the *central dogma of molecular biology*.

Background: The Central Dogma

The central dogma was proposed [13] as a compact description of how genetic information flows and is used in the cell. Despite occasional complaints [14], it remains a powerful description and simplification of this process. We first encountered genes in Chapter 5, but we did not take time to discuss how the information in those genes is used by the organism to carry out its various biological functions. It is this link that the central dogma provides.

Figure 7.3 gives a generalized overview of this information flow process. The genes are of course encoded in sections of the DNA double helix. To use the genetic information, that helix first opens to allow a complex group of proteins that we will call the *RNA polymerase* to bind to one-half of that DNA helix. This polymerase is an enzyme that can *transcribe* the DNA of the gene into another four-letter nucleotide code, namely *ribonucleic acid*, or RNA. Because the cell does not want to transcribe all of its DNA, the RNA polymerase binds specifically to a particular part of the beginning of the gene, called the *promoter*. The polymerase then synthesizes an RNA molecule that is complementary to the DNA template strand. Specifically, where the template DNA has a "G," the complementary RNA molecule will have a "C," while places where the DNA has an "A" will result in an RNA molecule with the base uridine ("U"). These "U" bases are chemically similar to the "T" base in DNA [15].

Depending on the type of RNA molecule and the organism, several other processes such as splicing and nuclear export can occur [15], but for simplicity, we will not consider them in our model. Instead, the next process we are interested in is

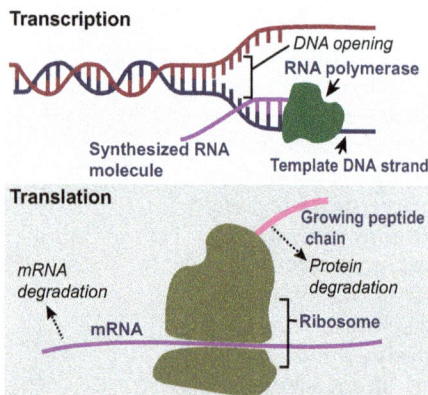

Figure 7.3 (Part of) the central dogma of molecular biology. The cell's DNA can relax its double helical structure sufficiently to allow an RNA polymerase complex to enter and synthesize a new RNA molecule using the information in the template DNA strand. We call this process *transcription*. If the RNA in question codes for a protein, we refer to it as a *messenger RNA* (mRNA). A complex molecular machine called a *ribosome*, which is composed of both RNA and protein, can then synthesize a protein from an mRNA, using that mRNA's sequence as well as transfer RNAs (tRNAs; not shown). This second process is called *translation*.

Second codon position

First codon position	T	C	A	G	Third codon position
T	TTT Phe [F] TTC Phe [F] TTA Leu [L] TTG Leu [L]	TCT Ser [S] TCC Ser [S] TCA Ser [S] TCG Ser [S]	TAT Tyr [Y] TAC Tyr [Y] TAA Stop TAG Stop	TGT Cys [C] TGC Cys [C] TGA Stop TGG Trp [W]	T C A G
C	CTT Leu [L] CTC Leu [L] CTA Leu [L] CTG Leu [L]	CCT Pro [P] CCC Pro [P] CCA Pro [P] CCG Pro [P]	CAT His [H] CAC His [H] CAA Gln [Q] CAG Gln [Q]	CGT Arg [R] CGC Arg [R] CGA Arg [R] CGG Arg [R]	T C A G
A	ATT Ile [I] ATC Ile [I] ATA Ile [I] ATG Met [M]	ACT Thr [T] ACC Thr [T] ACA Thr [T] ACG Thr [T]	AAT Asn [N] AAC Asn [N] AAA Lys [K] AAG Lys [K]	AGT Ser [S] AGC Ser [S] AGA Arg [R] AGG Arg [R]	T C A G
G	GTT Val [V] GTC Val [V] GTA Val [V] GTG Val [V]	GCT Ala [A] GCC Ala [A] GCA Ala [A] GCG Ala [A]	GAT Asp [D] GAC Asp [D] GAA Glu [E] GAG Glu [E]	GGT Gly [G] GGC Gly [G] GGA Gly [G] GGG Gly [G]	T C A G

Figure 7.4 The universal genetic code, used by most organisms to translate the 4-letter alphabet of DNA and RNA into the 20-letter alphabet used for proteins [15]. Because there are 20 possible amino acids, one or two bases of DNA/RNA are not enough to uniquely specify a given amino acid. Instead, amino acids are encoded by one of 61 three-base *codons* (the other three possible codons are stop codons to indicate the end of the protein). Of course, since 61 > 20, there are often several codons that encode the same amino acid: these cases are illustrated here with colored blocks.

the binding of the translational machinery, known as the *ribosome*, to the just-synthesized RNA molecule. That new RNA molecule is now known as a *messenger RNA*, or mRNA, because it carries the genetic "message" from the nucleus describing the protein to be synthesized. The ribosome, with the help of other RNA molecules called tRNAs (or transfer RNAs), reads the mRNA message in the four-letter DNA/RNA code. It then converts it into the 20-letter amino acid code using the scheme shown in Figure 7.4. The result of this complex process will be the synthesis of a new protein molecule based on the information in the DNA/RNA.

Proteins and Cellular Function

Before continuing, we should remind ourselves why this process is the *central dogma*. Figure 7.5 gives examples of just three protein structures from humans: each plays a key role in how our bodies function. For instance, the actin and myosin proteins in Figure 7.5B are the molecular source of the movements in your muscles. Likewise, all the enzymes we have been discussing in Chapters 5 and 6 were also proteins. In fact, almost any interesting biology you might think of has proteins at its heart, from the detection of light in the eye (rhodopsin) to fighting off infections with protein antibodies [16]. Hence, to understand the central dogma means to understand the key molecular processes by which the genetic information drives biological function.

Modeling the Central Dogma

We can create a model of this process in COPASI, but to do so we will need to add a couple of new details to the process we just described. The first involves *when* the RNA polymerase is able to bind to the gene's promoter and produce a transcript.

In organisms like humans, the proportion of the DNA in the genome that is directly or indirectly responsible for coding for the mRNA molecules (in other words, that is in *genes*) is only about 10% [17]. And even within that 10%, in any particular cell, most of the genes are not needed most of the time. For instance, your genome contains several slightly different versions of the genes coding for the subunits of hemoglobin (see Figure 7.5A), the protein in your blood responsible for carrying oxygen to the rest of your body [18]. Right now, more than 97% of the mRNAs from which the peptide subunits of hemoglobin are translated come from just three of those copies [19]. However, before you were born, your body produced different forms of hemoglobin. Those forms were encoded by duplicate

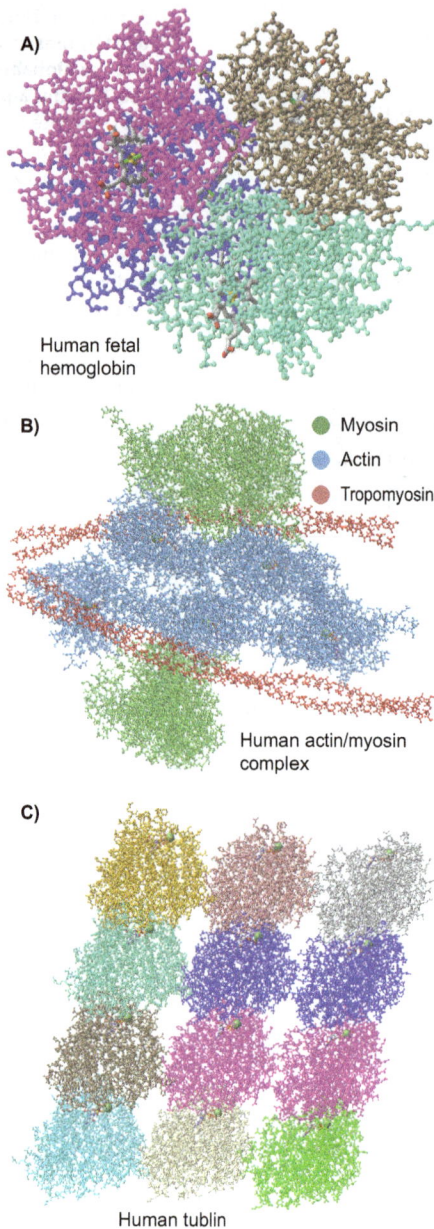

Figure 7.5 **Proteins are the (main) functional molecules in cells.** Shown are the three-dimensional shapes of three important proteins in your cells. (**A**) Human fetal hemoglobin tetramer. Each of the four hemoglobin peptides has an associated heme molecule in the center that can bind oxygen [16]; PDB structure [39] 4MQJ. (**B**) The human actin/myosin complex [40]. The movement of actin along myosin proteins allows your muscles to contract [41]; PDB structure 5JLH. (**C**) The cyto-skeletal molecule tubulin, which forms fibers through the association of repeated units of the same protein molecule [42]; PDB structure 5JCO.

copies of hemoglobin genes (see the previous chapter) that are similar but not identical to the ones you are using now [20]. Those sequence differences were large enough that your embryonic hemoglobin was able to pull oxygen away from the adult hemoglobin in your mother's blood and deliver it to you as you developed in her uterus [18]. Soon after birth, these genes for producing fetal hemoglobin were turned off [19], and as far as we know, they will not be used again in your lifetime.

The details of *how* genes are turned on and off are complex (see Chapter 14), but part of the process involves modifications to the packaging of the DNA. Genes that are not needed by a particular cell can be turned off by attaching methyl (CH_3) groups to their DNA or by modifications to the histone proteins around which that DNA wraps [15, 21]. Genes may be turned off in these ways either because they function at a different developmental stage (like the example of the fetal hemoglobins just given) or because they are not needed in the tissue in question. For instance, the genes coding for visual pigments like rhodopsin would be turned off in liver cells.

The Global Rules of Gene Expression

In this chapter we are interested in the set of genes that are turned on in a given cell: in other words, in the set of genes for which DNA methylation and histone modifications do not prevent transcription. Even here, however, the genes are not simply ready and waiting for an RNA polymerase to bind to their promoters and produce an mRNA. Instead, even transcriptionally active genes spend most of their time in a transcriptionally inaccessible state. In that state, the transcriptional machinery is unable to bind and start transcription [22, 23]. However, the transition between these inactive periods and the active ones is rather fast, on the order of seconds or minutes. As a result, gene expression occurs in bursts. In other words, when the DNA enters the transcribable "open" state, several mRNAs may be synthesized before it transitions back to the "closed" form [24, 25].

Using these various facts, we can now build a simple model of this process of the central dogma, as illustrated in **Figure 7.6**. Our model leaves out many features of real gene expression in eukaryotes, including the splicing out of the introns and the export from the nucleus [15] as well as the complex process of the initiation of transcription [16]. Instead, the model focuses on five key processes. Three of these we have already discussed: the opening and closing of the DNA (reaction #1 in Figure 7.6) and the processes of transcription (reactions #3 and #4 in Figure 7.6) and translation (reactions #5 and #6 in Figure 7.6).

The remaining two processes are the degradation of the synthesized mRNA and of protein molecules. It is probably obvious that cells would not want to keep mRNA molecules around indefinitely. If they did, the cell would have difficulty responding to changing conditions because it would be full of old mRNA molecules produced in response the previous conditions. In fact, the half-life of an mRNA varies from approximately tens of minutes in yeast cells to a few hours in mammalian cells [26, 27]. Protein half-lives in human cells are somewhat longer [28], partly for reasons we are about to explore.

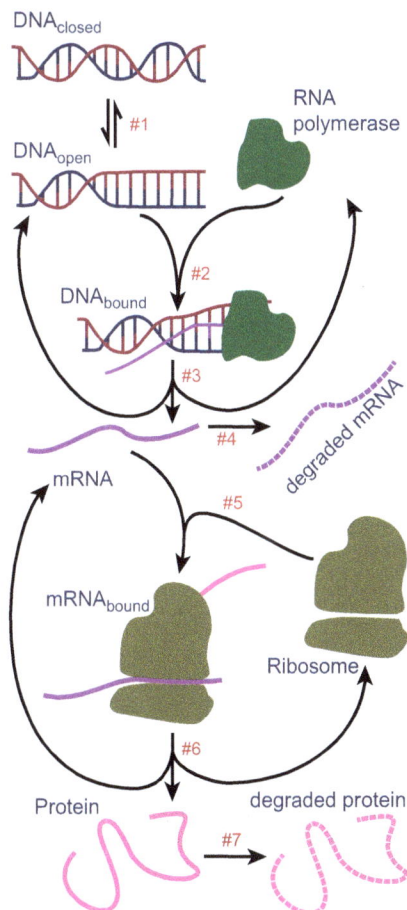

Figure 7.6 A simplified model of noisy gene expression showing the 10 species in our model (*blue*) as well as the 7 reactions (*red*). The promoter of the gene in question is assumed to exist in two states: "closed" and "open." Only the open state is accessible to the RNA polymerase to allow transcription. Once the polymerase binds, it is assumed to produce a new mRNA and then release from the DNA. The DNA remains in the open state, allowing for the binding of another polymerase. The new mRNA can either degrade (by analogy to radioactive decay) or meet with a ribosome and undergo translation, after which it is released to be potentially translated again. The resulting protein also has a half-life, which will generally be greater than that of the mRNA.

A Model of the Central Dogma

We can now see how our model of the system behaves in **Figure 7.7** (Exercise 7.2). For computational simplicity, we will use values of the mRNA and protein half-lives that are on the lower end of observed values but are still biologically plausible

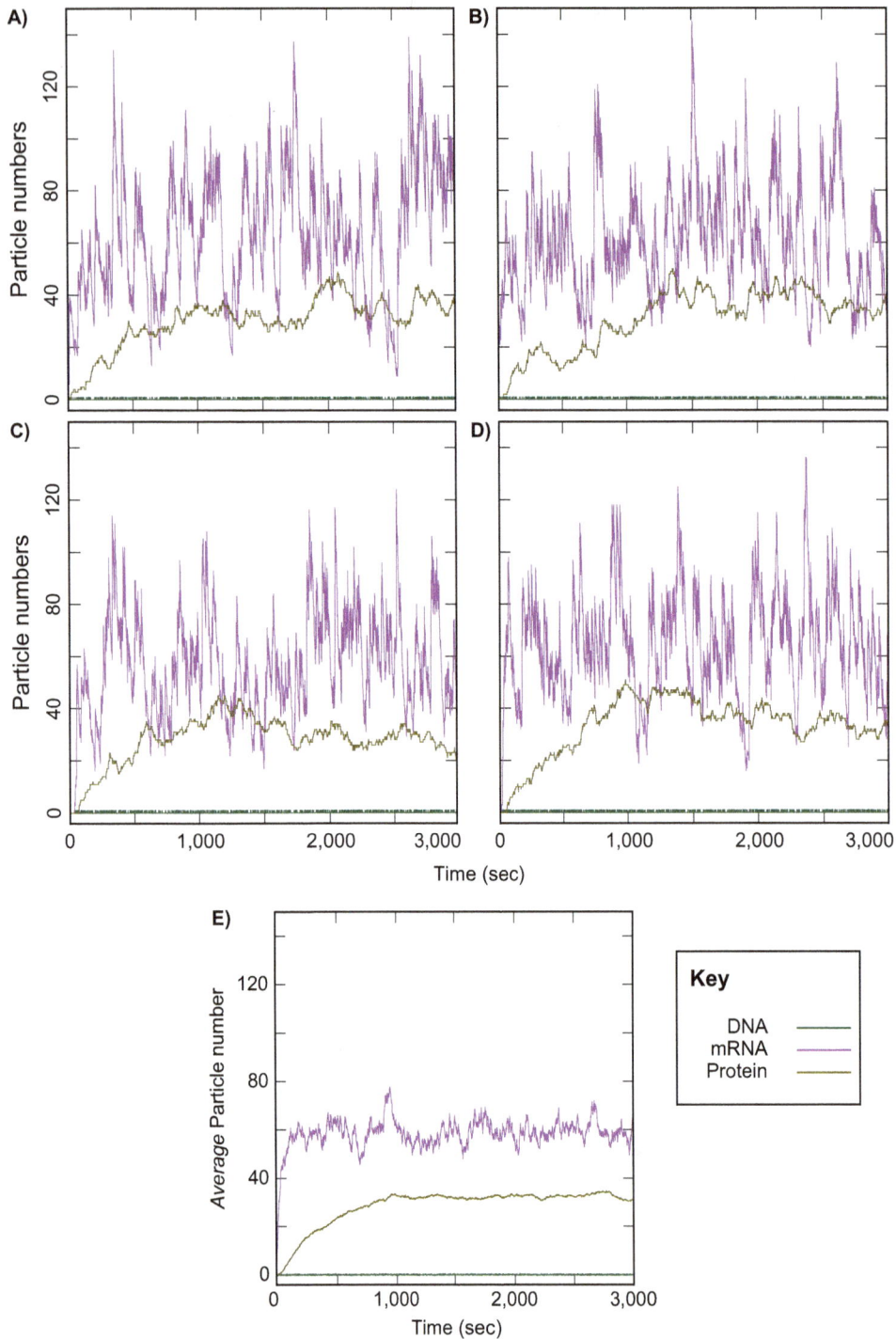

Figure 7.7 Simulations of the central dogma using stochastic models. (A–D) These panels show the results of modeling the system in Exercise 7.2 four times. In other words, these are outcomes for four initially identical cells but with random differences in DNA opening/closing and transcription and translation. The counts of open DNA are shown in green, while the counts of mRNA and protein molecules are given in purple and tan, respectively. **(E)** This panel shows the *average* particle numbers for each type of molecule across 25 simulations with these same initial conditions.

for organisms like yeast. With this model in hand, we will use the stochastic simulation approach we employed in Figure 7.2 to analyze it.

Exercise 7.2

Using COPASI, model the central dogma of biology using the model in Figure 7.5. We will split the reversible conversion between DNA_{closed} and DNA_{open} into two irreversible reactions to allow the stochastic simulation, giving us a total of eight reactions. The model has 10 chemical species (DNA_{closed}, DNA_{open}, DNA_{bound}, RNA polymerases, mRNA, ribosomes, proteins, and mRNA and protein degradation products). Make sure COPASI is set to use "Particle Numbers" at the top. We will use a compartment concentration of 10^{-14} ml to approximate the cell nucleus, a single molecule of DNA (initially in the closed state), 2,000 molecules of the RNA polymerase, and 100 ribosomes. All the other species have their initial concentration set to 0. The DNA is assumed to move from closed to open at a rate of 0.25 times per second and back to closed at a rate of 2 times per second. The reaction constant for the binding of the RNA polymerase to the open DNA (reaction #2) is set to 30,000 ml/(mmol·s) while the resulting rate of mRNA production (reaction #3) is 0.0005 mmol/(ml·s). The mRNA degrades at a rate of 0.03 times per second (reaction #4) and the protein (reaction #7) at a one-tenth of that rate (0.003/s). The ribosomes bind mRNA (reaction #5) at a rate of 100 ml/(mmol·s), and translation (reaction #6) occurs at a rate of 1 time per second. Create a plot showing the number of mRNA and protein molecules over time, as well as the DNA state (open, closed, or bound). Run several "Stochastic τ-Leap" simulations and compare their plots (c.f., Figure 7.7). The appropriate COPASI file is also available from the QBio website (http://qbio.statgen.ncsu.edu/data/CentralDogma/ NoisyGeneExpression.cps).

NOISY GENE EXPRESSION

Figure 7.7 shows the results of simulations from this model. The first four panels are from four simulations with identical initial conditions and model parameters. They show just how different such initially identical cells can become due to the random effects in our model. This molecular variation seen between cells is termed *noise* in processes of transcription and translation [29]. Although all the random events described in the model contribute to this noise, analyses suggest that the strongest contributor is the random opening and closing of the DNA that gives rise to the bursts in mRNA synthesis [22, 23]. Such noise also represents exactly the sort of molecular stochasticity Schrödinger was referring to in this chapter's epigraph.

However, the noise in Figure 7.7 is not simply the result of the small number of molecules involved. One of the notable features of the noise in our model is that the mRNA and protein noise levels are not the same. Why might this be?

Rather than trying to understand the pattern using our model, it is helpful to think more abstractly about how evolution might see noise. Biochemical noise is potentially harmful to an organism because that noise can prevent the organism from responding predictably and correctly to its environment [29, 30]. Consider the case of an idealized world without such noise. In that case, we can imagine that there is an evolutionary optimum level of protein made from each gene. This idea is easiest to understand in the context of an enzyme for metabolizing a sugar like galactose. In the presence of galactose, making too little of that enzyme is harmful because this prevents the cell from using the galactose to grow as quickly as it might otherwise. On the other hand, each molecule of the enzyme synthesized incurs a cost to the cell. That cost stems from the energy and building blocks used for synthesizing that enzyme, which are thus not available to other cellular processes [31, 32]. While we might struggle to find it, we can imagine that there is a level of protein production that most effectively balances these competing costs of metabolic need and enzymatic expense. In the absence of biochemical noise, natural selection, will, for many organisms, have had eons to explore the possibilities and find the required balance, just as we saw with the experimental example of glycerol in *E. coli* from the previous chapter [33].

What happens when we add noise, then? Essentially, we dilute the power of evolution to find the ideal protein level. The reason is that, even in cells programmed with the genetics that perfectly yields the needed level, biochemical noise will drive the cell away from that optimum. If our view of noise is correct, then we would hypothesize that organisms have evolved strategies for limiting noise in their machinery for synthesizing proteins. I write "protein" rather than "mRNA and protein" because the costs of noise are primarily in the realm of insufficient or excess numbers of protein molecules, since they, and not mRNAs, are the functional molecules.

Cellular Mechanisms for Noise Reduction

If we return to Figure 7.7, we see something rather striking. The protein levels are less noisy than are the mRNA levels, exactly as we might predict were natural selection to have controlled noise in protein levels. This difference in noise levels between the two types of molecule results from the particular combination of mRNA and protein synthesis and decay rates in our model. By synthesizing a large number of short-lived mRNA molecules and translating them relatively infrequently, the cell can reduce the noise in the protein expression levels relative to a strategy of making fewer, longer-lived mRNAs [34, 35]. We will better understand *why* this strategy works by the end of Chapter 9. For now, we can see evidence for it in the fact that, in yeast cells, genes whose absence is lethal to the cell (termed *essential* genes) are more likely to use this strategy of high mRNA production/ degradation than are nonessential genes [34]. That observation is consistent with essential genes being essential because a lack of their products is costly to the cell, meaning that noisy expression that perturbs those protein levels is especially costly for this type of gene. In Chapter 14, we will see another evolutionary strategy for reducing gene expression noise.

The presence of biochemical noise in these systems should also not be interpreted as meaning that the mRNA and protein levels are completely arbitrary. Figure 7.7E shows the results of averaging the DNA, protein, and mRNA molecule numbers over 25 such simulations. We can see there that our simulation starts with no protein or mRNA present; the levels of the two then climb, rapidly for the mRNA and more slowly for the protein, until they reach a rough steady state. At a biological level, these results imply that our earlier assumptions were correct: there is an underlying gene expression program that evolved to achieve the required protein levels, and we are seeing this program in the average expression levels. These results also indicate something else: if we were to study the levels of mRNA in cells by pooling a number of those cells, extracting RNA from them, and then sequencing it (i.e., RNASeq experiments [36]), we would not observe the biochemical noise in the system. The reason we would miss it is that by pooling our cells we created exactly the sort of averaging behavior that Schrödinger was explaining in the opening quotation! One of the most active areas of research in cellular biology today is in techniques for assessing the mRNA levels of single cells so that researchers can explore phenomena such as expression noise [37].

REFLECTIONS AND PREVIEWS

Multiple Models

In this chapter we have begun to explore biological models that include randomness. The first thing to notice about these analyses is that they are actually the *third* model we have made of biochemical reactions, adding to the differential equation models of Chapter 5 and the linear algebra models of Chapter 6. Having three different mathematical models of the same process may feel odd at first. The reason for this feeling is probably that many of us first encounter mathematical models in physics, where there is sometimes the unstated assumption that the model *is* the reality. In fact, even models in physics most often are useful and valid only in some contexts. For instance, Newton's model of universal gravitation does not explicitly include the fact that objects' masses increase as they approach the speed of light [38]. As we have seen, models are more often *complementary*: in other words, they represent the same underlying system in ways that are useful for different problems. In the last chapter we saw that the differential equation models of metabolism were powerful but hard to apply for large problems due to the number of parameters involved: we will revisit their usefulness in Chapter 15. The flux-balance models, on the other hand, sacrificed some details but could be applied across many genomes. And, as we just saw, stochastic models are unnecessary most of the time but are occasionally essential: we will reuse *them* in Chapter 14.

Randomness and Predictability

Returning to the question of randomness, we both started and ended the chapter with statements that hide an important puzzle: Why is it that pooling many individually random events produces predictability? In other words, what is the source of Schrödinger's claim that radioactive decay is predictable at the macroscopic level despite being unpredictable at the microscopic one?

In the next chapter, we will begin to explore this problem and start to develop some quantitative machinery for studying systems that include random behavior. We will find that these noise models are just one example of how the behavior of large groups of random entities can give rise to highly predictable systems – in other words, to *statistics*.

REFERENCES

1. Schrödinger E (1944) *What is life?* (Cambridge University Press, Cambridge).

2. Sagan C (1980) Chapter 8: travels in space and time. *Cosmos* (Random House, New York).

3. Levine IN (1995) *Physical chemistry*. 4th Edition (McGraw-Hill, New York), p. 901.

4. Lide DR ed (1994) *CRC handbook of chemisty and physics*. 75th Edition (CRC Press, Inc., Boca Raton).

5. Ade PA, Aghanim N, Arnaud M, Ashdown M, Aumont J, Baccigalupi C, Banday A, Barreiro R, Bartlett J, & Bartolo N (2016) Planck 2015 results-xiii. cosmological parameters. *Astronomy & Astrophysics* 594:A13.

6. Milo R & Phillips R (2015) *Cell biology by the numbers* (Garland Science).

7. Berry RJ & Hallam A eds (1986) *The Collins encyclopedia of animal evolution* (Willam Collins Sons and Co. Ltd, London), p. 144.

8. Pielak GJ (2005) A model of intracellular organization. *Proceedings of the National Academy of Sciences of the USA* 102(17):5901–5902.

9. Ellis RJ (2001) Macromolecular crowding: obvious but underappreciated. *Trends in Biochemical Sciences* 26(10):597–604.

10. Hudder A, Nathanson L, & Deutscher MP (2003) Organization of mammalian cytoplasm. *Molecular and Cellular Biology* 23(24):9318–9326.

11. Hoops S, Sahle S, Gauges R, Lee C, Pahle J, Simus N, Singhal M, Xu L, Mendes P, & Kummer U (2006) COPASI--a complex pathway simulator. *Bioinformatics* 22(24):3067–3074.

12. Gillespie DT (1977) Exact stochastic simulation of coupled chemical reactions. *The Journal of Physical Chemistry* 81(25): 2340–2361.

13. Crick FH (1958) On protein synthesis. *Symposia of the Society for Experimental Biology* 12:138–163.

14. Crick F (1970) Central dogma of molecular biology. *Nature* 227(5258):561–563.

15. Russell PJ (1994) *Fundamentals of genetics* (Harper Collins, New York).

16. Lodish H, Baltimore D, Berk A, Zipursky SL, & Matsudaira P (1995) *Molecular cell biology*. 3rd Edition (Scientifc American Books, New York), p 1344.

17. Lander ES, Linton LM, Birren B, Nusbaum C, Zody MC et al., (2001) Initial sequencing and analysis of the human genome. *Nature* 409(6822):860–921.

18. Hardison RC (2012) Evolution of hemoglobin and its genes. *Cold Spring Harbor Perspectives in Medicine* 2(12):a011627.

19. Giambona A, Passarello C, Renda D, & Maggio A (2009) The significance of the hemoglobin A2 value in screening for hemoglobinopathies. *Clinical Biochemistry* 42(18):1786–1796.

20. Stamatakis A (2014) RAxML version 8: a tool for phylogenetic analysis and post-analysis of large phylogenies. *Bioinformatics* 30(9):1312–1313.

21. Gilbert S (1991) *Developmental biology*. 3rd Edition (Sinauer, MA).

22. Raj A, Peskin CS, Tranchina D, Vargas DY, & Tyagi S (2006) Stochastic mRNA synthesis in mammalian cells. *PLoS Biology* 4(10):e309.

23. Becskei A, Kaufmann BB, & van Oudenaarden A (2005) Contributions of low molecule number and chromosomal positioning to stochastic gene expression. *Nature Genetics* 37(9):937–944.

24. Raser JM & O'Shea EK (2004) Control of stochasticity in eukaryotic gene expression. *Science* 304(5678):1811–1814.

25. Ko MS (1991) A stochastic model for gene induction. *Journal of Theoretical Biology* 153(2):181–194.

26. Wang H-F, Feng L, & Niu D-K (2007) Relationship between mRNA stability and intron presence. *Biochemical and Biophysical Research Communications* 354(1):203–208.

27. Sharova LV, Sharov AA, Nedorezov T, Piao Y, Shaik N, & Ko MS (2009) Database for mRNA half-life of 19 977 genes obtained by DNA microarray analysis of pluripotent and differentiating mouse embryonic stem cells. *DNA Research* 16(1):45–58.

28. Eden E, Geva-Zatorsky N, Issaeva I, Cohen A, Dekel E, Danon T, Cohen L, Mayo A, & Alon U (2011) Proteome half-life dynamics in living human cells. *Science* 331(6018):764–768.

29. Raser JM & O'Shea EK (2005) Noise in gene expression: origins, consequences, and control. *Science* 309(5743):2010–2013.

30. Barkai N & Leibler S (2000) Biological rhythms: circadian clocks limited by noise. *Nature* 403(6767):267–268.

31. Wagner A (2007) Energy costs constrain the evolution of gene expression. *Journal of Experimental Zoology Part B: Molecular and Developmental Evolution* 308(3):322–324.

32. Stoebel DM, Dean AM, & Dykhuizen DE (2008) The cost of expression of Escherichia coli lac operon proteins is in the process, not in the products. *Genetics* 178(3):1653–1660.

33. Ibarra RU, Edwards JS, & Palsson BO (2002) Escherichia coli K-12 undergoes adaptive evolution to achieve in silico predicted optimal growth. *Nature* 420(6912):186.

34. Fraser HB, Hirsh AE, Giaever G, Kumm J, & Eisen MB (2004) Noise minimization in eukaryotic gene expression. *PLoS Biology* 2:834–838.

35. Pires JC & Conant GC (2016) Robust yet fragile: expression noise, protein misfolding and gene dosage in the evolution of genomes. *Annual Review of Genetics* 50(1):113–131.

36. Marguerat S & Bähler J (2010) RNA-seq: from technology to biology. *Cellular and Molecular Life Sciences* 67(4):569–579.

37. Aldridge S & Teichmann SA (2020) Single cell transcriptomics comes of age. *Nature Communications* 11(1):4307.

38. Feynman RP, Leighton RB, & Sands M (2011) *The Feynman lectures on physics, Vol. I: The new millennium edition: mainly mechanics, radiation, and heat* (Basic Books).

39. Berman HM, Westbrook J, Feng Z, Gilliland G, Bhat TN, Weissig H, Shindyalov IN, & Bourne PE (2000) The protein data bank. *Nucleic Acids Research* 28(1):235–242.

40. von der Ecken J, Heissler SM, Pathan-Chhatbar S, Manstein DJ, & Raunser S (2016) Cryo-EM structure of a human cytoplasmic actomyosin complex at near-atomic resolution. *Nature* 534(7609):724–728.

41. Marieb EN (1991) *Human anatomy and physiology* (The Benjamin/Cummings Publishing Company, Redwood City, CA), p. 1040.

42. Vemu A, Atherton J, Spector JO, Szyk A, Moores CA, & Roll-Mecak A (2016) Structure and dynamics of single-isoform recombinant neuronal human tubulin. *Journal of Biological Chemistry* 291(25):12907–12915.

Time and Chance
Probability and Random Variables

8

> *"I returned, and saw under the sun, that the race is not to the swift,*
> *nor the battle to the strong, neither yet bread to the wise, nor yet*
> *riches to men of understanding, nor yet favour to men of skill;*
> *but time and chance happeneth to them all."*
> —*Ecclesiastes 9:11*

WHAT IS "RANDOM"?

The word "random" is one of those slippery words that we seem to understand less the more we think about it. When I was in high school, it had a brief vogue as expressing the idea of something that was unexpected in an amusing or charming way, such as an especially loud and colorful sweater. But even if we limit ourselves to the domain of science and mathematics, different fields think of what it means to be random in quite different ways.

In Figure 8.1A I show the variability in adult height for a group of 928 children, taken from measurements collected by the 19th-century geneticist and statistician Francis Galton [1]. His goal in collecting these data was to compare the children's heights to that of their parents, but our interest is actually simpler. In Figure 8.1B I show data from our simulations of noisy mRNA production in the previous chapter (see Figure 7.7).

It is at least curious that these two very different types of data appear to follow a common pattern in their variation: there is a central tendency, corresponding to the peak in each chart, and then a relatively symmetrical falling of observations on either side of that peak. Now, Galton showed very clearly that the data he collected were *not* random: one could predict with reasonable accuracy a child's height if you knew the heights of her or his parents [1]. On the other hand, our mRNA traces from Chapter 7 are, within the limits of computer simulation, truly random. There is no "reason" that one simulation has 84 mRNA molecules at a particular time and another has 63.

And yet, that distinction has less in it than it might appear. The mRNA simulations we made were capturing a *history* of cellular activities: of DNA openings, polymerase bindings, and mRNA decays. Were we to know that history, to know when the DNA was in the open state, when the polymerase bound, and so forth, we would know, from that information, the number of mRNA molecules present at a particular time. In some sense, then, for both the mRNA counts and for the height data, we can view the randomness as being a placeholder for a complex and unknown history. Now, accepting that argument, we still have a partial mystery. Galton's data were taken across more than 100 different parents: so, beyond history, we are observing something about the overall pattern of height in Victorian England. Why is it that the heights (and the mRNA counts) take on this bell-like curve?

DOI: 10.1201/9781003687504-8

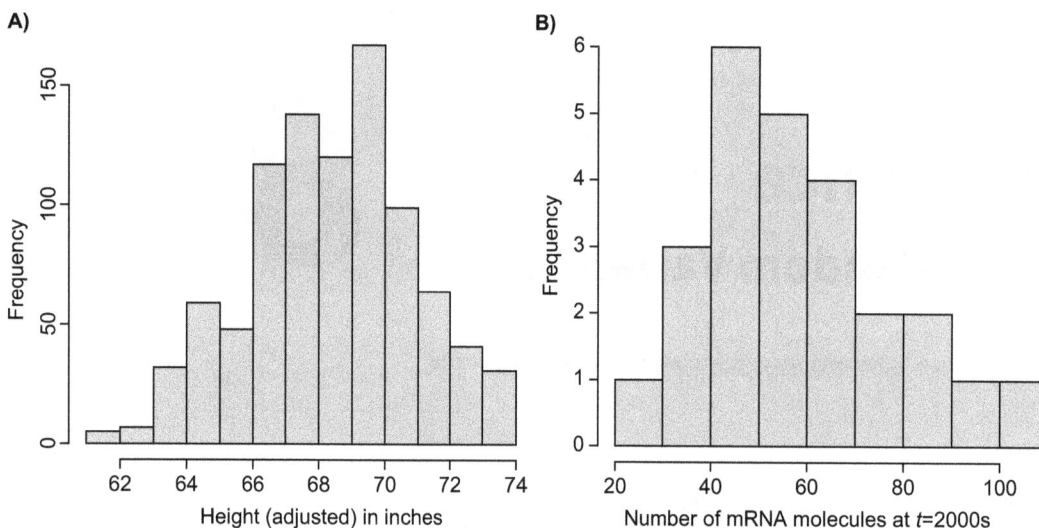

Figure 8.1 (Random) variation of a population about a central tendency in two very different systems. (A) The distribution of heights of 928 children in Victorian England, collected by Francis Galton [1]. The heights of the female children in the dataset were adjusted by a factor of 1.08 to be comparable to those of the males. **(B)** The distribution of the number of mRNA molecules per cell in our simulations of noisy gene expression from Chapter 7, taken at the (arbitrary) time $t = 2,000$ s.

MATHEMATICAL DESCRIPTIONS OF RANDOMNESS

To answer the question of why the plots in Figure 8.1 are bell-shaped will require us first to develop some mathematical machinery for describing randomness. That machinery will take us the remainder of this chapter, leaving the question of the bell shape until the beginning of Chapter 9. Hence, this chapter will have somewhat less biology than other chapters because we will focus on building these mathematical tools.

How can we get a grip on this concept of randomness? One place to start is at the subatomic level. There are very good reasons to believe that the underlying laws governing atomic and subatomic physics are probabilistic in nature, meaning that at some fundamental level we can only predict the *probabilities* of various outcomes or events. These laws are of course those of quantum mechanics [2], and they govern the radioactive decay described by one of their inventors, Schrödinger, in the previous chapter's epigraph.

Radioactivity, and therefore quantum mechanics, can mutate DNA sequences and is, as a result, of considerable importance in biology. In fact, we will use a toy example of radioactive decay and mutation in this chapter as an introduction to the mathematical treatment of *continuous random variables*. And it is quite likely that both human height and gene expression are at least influenced by events at the subatomic level. This idea is most obvious for gene expression, since behaviors such as DNA breathing are clearly tied to the movements and binding of atoms. But inheritance is also at its heart a molecular phenomenon, and an animal's height, as Galton found, is strongly influenced by its genes.

So, should we conclude that Figure 8.1 shows quantum effects? In all honesty, I do not know, or rather I do not know how much of the variation we see could be so attributed. Instead, let us consider another framing of the problem, one that draws attention to an underappreciated difference between biology and (much of) physics and chemistry. Biology is, at least in part, a *historical* science [3]. Darwin deeply appreciated this fact; his famous closing sentence for *The Origin of Species* makes the point better than I can [4]:

> There is a grandeur in this view of life, with its several powers, having originally breathed by the Creator into a few forms or only one; and that, whilst this planet has gone cycling on according to the fixed law of gravity, from so simple a beginning endless forms most beautiful and most wonderful have been, and are being evolved.

If we step back for a moment, we can see importance of history throughout biology, in the development of the adult animal from the embryo, in the similarity of parents (and grandparents) and their offspring, in the evolution of life across millions of years, and even in the behavior of individual cells.

Randomness as a Substitute for History

Take our gene expression model. If we want to know the number of mRNA molecules in the cell right now, knowing how many molecules there were a moment ago is a very good start, since whatever new mRNAs have been synthesized or old ones destroyed are added to the base of mRNA molecules that were present just a moment ago. In fact, the power of the simulations we did in the last chapter is now apparent: unlike in a real cell, we can *know* the whole history of our simulated cells and see how the overall pool of mRNAs increases or decreases at each moment.

DIGRESSION: BROWNIAN MOTION

To understand the importance of this idea, we will take a brief detour to the world that van Leeuwenhoek's microscope unveiled. When the early biologists looked into their microscopes, they saw motion and realized that there was an entire world of life that had been missed because it was too small to see with the unaided human eye.

However, this abundance of life masked something else important about the view through those microscopes. Ironically, it was a biologist, a botanist, who, having seen pollen grains moving under the microscope, had the foresight to examine inorganic substances of the same size under similar conditions [5]. When Robert Brown did this, he saw that these inanimate particles also moved, demonstrating that at least some of the motion scientists were seeing under their microscopes was not due to life, but to something else.

What might that phenomenon be? We know from the previous chapter that even a very small drop of water is packed full of molecules. Moreover, unless that drop is extremely cold indeed, the molecules in it are moving around at significant speed. Even under a microscope, of course, the molecules are too small to see. However, if we are looking at a particle that is still very small but big enough to see under the microscope, we can imagine that the water molecules are constantly bumping against it. And again, if the object is fairly small, the number of molecules colliding with it on one side might be slightly greater than the number on the other side, causing it to move away from that side. Of course, this motion is not likely to last too long, as in the next moment the situation could be reversed, moving it back in the other direction [2].

Writing an equation for this behavior will be tricky: How do we represent these little random "pushes?" As it turns out, if we do a bit more exploration first, there is a workable approach, and that approach has a very interesting link to the other idea of randomness that we were discussing with Galton's height data.

Applying the Concept of Brownian Motion More Broadly

To see the linkage, notice first that this *Brownian* motion that Brown saw does not *require* any truly random inputs. What I mean by that is that one can imagine a purely deterministic system of many, many particles. If we knew all the initial positions and velocities of those particles, we could model the system with tools from classical physics: force, velocity, momentum, mass. However, if we plead a certain ignorance of these initial conditions, we will find that we can make very accurate predictions about their behavior using models that treat them as random.

And of course, we can make the same argument about the heights of the children that Galton measured: if we were to add more and more information about their increasingly distant ancestors, we would gain at least a bit of predictive information about their heights [1]. But what if we do not, or cannot, know the full history?

In fact, we can almost never know the full history of a system. We cannot track all the molecules in a real cell, nor can we know the genomes of all our ancestors. Which is what brings us back to Figure 8.1: we may not need to know the history—we may be able to approximate it using tools from statistics. The reason was provided by the quotation from Schrödinger in Chapter 7. The unknown history of a single individual molecule or human being may make their behavior (or height) hard to predict, but if we bring together enough of those unknown histories, our powers of aggregate prediction can be considerable. I refer to this type of randomness as *thermodynamic* randomness to imply that it represents a substitute for complete knowledge of the history of a large group of entities. An equally good name might be "population randomness." In both cases, we can make useful predictions about such systems even without knowing their full history. But to do so, we need a little bit of mathematical and statistical machinery.

PROBABILITY AND RANDOM VARIABLES

From here on, we will be framing our discussions in terms of probabilities. We all have a reasonable intuition of what a probability is, but it is as well to give a quick reminder before moving on. To keep things simple, let us just say that we are interested in, over a large number of trials, how often a particular event occurs, like rolling a 5 on a die. For a fair die, we would expect one out of six rolls to give a 5, and hence the probability of a 5 is 1/6, or about 0.1666. A probability, therefore, is a number between 0 and 1 and represents either the chance of the event occurring in a single trial or the proportion of the time the event occurs over many trials.

Dice are a nice example to play with because we know that there are only six outcomes, each of which occurs with equal frequency. And of course, if we add the probability of each of those six equally probable outcomes, we get:

$$P = \sum_{i=1}^{6} \frac{1}{6} = 1.0 \tag{8.1}$$

These probabilities add to 1 because only six outcomes are possible and at least one of them must occur. Even when we move on to much more complex probabilistic models, we will still need to be sure that we have accounted for all possible outcomes and that the total probability of *something* happening in a trial is 1.0. Notice that rolling a die is mostly another example of thermodynamic randomness. If we knew the starting position of the die, the forces applied to it, the gravity acting on it, and so forth, we could probably predict its final state. But an equal chance of all six faces landing upward is a simpler and yet often sufficient model.

We can express this idea of event with an unknown random outcome as being a *random variable*. As a metaphor, **Figure 8.2** represents random variables as "black boxes" that generate numbers through some unseen internal process. Thus, Figure 8.2 gives some potential generators of random variables and shows the outcomes that are possible and their probabilities.

An Example of a Discrete Probability Distribution: The Binomial Distribution

The coin and the die in Figure 8.2 are probably quite instinctive to many of us, but the third example, flipping many coins and counting the heads, needs some examination. First, notice that we do not care about the *order* in which the heads arrive, simply about their number. Figure 8.2C assumes a fair coin, so it is not terribly surprising that the most likely outcome of flipping a coin 20 times is to have 10 heads and 10 tails. What might be more surprising is that other outcomes are observed a reasonable frequency of the time.

In Exercise 8.1 you will recreate the data of Figure 8.2C, performing a total of 60 trials rather than 40. You will probably see a curve that looks more similar to those in **Figure 8.3** than does Figure 8.2C, but of course, with random events, you never can tell.

A)

B)

C)

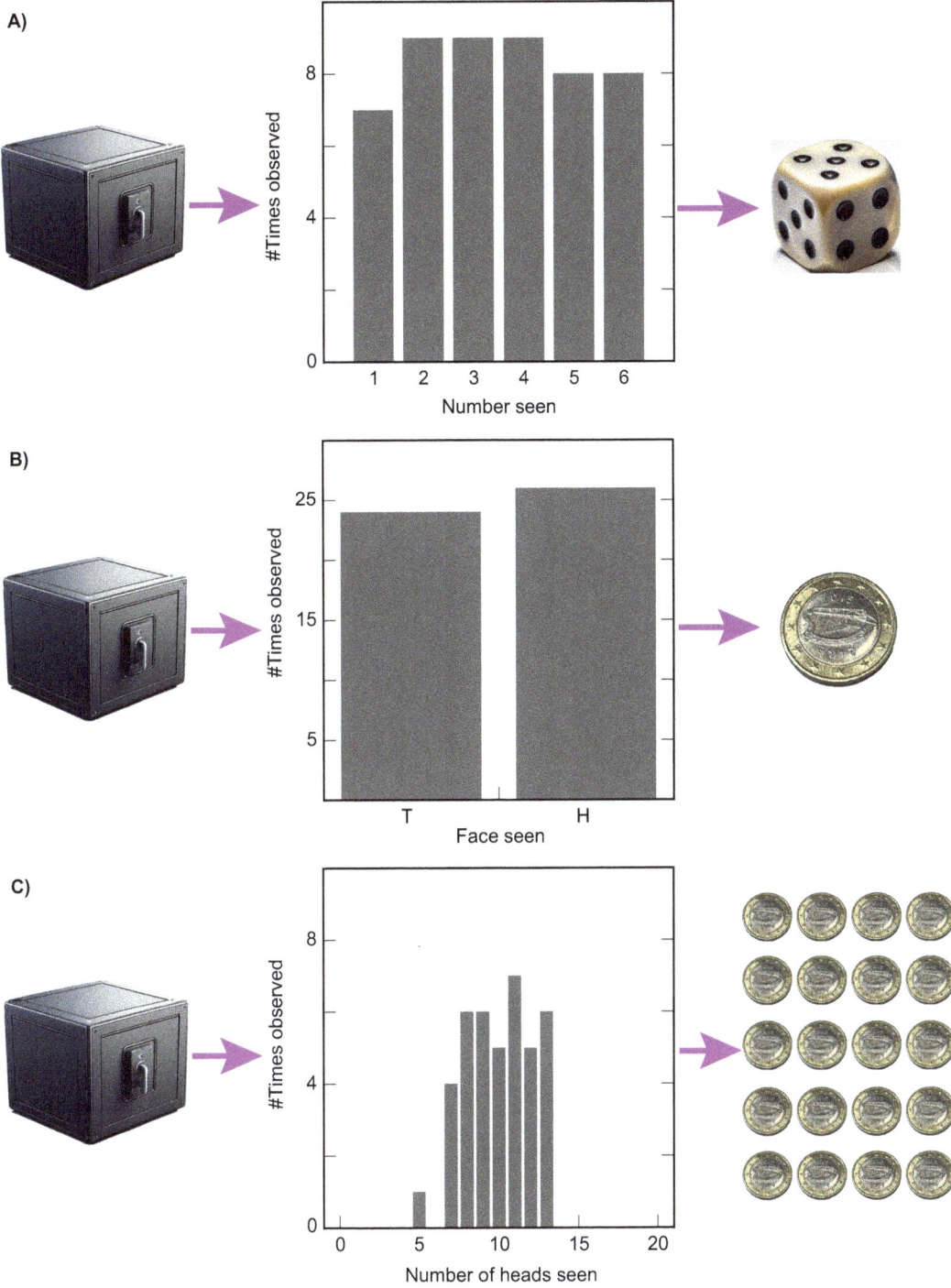

Figure 8.2 Three examples of discrete random variables. (**A**) Simulation of rolling a fair die 50 times. (**B**) Simulation of flipping a fair coin 50 times. (**C**) Simulation of flipping a coin 20 times and counting the number of heads, repeated 40 times.

Exercise 8.1

Using the website http://qbio.statgen.ncsu.edu/CoinFlip/, create a table of the number of heads observed when flipping 20 coins across 60 trials (in other words, run the program 60 times and record the results). Make a histogram of these results.

A)

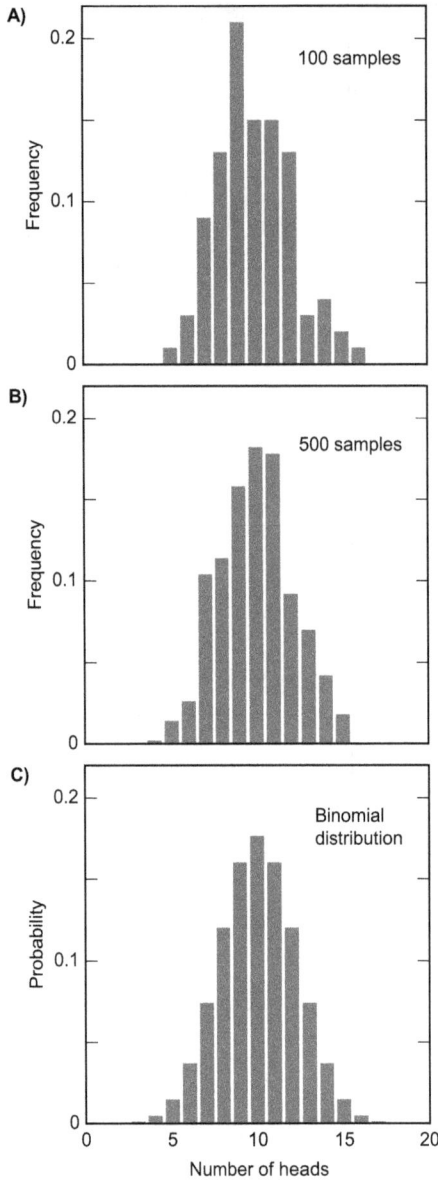

B)

C)

Figure 8.3 The number of heads (n, x − axis) in 20 coin flips, seen from different numbers of samples. (**A**) The frequency of *n* heads seen across 100 samples of 20 flips. (**B**) The frequency of *n* heads seen across 500 samples of 20 flips. (**C**) The probability of each number of heads computed with the binomial distribution (Equation 8.5).

In **Figure 8.3** I show the same chart of the outcomes (which we will call a *histogram* from now on) as in Figure 8.2, except that I have increased the number of samples, first to 100 (**Figure 8.3A**) and then to 500 (**Figure 8.3B**). These increases show a trend consistent with our intuition: as the number of samples increases, the histograms become more symmetric (because the coin is fair) and the peak at 10 heads becomes more defined. **Figure 8.3C** shows what the histogram should look like in the limit of a very large number of samples. These probabilities are given by the *binomial* distribution [6], which we will now construct for ourselves.

To begin, we need to consider the process of a single coin flip, which has a probability p of being a head. In our case, $p = \frac{1}{2}$ because the coin is fair. If we wanted to compute the probability of a particular series of flips, say HTTH, that probability would be given by:

$$p \cdot (1-p) \cdot (1-p) \cdot p = \left(\frac{1}{2}\right)^4 \tag{8.2}$$

because the probability of a tail is $1 - p$ and $p = (1 - p)$ for our fair coin. We can thus generalize Equation 8.2 to any number of flips n: the probability of any particular sequence of heads and tails will simply be p^n.

Numbers of Combinations

However, we are not interested in the precise sequence of flips, merely the total number of heads. Equation 8.2 shows that every *particular sequence of flips* has the same probability. For 20 flips, that value is always $\left(\frac{1}{2}\right)^{20}$, or about 1 in a million. The reason that 10 heads is more common than 1 head therefore lies instead in the number of different possible sequences that have 10 heads in them. That number is much larger than the number that have only a single head. In fact, there are only 20 sequences with a single head, because that single head occurs either on the first flip, or on the second, and so forth. So how many sequences have k heads and $n-k$ tails?

To solve this problem, it is helpful to imagine a group of *n* boxes into which we will place our *n* coins (**Figure 8.4**), which we have already flipped and for which we know the number of heads. How many different orders could we put these boxes in? Well, we have 20 boxes to choose from for the first, 19 for the second, and so forth. For all 20 coins, the number of color orders is $20 \cdot 19 \cdot 18 \ldots 3 \cdot 2 \cdot 1$, a computation that you may recall is denoted as 20! or 20 *factorial*. This number is of course very large. However, we know that *k* of the boxes have coins showing heads and $n-k$ have ones showing tails. In each case, all the boxes with heads and all the boxes with tails are indistinguishable from each other. Just looking at the boxes with heads, there are *k!* arrangements of those boxes that are identical from our perspective. Similarly, there are $(n-k)!$ orderings of the tails that are indistinguishable (see Figure 8.4B). Because each of the *k!* head orders we don't care about can be combined with any of the $(n-k)!$ tail orders we don't care about, there are

$$k! \cdot (n-k)! \tag{8.3}$$

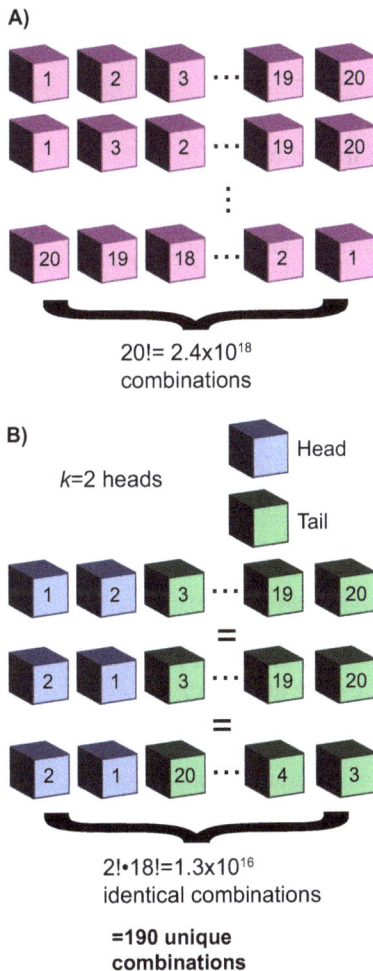

Figure 8.4 A cartoon showing how to compute the number ways to flip *k* heads in *n* flips. (A) This panel shows that if we treat each flip as distinguishable, we could flip the coins in any one of 20! ways. For $n=20$, n! works out to approximately 2.4 million trillion possibilities (2.4×10^{18}). However, we would like to treat all the heads as being equivalent, as well as all the tails. **(B)** This panel illustrates how this simplification reduces the number of possibilities dramatically. For illustration, I use the example of flipping $k=2$ heads. Any order of those two flips is equivalent to any other, and there are 2! = 2 such orders. Similarly, there are 18! orders of the tails that are equivalent. When we remove these 2!•18! orders from the original 20!, we are left with only 190 unique combinations (see Equation 8.4).

total orders that are identical for our purposes. We can remove these orders from the total set of $n!$ by dividing that n total number of orders by Equation 8.3, giving us the number of distinct orders in which we might see k heads and $n-k$ tails:

$$\frac{n!}{k! \cdot (n-k)!} \tag{8.4}$$

Writing the Binomial Distribution

When we combine this expression for the number of possible orders of the heads and tails with the probability of k heads and $n-k$ tails from Equation 8.2, we find that

$$\frac{n!}{k! \cdot (n-k)!} \cdot k^p \cdot (n-k)^{(1-p)} \tag{8.5}$$

gives the probability of k heads in n flips for a coin with the arbitrary probability of a head given by p. This equation gives the binomial distribution. Using the probabilities from it (i.e., Equation 8.5), we can compute the probability of any number of heads, allowing us to produce Figure 8.3C.

We call this binomial distribution a *discrete probability distribution* because the outcomes can be mapped onto integers. The die we considered earlier in the chapter would generate another discrete probability distribution. Likewise, the *geometric distribution* gives us the probability of observing the first head in our series of flips on the nth flip. However, rather than explore more discrete distributions, we are now going to see if we can extend this idea of a probability distribution to *continuous* outcomes.

CONTINUOUS RANDOM VARIABLES

Unlike our example random variables so far, a person's height in Figure 8.1A is measured on a continuous scale. And if we think a bit more, we realize that we might want to analyze random effects in many areas whether the measure of interest is continuous: the distance traveled by a particle under Brownian motion, the time until a cell mutates, or the height of an individual. We will start exploring this concept with an example from the treatment of cancers, although we will have to make it a slightly contrived one for clarity.

An Example of Continuous Probabilities from Cancer Treatment

Recalling our work in Chapter 4, we can reflect that treatments for cancer generally exploit the fact that a key danger of cancer is that the cancerous cells are rapidly dividing [7]. As such, drugs or treatments that are harmful to dividing cells can potentially slow or stop cancer progression [7]. Although cancer cells may be more sensitive to such treatments than normal cells are, the potential for side effects in healthy cells that are also dividing is a serious concern. Radiation, one such treatment, can damage the DNA of the cancer cells, but of course, the dose necessary to do so is unlikely to be good for the neighboring healthy cells either. One of the revolutions in cancer treatment we have seen in the past two decades has been radiation treatments that use the precise targeting of a number of different radiation beams to kill tumor cells while limiting the dose absorbed by the normal cells. These different beams are guided to intersect such that the high doses of radiation are seen only by the tumor cells (**Figure 8.5**), sparing the rest of the body [8, 9].

We will model such an approach, but, for simplicity, we will consider the effect of just one such beam on a hypothetical single cancer cell. We will assume a slightly abstract version of this system where the beam is approximated by a series

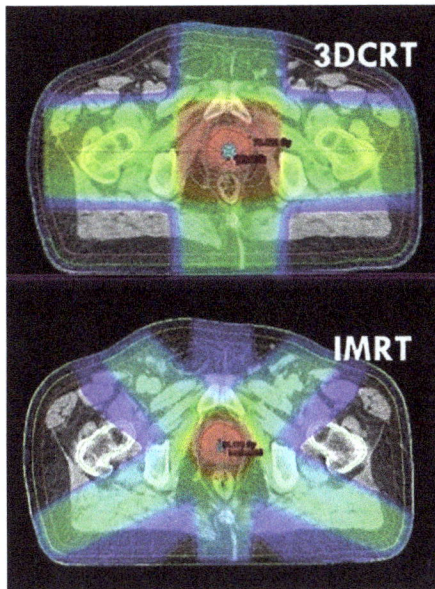

Figure 8.5 Three-dimensional conformal radiotherapy (3DCRT [top] and image-guided intensity-modulated radiotherapy (IMRT [bottom]). The patient is placed so that several beams of radiation can be precisely directed to intersect only in the tumor. The result is a low radiation dose for all parts of the body *except* the tumor. (From Sveistrup J et al. [2014] *Radiation Oncology* 9:1–8. With permission from Springer Nature.)

of randomly fired high-energy particles. Each of these particles has some very small chance of hitting the cell in a critical area and killing it. We would then like to estimate how long the cell will live under treatment with this beam. We define λ to be the *firing rate* of the radiation gun in some arbitrary units that we will not worry about. Now, clearly the random variables we have used so far are inappropriate for this question because time is a continuous quantity. If we assume that we have an arbitrarily accurate stopwatch, the chance that the cell dies at any *exact* instant is effectively zero. Instead, we will need to offer our predictions in terms of a range: "there is x% chance the cell will die between y and z minutes," for instance.

The first question to ask ourselves is if there is any coupling between the gun and the cell: in other words, does the cell somehow "know" something about the gun, or vice versa? The answer to that question is no: the two are independent. The gun fires at a constant rate throughout the experiment, and the cell cannot remember the gun's misses. This condition means that, given the fact that the cell is still alive at any given point in time, the chance of it dying in the next (very) short interval is constant and, very importantly, independent of the amount of time it has lived so far.

This last point is key and somewhat contrary to our intuition. We are used to thinking about objects like cars or animals with more or less defined lifespans. In those cases, we can roughly imagine that a certain amount of wear and tear is accumulating for each year the car is used. At some point, then, the cumulative damage is such that the car is no longer usable.

This intuition is not wholly wrong for cells (but see the next chapter). However, we will assume that the average natural lifespan of the cell is much longer than our experiment, so that the chances of a natural death over the course of the experiment are negligible. The cell also, as we already assumed, has no memory of any prior failed shots. With these assumptions, the chance of the cell dying in a given short interval is only a function of whether a radiation hit lands during that interval.

BACK TO THE INVERSE PROBLEM: PROBABILITY DISTRIBUTIONS

This framing of the problem brings us back to the ideas with which we started the book: we do not know the survival probability of the cell, but we may be able to work out how that probability changes in time. In other words, we can work out the *derivative* of the survival probability. How might we do this?

We can express the idea of a small, constant chance of dying c in a short interval with the equation:

$$P(t+\Delta t)=(1-c)\cdot P(t) \tag{8.6}$$

Here P represents the probability of the cell being alive at time t. This equation will be misleading if we let Δt become too large, because then there is a chance of two radiation shots landing in the same interval. But assuming Δt is small, we can expand about $(1-c)$ to give:

$$P(t+\Delta t)\cong P(t)-c\cdot P(t) \tag{8.7}$$

And if we move the term only in $P(t)$ to the other side, we have:

$$P(t+\Delta t)-P(t)\cong -c\cdot P(t) \tag{8.8}$$

The left-hand side of Equation 8.8 looks very much like the amount of change seen in $P(t)$ over a short interval. If we write:

$$\Delta P\cong -c\cdot P(t) \tag{8.9}$$

we can then let $c=\Delta t\cdot\lambda$, giving:

$$\Delta P\cong -\Delta t\cdot\lambda\cdot P(t) \tag{8.10}$$

which, when we divide both sides by Δt, gives:

$$\frac{\Delta P}{\Delta t}\cong -\lambda\cdot P(t) \tag{8.11}$$

If we take the limit at $\Delta t\to 0$, we have an expression for the derivative of our probability function:

$$\frac{dP}{dt}=-\lambda\cdot P(t) \tag{8.12}$$

If we look back at Tables 1.2 and 1.3 in Chapter 1, we see that functions of e^x have derivatives of the form e^x. That fact in turn reminds us of our model of rabbit growth and, particularly, of fox population decay in Chapter 3. There we saw that functions of the form:

$$P(t)=-e^{-\lambda t}+C \tag{8.13}$$

have derivatives that match Equation 8.12 (You can verify this point in Exercise 8.2). It seems fair, therefore, to argue that the probability of surviving to a particular time in this framework is some kind of exponential function. But how exactly would we use Equation 8.13? To answer that, we will first complete Equation 8.12 and write it as:

$$\frac{dP}{dt}=-\lambda\cdot e^{-\lambda t} \tag{8.14}$$

We will start with a quick definition. **Figure 8.6A** shows Equation 8.14 for the case of $\lambda=2$. In statistics and probability theory, this derivative function has a special name: the *probability density function* (PDF; [10]). Now, when it comes to Equation 8.14, our instinct is to look at Table 1.3 and find its symbolic integral. However, we are about to find that we are actually interested in the *definite* integral of Equation 8.14. The definite integral in this context is the area under the derivative curve between two values of t, t_1 and t_2, where $t_1<t_2$ (e.g., **Figure 8.6B**).

A)

B)

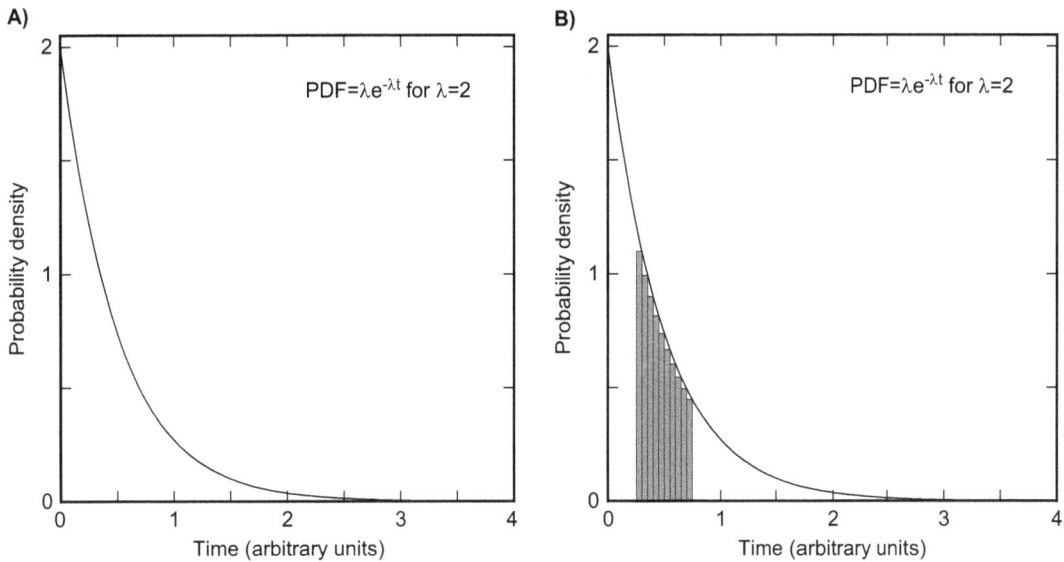

Figure 8.6 The exponential probability distribution, expressed in terms of its *probability density function* (PDF). (A) The PDF for an exponential distribution with $\lambda = 2$. **(B)** Definite integration of the PDF from $t_1 = 0.25$ to $t_2 = 0.75$ gives the probability of cell death occurring in this time interval.

To understand why we desire this quantity, first look back at Equation 8.12. If we are interested in the probability of the cell dying in a short interval Δt, we could approximate that probability with $\Delta t \cdot dP / dt$. In other words, we would apply the idea of approximating a function by summing up short intervals of its derivative from Chapter 1. As our time interval grows longer, we will need to use more than one Δt, corresponding to summing up many rectangles as shown in Figure 8.6B. If we want to know the probability that the cell dies between time t_1 and t_2, those rectangles will start at t_1 and end at t_2. Summing up the area between t_1 and t_2 gives the probability we were in search of—that of cell death between these two intervals.

Using the Exponential Distribution

Although we did not discuss this idea in Chapter 1, calculating the area under *part* of the derivative curve is referred to as a *definite integral*. We write that integral as:

$$P\left(t_1 \le t \le t_2\right) = \int_{t_1}^{t_2} \lambda \cdot e^{-\lambda t} dt \tag{8.15}$$

In the case of the exponential probability distribution, there is actually a symbolic antiderivative (i.e., Equation 8.13), and so we should be able to use it to evaluate the integral in Equation 8.15. However, before we can do that, we need think for a moment about the constant C and what it means to be a probability distribution function.

The rules of probability above say that the sum of the probability of all possible events must equal 1.0. What would the equivalent of this rule be in continuous space or time? Simply put, it means that the probability that the cell dies between time 0 and the limit of infinite time must be 1.0. Using the notation of Equation 8.15, we would express this rule as:

$$P(0 \le t < \infty) = \int_{0}^{\infty} \lambda \cdot e^{-\lambda t} dt = 1 \tag{8.16}$$

In other words, we need a value of C from Equation 8.13 that will turn that equation into what is called a *cumulative distribution function*: the probability that cell death *has* occurred by time t. Using the symbolic antiderivative from Equation 8.13 to evaluate Equation 8.16 thus corresponds to evaluating Equation 8.16 at $t = 0$ and as $t \to \infty$. We can write that computation using the rules for writing the definite integral for a function with a known antiderivative as [11]:

$$\int_{t_1}^{t_2} \lambda \cdot e^{-\lambda t} dt = C - e^{-\lambda t} \big|_{t_1}^{t_2} \tag{8.17}$$

This notation simply tells us to evaluate the symbolic antiderivative at time t_2 and subtract from that the value at t_1:

$$C - e^{-\lambda t} \big|_{t_1}^{t_2} = \left(C - e^{-\lambda t_2} \right) - \left(C - e^{-\lambda t_1} \right) \tag{8.18}$$

We can now evaluate Equation 8.18 using $t_1 = 0$ and as $t_2 \to \infty$:

$$\int_0^\infty \lambda \cdot e^{-\lambda t} dt = C - e^{-\lambda t} \big|_0^\infty \tag{8.19}$$

Since $e^{-\lambda \cdot 0} = 1$ and $\lim_{t \to \infty} e^{-\lambda t} = 0$, Equation 8.19 evaluates to 1 for any value of C. However, we also need that $P(0) = 1$ and $\lim_{t \to \infty} P(t) = 0$. Letting $C = 1$ in Equation 8.13 allows both of those conditions to be true, such that our final cumulative probability distribution function [10] is:

$$P(t) = 1 - e^{-\lambda t} \tag{8.20}$$

To see that Equation 8.20 obeys the rule that the total probability of death sometime between $t = 0$ and $t \to \infty$ sums to 1.0, we can test it using Equation 8.17. Unsurprisingly we find that the total probability works out to 1.0.

We can also simplify Equation 8.20 to:

$$1 - e^{-\lambda t} \big|_{t_1}^{t_2} = e^{-\lambda t_1} - e^{-\lambda t_2} \tag{8.21}$$

which will be our probability of interest.

Returning to the original problem of finite values for t_1 and t_2, we simply plug the respective values into Equation 8.21. In the case of **Figure 8.7**, we have $e^{-2 \cdot 0.25} - e^{-2 \cdot 0.75} = 0.383$.

Figure 8.7 reminds us again of the relationship between the probability density function and the cumulative distribution function: integrating between t_1 and t_2 in Equation 8.14 is equivalent to taking the difference in the values of Equation 8.20 at t_2 and at t_1 and computing the area under the curve of Equation 8.12 between t_1 and t_2.

With all this machinery in place, it is reasonable to go back and consider the effects of λ: the firing rate of our radiation gun. **Figure 8.8** shows how increasing this intensity from $\lambda = 2$ to $\lambda = 3$ changes the chances of survival. Consider $t = 1.4979$. At that time, with $\lambda = 2$, there is a 95% chance that the cell is already dead. How does this probability change with $\lambda = 3$? Interestingly, even though the beam intensity has increased by 50%, the probability of cell death over the interval from 0 to 1.4979 only increases to roughly 99%.

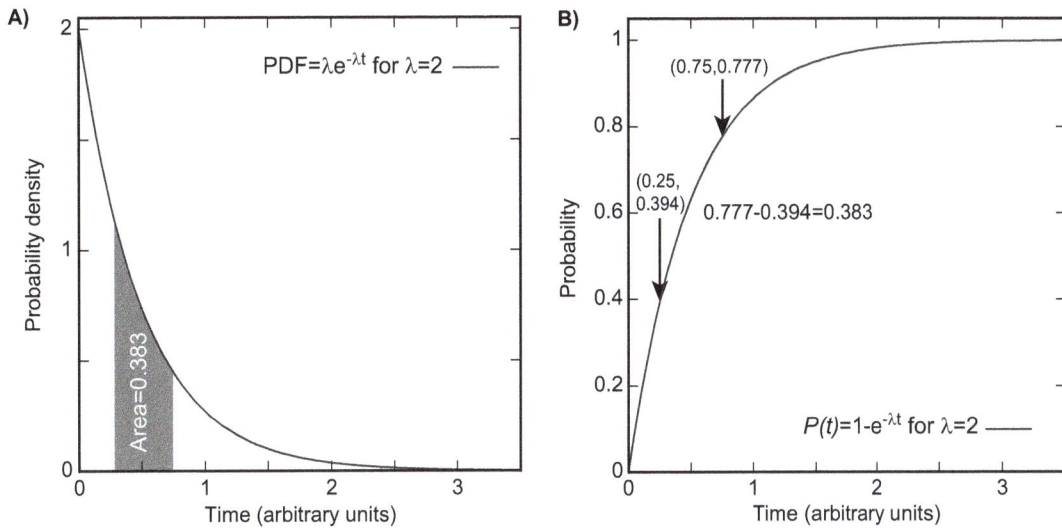

Figure 8.7 Computing the probability of cell death between 0.25 and 0.75 time units. (**A**) Using the probability density function, this computation is equivalent to computing the area under the curve between 0.25 and 0.75 (*shaded*). (**B**) For the exponential distribution, we can find a symbolic antiderivative of Equation 8.14, which, when we constrain it so that the total area under the curve from $t = 0$ to infinity is 1.0 by the rules of probability, gives Equation 8.20 (*purple curve*). Evaluating Equation 8.20 at 0.25 and 0.75 and taking the difference gives us the same probability we obtained by integration in **A**.

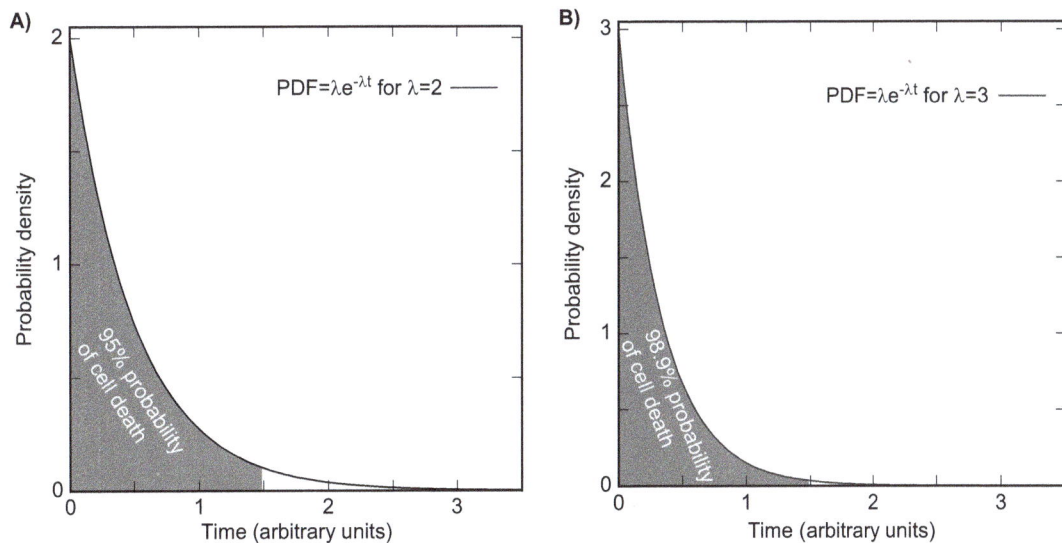

Figure 8.8 Comparing survival probabilities for a beam intensity of $\lambda = 2$ (A) versus $\lambda = 3$ (B). Solving Equation 8.20 for the value of t that gives $P(t) = 0.95$ (i.e., a 95% chance of killing the cell) yields $t = 1.4979$. Plugging this same t into Equation 8.20 with $\lambda = 3$ gives a 98.9% chance of cell death having occurred prior to $t = 1.4979$.

OTHER CONTINUOUS PROBABILITY DISTRIBUTIONS

As you might imagine from our discussion of discrete probability distributions, there are many different continuous probability distributions that describe differing types of phenomena. We will only consider three of these in this book and will leave an extended discussion of them until the next chapter. However, **Figure 8.9** presents the two new examples and illustrates the shapes of their density functions compared to that of the exponential distribution. The uniform distribution behaves very similarly to the discrete uniform distribution we saw for dice and coins. Just like those distributions, it is scaled by the range of potential outcomes, such that a continuous uniform distribution ranging over [0..2] would have density function of $dP/dt = 0.5$, while the one ranging over [0..1] would have a density function of $dP/dt = 1.0$.

A) Exponential distribution

$$PDF = \lambda e^{-\lambda t} \text{ for } \lambda = 2$$

B) Uniform distribution

$$PDF = 0.5$$

C) Normal distribution

$$PDF = \frac{1}{\sqrt{2\pi\sigma^2}} \cdot e^{-\frac{(x-\mu)^2}{2\sigma^2}}$$

for
$\mu = 0, \sigma = 2$

Value

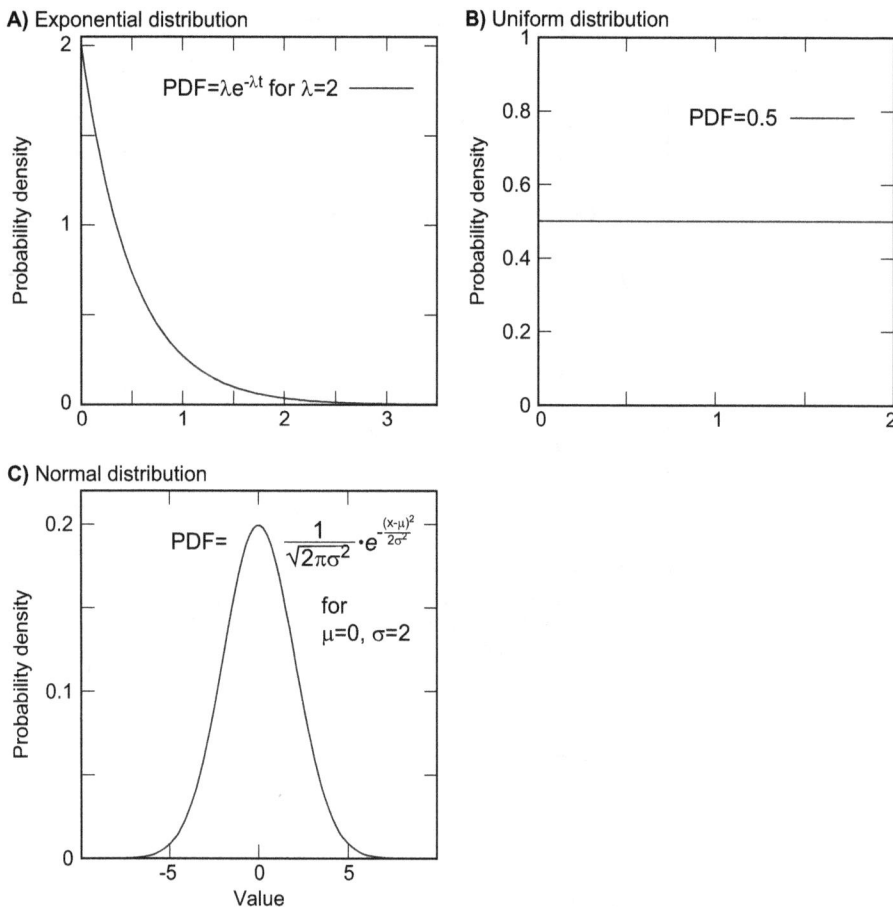

Figure 8.9 Three different continuous probability distributions. (**A**) The exponential distribution that we have discussed extensively. (**B**) The uniform distribution, which is the continuous analog of the die and coin distributions we saw in Figure 8.2. (**C**) The normal distribution, which has two parameters: a mean μ and a measure of variability σ. Because we have picked $\mu = 0$, this normal distribution is centered at $x = 0$.

Figure 8.9C is of particular interest, because if we look back to Galton's data from Figure 8.1, we see that it very likely follows this *normal* distribution, since height is measured on a continuous scale. The next chapter will consider the origins and uses of normal distributions.

REFLECTIONS AND PREVIEWS

In this chapter we developed the mathematical tools needed to describe random events on both discrete and continuous scales. We found that some of the differential and integral calculus we considered Chapter 1 is very useful in the case of continuous probability distributions. In the next chapter we will develop the mathematical theory for why both the height and mRNA count data of Figure 8.1 have their distinctive bell shape. To do so, we will return to Brownian motion and describe it in terms of *random walks*. We will then see that not only is the normal distribution of Figure 8.9C expected in a wide variety of cases, but that the normal distribution gives us a powerful set of tools for comparing samples from populations that are governed by more or less *any* of the probability distributions discussed in this chapter.

Exercise 8.2

Using the rules of differentiation we used in Chapter 3, verify that the derivative of Equation 8.13 is in fact Equation 8.12.

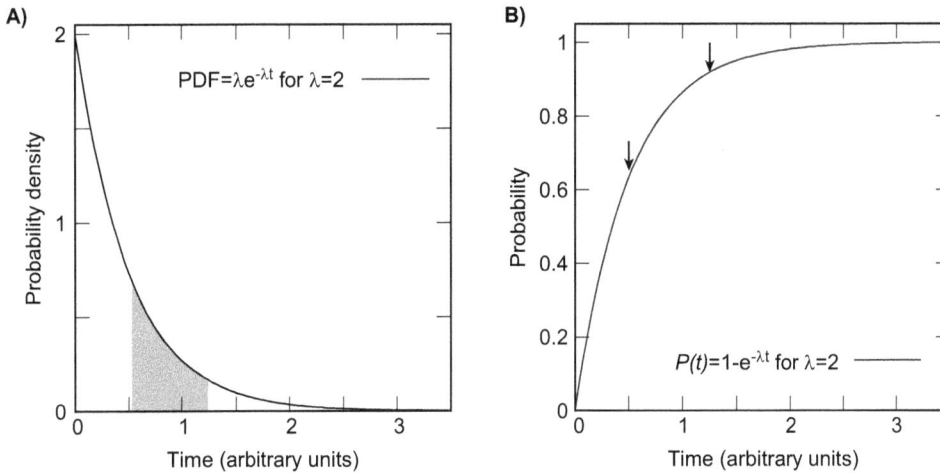

Figure 8.10 **Computing the probability that cell death occurs between** $t = 0.55$ **and** 1.25 **time units.** See Exercise 8.3.

Exercise 8.3

Using the rules of differentiation we used in Chapter 3, verify that the derivative of Equation 8.20 is Equation 8.14. Then repeat the calculation of the survival probability shown in Figure 8.7 for $t_1 = 0.55$ and $t_2 = 1.25$, consulting **Figure 8.10** for reference.

REFERENCES

1. Galton F (1886) Regression towards mediocrity in hereditary stature. *The Journal of the Anthropological Institute of Great Britain and Ireland* 15:246–263.
2. Feynman RP, Leighton RB, & Sands M (2011) *The Feynman lectures on physics, Vol. I: The new millennium edition: mainly mechanics, radiation, and heat* (Basic Books).
3. Gould SJ (1989) *Wonderful life: the burgess shale and the nature of history* (W. W. Norton, New York).
4. Darwin C (1859) *The origin of species by means of natural selection* (John Murry, London).
5. Haw MD (2002) Colloidal suspensions, Brownian motion, molecular reality: a short history. *Journal of Physics: Condensed Matter* 14(33):7769.
6. Sokal RR & Rohlf FJ (1995) *Biometry*. 3rd Edition (W. H. Freeman and Company, New York).
7. Weinburg RA (2007) *The biology of cancer* (Garland Science, New York).
8. Verhey LJ (1999) Comparison of three-dimensional conformal radiation therapy and intensity-modulated radiation therapy systems. *Seminars in radiation oncology*, (Elsevier), pp. 78–98.
9. Sveistrup J, Rosenschöld PM, Deasy JO, Oh JH, Pommer T, Petersen PM, & Engelholm SA (2014) Improvement in toxicity in high risk prostate cancer patients treated with image-guided intensity-modulated radiotherapy compared to 3D conformal radiotherapy without daily image guidance. *Radiation Oncology* 9:1–8.
10. Grinstead CM & Snell JL (2006) *Grinstead and Snell's introduction to probability: the chance project* (American Mathematical Society).
11. Larson RE & Hostetler RP (1986) *Calculus with analytic geometry*. 3rd Edition (D. C. Heath and Company, Lexington, MA), p. 1013.

Is It Normal?

Sampling, Statistics, and the Central Limit Theorem

9

> *"Not all those who wander are lost."*
> —J. R. R. Tolkien, "The Fellowship of the Ring"

AN EXPERIMENT IN SAMPLING

We left the preceding chapter with just the slightest suggestion of a cliffhanger: Why do our samples of mRNA abundances and children's heights both have a shape that is rather like that of a normal probability density function? (For reference, I have reproduced this figure as **Figure 9.1**.) The answer to that question gives us one of the most important tools in statistics, making it worth spending a bit of time on discovering, or at least illustrating, the answer for ourselves.

To begin with, we will use some of the new machinery from the last chapter to prove that the effect we believe we observed is real. In Exercise 9.1, you will take a series of samples from the exponential distribution we explored in the previous chapter. **Figure 9.2** shows my results from taking two quite large ($n = 200$) samples from such a distribution. As you can see, when I make histograms of these two sets of 200 values, they match fairly well with the true probability density function (PDF) of an exponential distribution. Moreover, both the PDF and the two histograms quite clearly have asymmetrical shapes that are nothing like the bell-like curves seen for the children's heights and mRNA abundances in Figure 9.1.

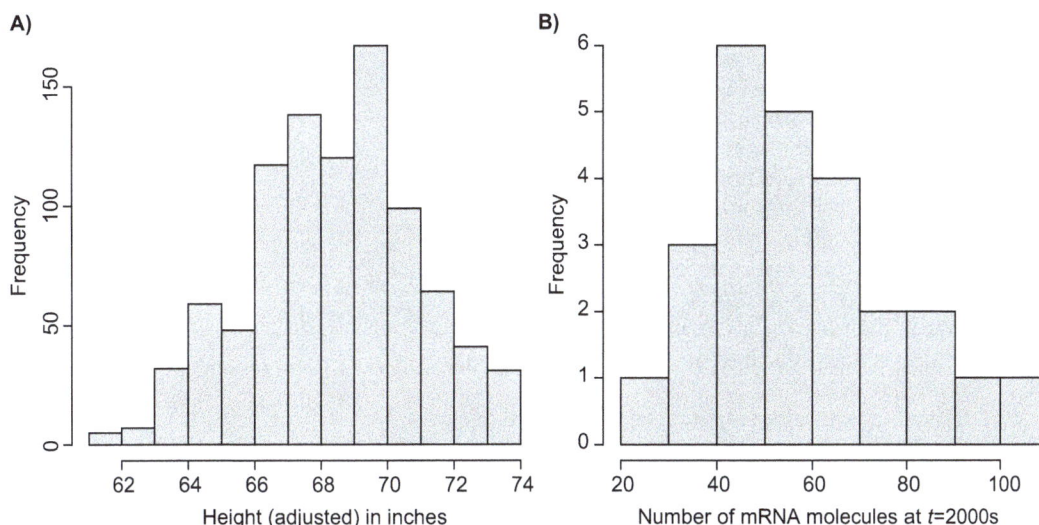

Figure 9.1 Reminding ourselves of the random variation about a central tendency seen in two very different systems from Chapter 8. (**A**) The distribution of heights of 928 children in Victorian England, collected by Francis Galton [2]. The heights of the female children in the dataset were adjusted by a factor of 1.08 to be comparable to that of the males. (**B**) The distribution of number of mRNA molecules per cell in our simulations of noisy gene expression from Chapter 7, taken at the (arbitrary) time $t = 2,000$ s.

DOI: 10.1201/9781003687504-9

Exercise 9.1

Using the website (http://qbio.statgen.ncsu.edu/CLTsim/), take 50 samples, each of size 100 from an exponential distribution with a distribution parameter $\lambda = \frac{1}{2}$. Compute the mean of each sample and then plot the distribution of those means. Repeat the exercise for a uniform distribution with values between 0 and 4.

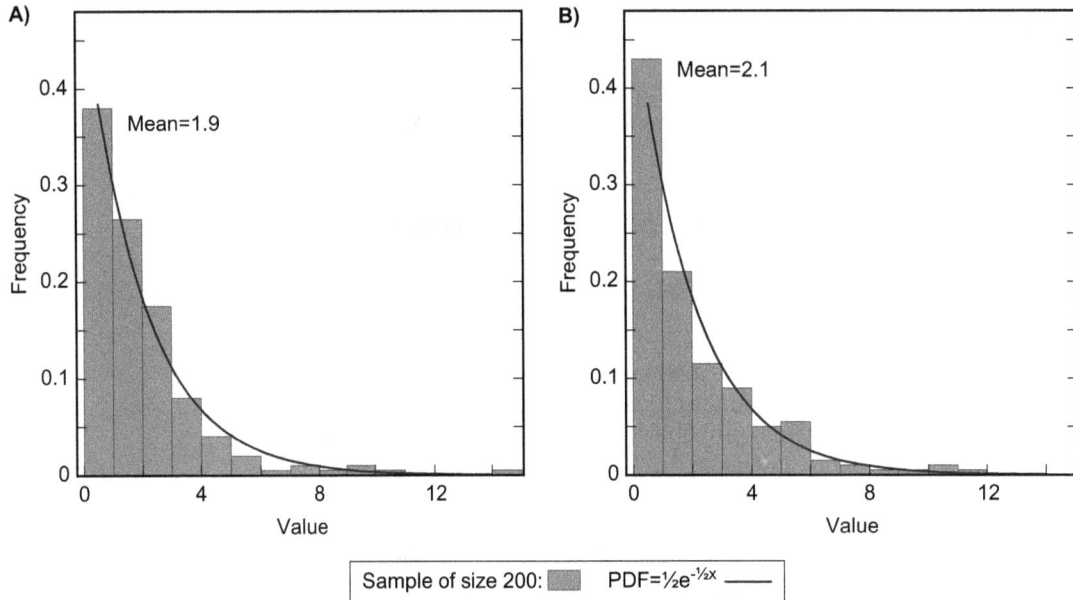

Figure 9.2 Histograms from two samples of 200 values (A and B) from an exponential distribution with $\lambda = \frac{1}{2}$. In each case, I plot these histograms alongside the probability density function for the exponential distribution. I also give the observed mean for each sample.

Behavior of Sample Averages

The magic, so to speak, happens when we stop considering single samples from a distribution and consider how the *average* of a sample behaves. In other words, what happens if we repeat this sampling procedure many times and consider the average of each sample? The result is rather surprising. As **Figure 9.3A** shows, the distribution of these *sample averages* looks suspiciously similar to Figure 9.1.

To show that I haven't cheated too extravagantly, **Figure 9.3B** shows the results of taking the sample average from 400 samples of size 200 from a uniform distribution on [0,4]. When I plot the distribution of those averages, they also have that distinctive bell shape, even though the PDF of a uniform distribution looks nothing like either a normal or an exponential distribution.

Just to reinforce the point, I have illustrated a normal probability density function on top of each panel in Figure 9.3. In each case, I used 2.0 as the mean of the illustrated normal distribution. This choice is appropriate because, although we did not discuss the fact in Chapter 8, the mean of an exponential distribution is $1/\lambda$ – this case, 2. Similarly, it is intuitively obvious that the mean of a uniform distribution on [0,4] is just the central value of 2.

We will not worry yet how I derived the variance (σ^2) of these two normal distributions in Figure 9.3. Instead, to understand how these two very different distributions could give rise to such similar behavior in their sample means, we will take a slight detour back to the idea of Brownian motion we briefly discussed in the previous chapter.

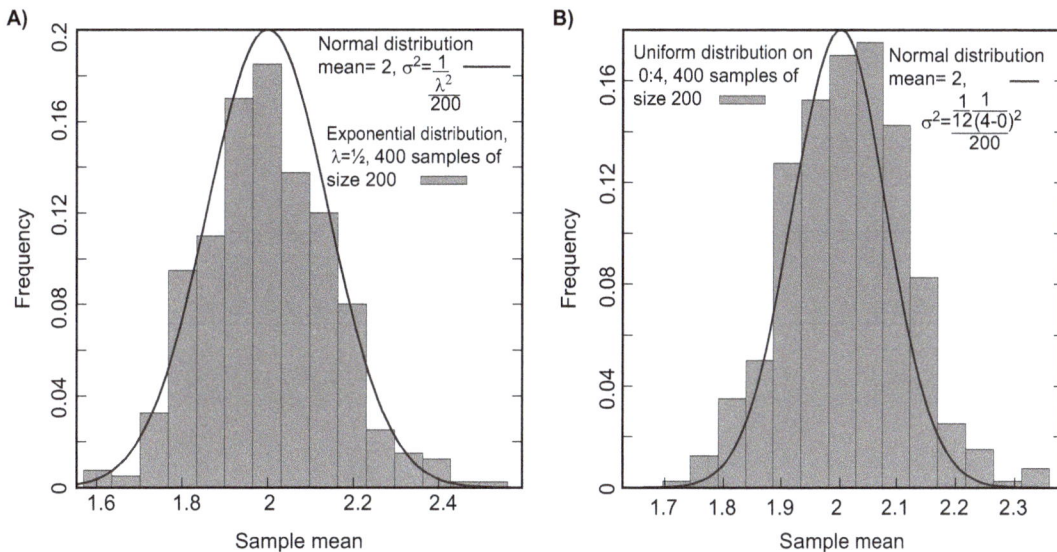

Figure 9.3 Samples of samples drawn from an exponential (A) and uniform (B) distributions. In each case, samples of size 200 were drawn 400 times and the *mean* of each resulting sample computed. The histograms show the distribution of these means, with a normal distribution with a mean of 2 shown for reference. Notice the similarity between the histograms and these normal curves. We will derive expressions for the variance later in this chapter [1].

DIGRESSION: RANDOM WALKS AND DIFFUSION

Before we consider a somewhat physically realistic system, let us consider the simplest case of diffusion: a one-dimensional random walk in discrete time with unit step sizes [3, 4]. While that description sounds slightly intimidating, all it means is that we start with a particle or person at position 0 on the x-axis, and, at each time step, the walker may either step forward one unit or backward one unit, a choice made by flipping a fair coin. **Figure 9.4A** shows three possible realizations of this process over 40 steps. Despite the fact that there is no bias in the step direction, we notice that the walker does not invariably end the walk at $x = 0$.

It is not hard to compute the chances that the walker will end their walk at any given value of x by using the binomial distribution we saw in Chapter 8. Recall that the binomial equation was:

$$\frac{n!}{k! \cdot (n-k)!} \cdot p^k \cdot (1-p)^{(n-k)} \tag{9.1}$$

This equation gives the probability of observing k heads in n coin flips. To get the probability of reaching position $x = k$ after n steps, we first arbitrarily assign heads to be the positive step direction. Then, we multiply the probability above by the corresponding step directions, giving us:

$$\left(\frac{n!}{k! \cdot (n-k)!} \cdot p^k \cdot (1-p)^{(n-k)} \right) \cdot \left(k - (n-k) \right) \tag{9.2}$$

Here, the factor of $k - (n-k)$ gives us the difference between the number of positive and negative steps. If $k = n-k$, we will end up at 0, while if $k > n-k$, our final position will be positive; if $k < n-k$, it will be negative.

Figures 9.4B and **9.4C** show the probabilities of net distances traveled after 20 and 40 steps, respectively. Clearly, as more steps are taken, the range of possible total distances traveled increases, but, because there is no bias to the walk, the most probable outcome remains no net progress $(x = 0)$.

A)

B)

C)

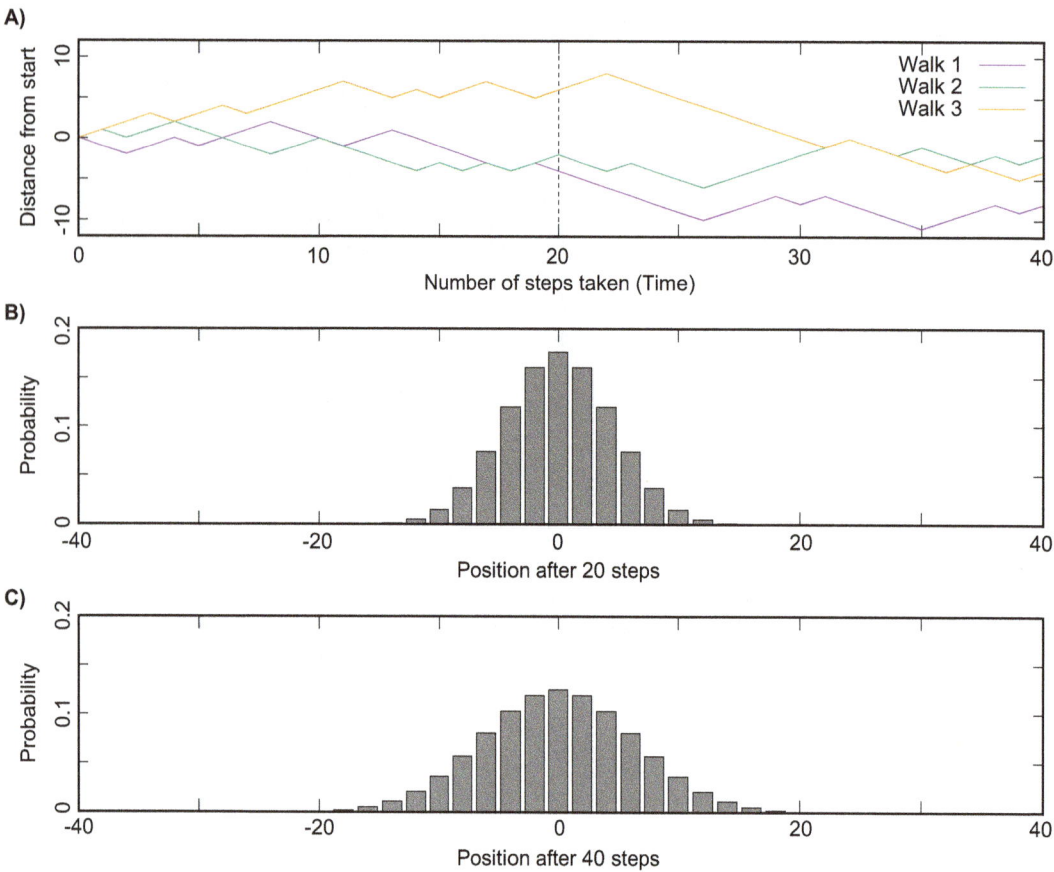

Figure 9.4 Random walks in one dimension. (A) three walks of length $n = 40$ are shown, with steps (time) on the x − axis and net distance (or position) on y. **(B)** In this panel I show the probability of ending up at position x for a walk of length $n = 20$. The x − axis scale runs from −40 to 40 for comparison with **C**, but the maximum ending positions for 20 steps are of course −20 and 20. **C** shows the probability of the possible final positions after 40 steps.

Continuous Random Walks

Figure 9.4 reminds us of the symmetrical distributions we saw in Figure 9.3, but it is a discrete model and so only gives intuition. The next step is to develop a continuous analog of this walk where our particle can take on any position along the x-axis. Although our goal is a probability of the particle's position, it will be helpful to start out by considering a large number of particles (Figure 9.5). When $t = 0$, these particles begin all piled up together on the y-axis ($x = 0$, Figure 9.5A). We then imagine watching them move along x as time increases (Figure 9.5B, C).

In Figure 9.5 we see the dependance of particle position on t: the amount of time there has been for diffusion to occur. Clearly, as t increases, more diffusion is seen. However, if we were to fix t, we could also consider how the concentration of particles is changing as we move along the x − axis. To understand this behavior, Figure 9.6 invites us to consider a small region of this system near $x = 0.05$, with $t = 0.25$. If we focus just on the region $x \pm \Delta x$, we can define Δc as the difference in particle concentration between x and $x + \Delta x$. If we assume that the direction (negative or positive) of each particle's motion is random, we will expect that the net movement of particles will be from regions of high density to low density. The reason for this behavior is just that there are more particles in that high-density region and half of them are moving in the direction of lower density. How fast are they moving? We will assume that the average speed of the particles is given by the diffusion constant D. Here D represents the rate at which particles cross the unit plane at x. Thus, the net flow rate of particles J is going to be given by the

A)

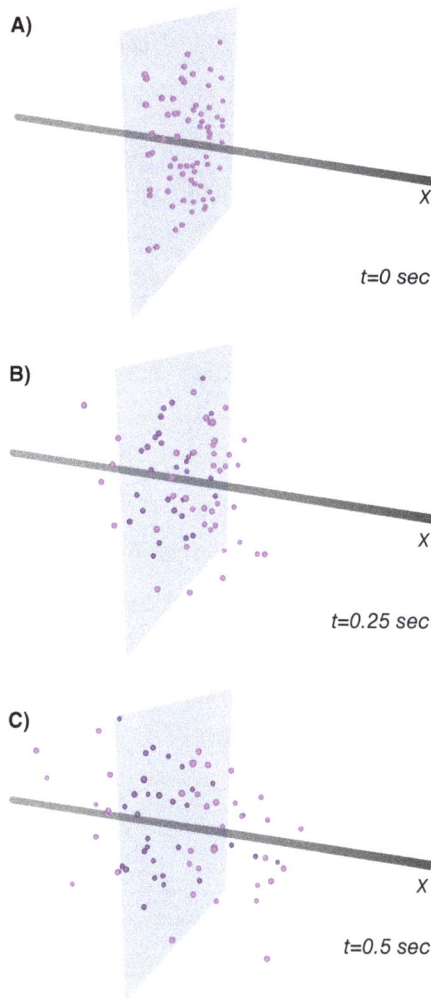

t=0 sec

B)

t=0.25 sec

C)

t=0.5 sec

Figure 9.5 Diffusion in one dimension (x) from an initial concentration of particles concentrated at x = 0. Each particle diffuses independently along the x – axis for time t with an average diffusion speed $D=1$. (Note that the y and z coordinates are arbitrary.) (**A**) $t=0$: no net diffusion. (**B**) $t=0.25$ s. (**C**) $t=0.5$ s.

combination of the local difference in particle density $(\partial c / \partial x)$ and D, allowing us to write:

$$J = -D \cdot \frac{\partial c}{\partial x} \tag{9.3}$$

which just tells us that the rate of flow J across x goes from higher to lower concentrations at relative rate D [3]. What we would like now is function of the form $c(x,t)$. That function will describe the particle concentration for all points along x at a particular time t.

Let us first approximate the number of particles in a local region around x at a particular time t as:

$$c(x,t) \cdot \Delta x \tag{9.4}$$

Equation 9.3 gives the time-dependent flow rate into the region around x, meaning that if we are willing to assume that no particles are being created or destroyed in this region, the change in particle number in the region around x in time could be written as:

$$\frac{\partial}{\partial t} c(x,t) \cdot \Delta x = J(x) - J(x+\Delta x) \tag{9.5}$$

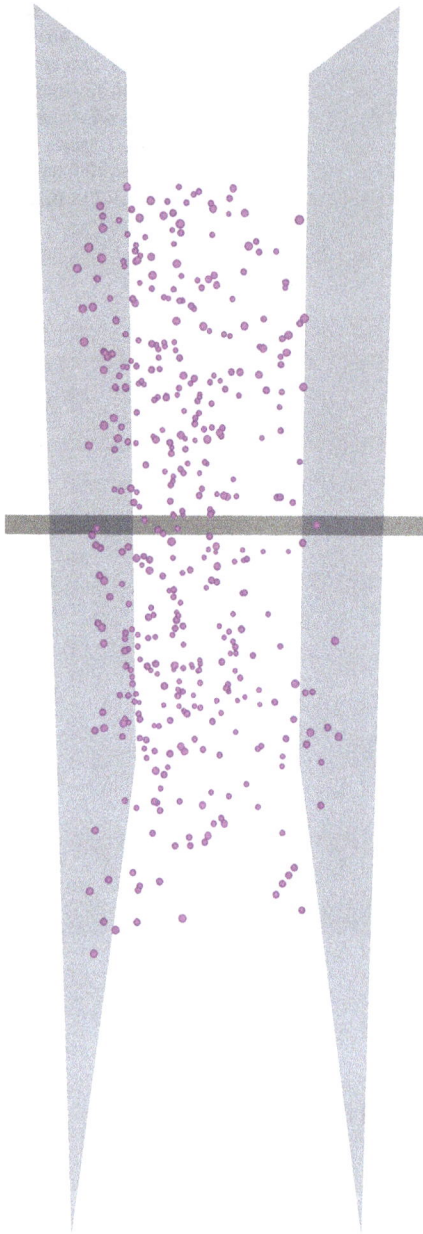

Figure 9.6 **"Concentration gradient" of particles near $x = 0.05$ for $D = 1$ and $t = 0.25$ s.** A large interval of $\Delta x = 0.7$ is used for illustrative purposes. As can be seen, the concentration is decreasing in the positive x direction, implying that $\partial c / \partial x < 0$ in this local region.

In other words, the change in concentration is simply the difference between all the particles coming in at x less those leaving at $x + \Delta x$. If we move the Δx to the other side, we get:

$$\frac{\partial}{\partial t} c(x,t) = \frac{J(x) - J(x + \Delta x)}{\Delta x} \tag{9.6}$$

We can see that by taking the limit as $\Delta x \to 0$, we can view the right side of this equation as the derivative of J with respect to x:

$$\frac{\partial}{\partial t} c(x,t) = \frac{\partial J}{\partial x} \tag{9.7}$$

Substituting for J from Equation 9.3 gives:

$$\frac{\partial}{\partial t} c(x,t) = \frac{\partial \left(D \frac{\partial c}{\partial x} \right)}{\partial x} \tag{9.8}$$

If we expand this second derivative operation, we can see that we end up with the second derivative of c with respect to x:

$$\frac{\partial c}{\partial t} = D \frac{\partial^2 c}{\partial x^2} \tag{9.9}$$

Equation 9.9 is referred to as the diffusion or *heat equation*. It is also what is known as a *partial differential equation*. That is just another way of saying that we have an equation that describes the relationship between the derivatives of a function of more than one dependent variable (here x and t). As you might suspect, in general it is difficult to find analytic solutions for such equations. However, this particular example is susceptible to a trick that gives us a solution. The mathematics of that trick are perhaps beyond the level of this book. But the result is interesting enough that, with some reservations, I am presenting them here. I will simply add that understanding the next few paragraphs is not essential to the key point of this chapter: they are just showing that "there is nothing behind my back."

Digression: Fourier Transformations and the Solution to the Heat Equation

What we will do to solve the heat equation is to apply what is called the *Fourier transformation* to the problem [5]. Fourier transforms, very roughly, take a function that is described in terms of positions and time (say x and t) and re-express those functions in terms of frequencies. One can think of those frequencies as being the characteristic frequencies of a *sine* or a *cosine* function [6]. The Fourier transformation of our desired solution equation $c(x,t)$ with respect to the position variable x is given by [5]:

$$\mathcal{F}\{c(x,t)\} = \int_{-\infty}^{\infty} e^{i\xi x} \cdot c(x,t) dx \tag{9.10}$$

We could say a lot more about this transformation and its power, but that would take us rather far off topic. Suffice it to notice that this $e^{i\xi t}$ term can be rewritten using Euler's formula:

$$e^{i\xi x} = \cos(\xi x) + i \cdot \sin(\xi x) \tag{9.11}$$

Here i is the imaginary number given by the square root of -1. In other words, the Fourier transformation is a way to represent a function as the sum of an arbitrary number of sine and cosine functions with different periods.

The value of this transformation is that it behaves in useful ways relative to the derivatives of $c(x,t)$. As shown in Appendix 9A, the derivative with respect to t of the Fourier transformation of $c(x,t)$ just the Fourier transformation of the derivative of $c(x,t)$ with respect to t:

$$\frac{\partial}{\partial t}\left(\mathcal{F}\{c(x,t)\}\right) = \mathcal{F}\left\{\frac{\partial c}{\partial t}\right\} \tag{9.12}$$

In addition, the second derivative of $c(x,t)$ with respect to x after transformation is just

$$\frac{\partial^2}{\partial x^2}\left(\mathcal{F}\{c(x,t)\}\right) = -\zeta^2 \mathcal{F}\{c(x,t)\} \tag{9.13}$$

(Again, see Appendix 9A.)

Using these two properties, we can now take the Fourier transform of both sides of Equation 9.9:

$$\mathcal{F}\left\{\frac{\partial c}{\partial t}\right\} = D \cdot \mathcal{F}\left\{\frac{\partial^2 c}{\partial x^2}\right\} \tag{9.14}$$

which, using 9.12 we will rewrite as:

$$\frac{\partial}{\partial t}\left(\mathcal{F}\{c(x,t)\}\right) = D \cdot \mathcal{F}\left\{\frac{\partial^2 c}{\partial x^2}\right\} \tag{9.15}$$

We can then use 9.13 to get:

$$\frac{\partial}{\partial t}\left(\mathcal{F}\{c(x,t)\}\right) = -D\xi^2 \mathcal{F}\{c(x,t)\} \tag{9.16}$$

Looking at this equation, notice that $\mathcal{F}\{c(x,t)\}$ is just "some function." If that is the case, Equation 9.16 is a differential equation saying that that function (call it $\hat{c}$) has a time derivative equal to a constant (m) times the function itself. We have encountered this differential equation several times already in this book:

$$\frac{d\hat{c}}{dt} = m \cdot \hat{c}(t) \tag{9.17}$$

As a result, we know that such equations have solutions of the form:

$$\hat{c}(t) = Ne^{mt} \tag{9.18}$$

where N is just another constant. Hence, the solution to Equation 9.9 in Fourier space is just an exponential function with an unknown constant N in front of it:

$$\mathcal{F}\{c(x,t)\} = Ne^{-D\xi^2 t} \tag{9.19}$$

To get the value of N, we need the initial conditions of our system. Under our assumption that our random walk starts at $x = 0$, we have:

$$c(x,0) = \begin{cases} 1 & \text{if } x = 0 \\ 0 & \text{otherwise} \end{cases} \tag{9.20}$$

All we are saying here is that before any time has elapsed, the probability of being anywhere *except* $x = 0$ is 0. Now we need take the Fourier transform this equation, which is known as Dirac's delta function [7]. Because the delta function is zero everywhere *except* $x = 0$, its Fourier transform is just $e^{i\xi x}$ evaluated at $x = 0$, namely $e^{i\xi 0}$, or 1. We therefore set $t = 0$ and plug $\mathcal{F}\{c(x,0)\} = 1$ into 9.19 to yield:

$$1 = Ne^{-D\xi^2 0} \tag{9.21}$$

Meaning that $N = 1$ and our complete solution to the random walk in Fourier space is just:

$$\mathcal{F}\{c(x,t)\} = e^{-D\xi^2 t} \tag{9.22}$$

Having now solved Equation 9.9 in Fourier space, it remains to see if we can convert that solution back to "normal" space. Given a Fourier-transformed function $\hat{c}(\xi,t)$, the *inverse* Fourier transformation is defined as [5]:

$$\mathcal{F}^{-1}\{\hat{c}(\xi,t)\} = \frac{1}{2\pi}\int_{-\infty}^{\infty} \hat{c}(\xi,t) \cdot e^{-i\xi x} d\xi \tag{9.23}$$

We could evaluate this integral for our expression for $\hat{c}(\xi,t)$ in Equation 9.22. However, a simpler route is to consult a table of functions and their Fourier transformations, where we find [5]:

$$\mathcal{F}\left\{e^{-ax^2}\right\} = \sqrt{\frac{\pi}{a}} e^{-\xi^2/4a} \tag{9.24}$$

In other words, there is a class of exponential functions in normal space that have Fourier transformations that look a lot like our solution in Equation 9.22. In fact, if we multiply 9.24 by $\sqrt{a/\pi}$, we get:

$$\sqrt{\frac{a}{\pi}} \cdot \mathcal{F}\left\{e^{-ax^2}\right\} = \sqrt{\frac{a}{\pi}} \cdot \sqrt{\frac{\pi}{a}} e^{-\xi^2/4a} \tag{9.25}$$

We can see from the definition of the Fourier transformation in Equation 9.10 that it the transformation of a function by a constant is just the original transformation multiplied by that constant. As a result, we can move the $\sqrt{a/\pi}$ term on the left *inside* the transformation. Because we can also cancel the $\sqrt{a/\pi}$ and $\sqrt{\pi/a}$ terms on the right, we have:

$$\mathcal{F}\left\{\sqrt{\frac{a}{\pi}} \cdot e^{-ax^2}\right\} = e^{-\xi^2/4a} \tag{9.26}$$

If we now propose that $a = 1/4Dt$, we find:

$$\mathcal{F}\left\{\frac{1}{\sqrt{4Dt\pi}} \cdot e^{\frac{-x^2}{4Dt}}\right\} = e^{-\xi^2 Dt} \tag{9.27}$$

meaning that we now have a Fourier transformed function that exactly matches our solution in Equation 9.22. We can therefore conclude that an *untransformed* $c(x,t)$ function of:

$$c(x,t) = \frac{1}{\sqrt{4Dt\pi}} \cdot e^{-x^2/4Dt} \tag{9.28}$$

is the solution to Equation 9.9 in "regular space." This equation applies to cases where $t > 0$, with Equation 9.20 applying when $t = 0$. We can check our solution by taking derivative of Equation 9.28 with respect to t. This computation is shown in Appendix 9B, where we find that we indeed regenerate Equation 9.9.

The Continuous Random Walk Is Described by a Normal Distribution

Having found that Equation 9.28 describes the continuous random walk, let us examine it a bit more closely. If we let $\sigma^2 = 2Dt$, we have:

$$c(x,t) = \frac{1}{\sqrt{2\pi\sigma^2}} \cdot e^{-x^2/2\sigma^2} \tag{9.29}$$

which, if we look back to Figure 8.10, is just the PDF for a normal distribution. In other words, at time t, the particle distribution for the random walk is a normal distribution with mean 0 and variance $\left(\sigma^2\right)$ of $2Dt$. Looking at Equation 9.28, we take comfort in the fact that, as time increases, the spread of the walk increases as well (**Figure 9.7**), just as our result suggests it should. What about our diffusion example from Figure 9.5? Well, Equation 9.28 can be seen equally well as describing one-dimensional Brownian motion with a diffusion rate D [4], meaning that the distribution of concentrations for these particles will follow this same normal density.

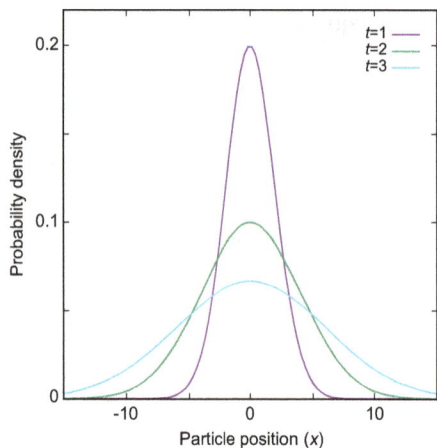

Figure 9.7 Probability density for a continuous random walk for times $t = 1, 2$ and 3 with a value of $D = 1$, taken from Equation 9.28.

RELATING THE RANDOM WALK TO THE BEHAVIOR OF SAMPLES

At this point, the natural question to ask is why we have just spent all this time on the topic of random walks. To answer that question, we should go back to Figures 9.1 and 9.2 and reimagine what we are seeing. In particular, we could consider representing a child's height as resulting from the *sum* of a series of small random events that I will refer to as *impulses*. For example, focusing only on the genetic component of height, we could argue that the child's height is the result of a series of these impulses, one from each gene, of which about half push to the right of the mean height of the whole population and about half to the left. (Since each child has a different combination of those genes, their heights will be slightly different.) Seen in this way, the children's heights can be seen as the result of a conceptual random walk. To be a bit more concrete, consider each value x_i from a particular sample in Figure 9.2. If we assume that our population mean is μ, we could represent these sample means as:

$$\sum_i \left(\mu - x_i\right) \tag{9.30}$$

By definition, half of these distances will be positive (greater than the mean) and half less than the mean. Essentially, we have a random walk centered at μ. Generalizing a bit, we can claim that the random walk provides an explanation of the ubiquity of the bell-like curves we have seen in Figures 9.1 and 9.2. To reinforce this idea, think back to our simulations of noisy gene expression in Chapter 7. Ignore the initial phases of the simulation for a moment and consider what happens after the expression levels have come to equilibrium. At any given moment, the mRNA level m will decrease a bit due to decay or increase a bit due to transcription. As a result, we will see the particle numbers "walk" back and forth around the mean: when we sum up those little steps, we will generate another normal distribution, centered on the mean mRNA level for the model.

DISTRIBUTION OF SAMPLE SUMS AND MEANS: THE CENTRAL LIMIT THEOREM

Beyond being conceptually satisfying, the fact that samples that depend on the sum of random inputs will follow a normal distribution is tremendously useful on a practical level because it allows us to make statements about the nature of an entire population using only a sample from it.

To understand how our random walk discussion enlightens us on the behavior of samples, rather than considering the sample mean, we will start with the sample *sum, S. S* is directly analogous to the distance traveled in the random walk: it is the sum of all the sampled values. Those values themselves correspond to the impulses in the random walk.

We will assume that S is computed from sampling n items from some underlying probability distribution F. F could take on many forms: normal, exponential, uniform, among others. The two rules F needs to follow are that it have a finite mean and that it have a finite and non-zero *variance* (8). We have not discussed how to compute the mean μ or the variance σ^2 of a continuous probability distribution, but the two computations are quite similar. Computing the mean involves taking the integral of the probability density function $f(x)$ of F multiplied by the value of the random variable itself [8]:

$$\mu = \int_{-\infty}^{\infty} x \cdot f(x) dx \qquad (9.31)$$

Effectively, we are multiplying each potential value x from the distribution by its proportional probability of occurring, namely $f(x)$.

Similarly, the variance of F is computed by taking an integral of $f(x)$ multiplied by the square of the difference between x and the distribution mean [8]:

$$\sigma^2 = \int_{-\infty}^{\infty} (x - \mu)^2 \cdot f(x) dx \qquad (9.32)$$

Again, we are weighting each value's squared distance to the mean, $(x - \mu)^2$, by the relative probability of that value (again given by $f(x)$). We see from these two equations that μ and σ^2 can be computed for many different distributions, including the exponential and uniform distributions: they are not just parameters for normal distributions.

We should now return to our sample sum S. If S is drawn from F and n is fairly large, we would expect to find that $S \approx n\mu$. However, because it is only a sample, S will not take on exactly this value. (Of course, in the real world, we also do not know what the population mean μ actually is; that is one of the reasons we have taken the sample. We will come back to this problem later.)

What our random walk system allows us to do is to quantify *how close* S is likely to be to $n\mu$. This question comes down to asking how variable S is. The first component of the answer is obvious: the larger the variability in our impulses, the larger the variability in S. In other words, the variability in S should be proportional to σ^2 from F. Looking back at the random walk, we are happy to see that, indeed, the variance of that distance traveled includes D: the diffusion rate that is the equivalent of the variance of the diffusion system.

But we also expect the variability of S to depend on n: as we add more draws from F, it becomes possible for S to be increasingly far from $n\mu$. The analog of n in our random walk is t, and, looking back, we see that the variance of the distance traveled in the walk is *also* proportional to t.

Statisticians have formalized this idea of the behavior of a sample sum as the *Central Limit Theorem* [8]. It states that as n increases, the distribution of S approaches a normal distribution with mean $n\mu$ and variance $n\sigma^2$ [1], which we can denote at $S \sim N(n\mu, n\sigma^2)$. (We would read this as "S is normally distributed with mean n times μ and variance n times σ^2.")

The Central Limit Theorem for Means

Although the association between the behavior of the sample sum and the random walk is neat and relatively easy to compute, it would be even more useful to say something about the behavior of $\bar{x}$: the sample *mean*. Since $\bar{x}$ for a particular sample is just S/n, it is clear that we expect $\bar{x}$ to approach μ, the mean of F. The more critical question is what the variance is.

We will compute this variance shortly, but first it is worth exploring our intuition a bit. Even if the exact value of the variance of the sample mean is not obvious, the general form of the distribution is somewhat instinctive. Given that we know that the distribution of S is normal and that $\bar{x}$ is just S divided by n, our visual imagination should suggest that multiplying a bell-shaped curve by a constant should yield a second bell-shaped curve. Since $n > 1$, we also know that $1/n < 1$, meaning that our new normal distribution for the sample mean should have a narrower shape (smaller variance) than that for the sample sum S.

To work out precisely what that variance is, we will first represent the distribution of S with a probability density function $s(y)$ (the derivative of S). Its values are given by the corresponding normal probability density function of:

$$s(y) = \frac{1}{\sqrt{2\pi n\sigma^2}} \cdot e^{-(y-n\mu)^2 / 2n\sigma^2} \tag{9.33}$$

(We are using y as our distribution variable to avoid confusion with $\bar{x}$, our sample mean.) Next, we will try to compute the new and unknown distribution of $\bar{x}$, which we will name X. Because X is a probability distribution, we can find the probability that the sample mean is less than a dummy variable y by evaluating the probability distribution function X at y: $X(y) = \text{Prob}(\bar{x} \leq y)$. Thinking back to Chapter 8, this computation is equivalent to integrating X's probability density function $x(y)$ between negative infinity and y. But we also know that that mean of X is just $\bar{x} = S/n$, meaning that the probability that $\bar{x} \leq y$ must be the same as the probability that $S \leq ny$. Or, in terms of the two distribution functions S and X:

$$S(ny) = X(y) \tag{9.34}$$

Next, we consider what Equation 9.34 represents in terms of the probability density functions by differentiating both sides with respect to y:

$$\frac{\partial}{\partial y}(S(ny)) = \frac{\partial}{\partial y}(X(y)) \tag{9.35}$$

The right side of this equation is just the definition of the PDF for X, namely what we are trying to compute. To compute $\frac{\partial}{\partial y}(S(ny))$, we apply the chain rule to get:

$$s(ny) \cdot \frac{\partial}{\partial y}(ny) = \frac{\partial}{\partial y}(X(y)) \tag{9.36}$$

Or, if we reverse and simplify:

$$x(y) = n \cdot s(ny) \tag{9.37}$$

meaning that the PDF of X is just n times the PDF of S evaluated at ny. So, plugging in ny for y in Equation 9.33 and multiplying by n, we have:

$$x(y) = n \cdot \frac{1}{\sqrt{2\pi n\sigma^2}} \cdot e^{-(ny-n\mu)^2 / 2n\sigma^2} \tag{9.38}$$

or

$$x(y) = \frac{n}{\sqrt{2\pi n\sigma^2}} \cdot e^{-n^2(y-\mu)^2 / 2n\sigma^2} \tag{9.39}$$

If we move the n in the first term into the denominator and the n^2 in the exponential similarly, we have:

$$x(y) = \frac{1}{\sqrt{2\pi(\sigma^2/n)}} \cdot e^{-(y-\mu)^2 / 2(\sigma^2/n)} \tag{9.40}$$

In other words, just as the *sum* of a series of items sampled from a distribution with finite mean and variance approaches a normal distribution with mean $n\mu$ and variance $n\sigma^2$ when n is large, the *mean* of that sample *also* approaches a normal distribution. In this case, that normal distribution X has a mean of μ and a variance of $\sigma^2/n : \bar{x} \sim N\left(\mu, \sigma^2/n\right)$. This result is the Central Limit Theorem for sample means [1].

Application of the Central Limit Theorem: Fruit Fly Lifespans

We will explore the usefulness of this new understanding of the behavior of samples drawn from a distribution with a simple example involving the lifespan of fruit flies.

The inevitability of death is so instinctive to us that the question of why organisms die is worth thinking about for a moment. First, we should realize that, from a cellular perspective, death is an entirely different phenomenon. Every cell in your body is a product of an unbroken chain of successful cell divisions stretching back to the origins of cellular life more than 3 billion years ago. Now, our own cells are an imperfect example here because, with the exception of a handful of eggs cells in women, they all will die one day (recall our discussions in Chapter 4). But consider instead a free-living bacterial cell. While such cells die all the time, in some sense they do not *need* to die: they are able to replace all of their molecular parts except their genes with new ones; moreover they are able to make (nearly) perfect molecular copies of themselves, resetting their aging clock to zero.

Hence, while one part of death is the accumulated wear and tear we are all familiar with from the degradation of our material possessions, this analogy is imperfect because, at a molecular level, many cells can repair most damage of this kind.

On the other hand, all organisms suffer from what we might term "extrinsic mortality," which just corresponds to death by accidents, predation, disease, and so forth. As a result, every organism has a finite lifespan. We are also, however, all *evolved* organisms. We might expect that, all other things being equal, evolution would seek genetic combinations that are long-lived. Such combinations would be good at repairing the intrinsic causes of aging, including wear and tear.

The difficulty with this idea lies in the extrinsic mortality, because Darwinian selection cannot evolve genetic combinations to extend lifespan indefinitely, since extrinsic mortality means that no individual organism will live that long, regardless of the perfection of its genes [9]. To oversimplify a bit, if every individual is dead due to extrinsic mortality by age 90, genetic combinations that extend life from 90 to 100 but have no effect before that age cannot be selected for.

Modern humans are a poor place to start reasoning about aging because our evolutionarily recent development of complex societies and their associated technology have increased our expected lifespans. As a result, for most of us, death from old age is a realistic and concerning prospect [10]. If, however, we consider our ancestors of a few millennia ago (or other animal populations), extrinsic mortality probably killed most individuals before aging was a significant concern [10]. Hence, we might expect to find that the age at which a wild organism starts to experience senescence would be greater than the age at which the majority of the individuals of that species have died due to extrinsic events.

This argument has even greater force when we consider the possibility that there might be a cost to the organism in investing in anti-aging defenses. In particular, such investment could use resources that could otherwise be applied to greater reproductive success [11]. In this view, resources devoted to extending lifespan might incur the cost of fewer offspring, making long-lived individuals paradoxically less successful evolutionarily.

As it turns out, we can test some of these ideas in an interesting way. We will see in Chapter 15 that an animal's energy usage per unit time (e.g., in calories burned per hour) is, not surprisingly, strongly associated with its body size [12]. Somewhat more unexpectedly, there is also a similar association between metabolic rate and lifespan, although the mechanisms behind that association are still being debated [13]. For our purposes, what matters is that we can usually make a fairly good prediction of an animal's expected lifespan from its body size. Where things get interesting is in the exceptions. Vertebrates such as birds (especially birds like the albatross) [14, 15], bats [15, 16], and naked mole rats [17] live much longer than would be expected given their size. What links this rather unlikely group of animals? A very intriguing answer is a relatively low rate of predation and hence of external mortality [14, 15]. All these animals have ecological niches (air, caves, burrows) that help shelter them from predation and potentially other sources of mortality. As a result, it appears that they *have* evolved protections from aging beyond those of other vertebrate precisely because their reduced rates of external mortality make such investments evolutionarily sensible in their cases. As you might expect given our own fear of death, the genomic changes that evolution used to extend lifespans in these animals are being intensively studied by scientists to help us learn how to extend human lives [17].

Experimentally Testing the Link between Mating Success and Lifespan

If the preceding discussion strikes you as quite speculative, you might be wondering if we could find a more direct approach to testing for an association of lifespan and reproduction. And, in fact, we can. In particular, we will look at one potential corollary to this line of thought: that organisms might partition their resources differently prior to and after reproduction, decreasing investment in lifespan once they have reproduced. If the organism is experiencing an environment where reproduction is not possible, perhaps because of a lack of mates, they will invest energy in lifespan extension, "hoping" that the reproductive environment improves. Once mating has occurred, the organism might instead invest in pursuing more mating opportunities in that currently good reproductive environment and invest less in extending lifespan.

As it happens, one of the most commonly used sample datasets in biological statistics addresses exactly this question. Partridge and Farquhar compared the lifespans of two groups of fruit flies, asking whether male fruit flies that have mated several times have a shorter remaining lifespan than those of the same age that have mated only once [18].

If we make a histogram of the lifespan of these 125 flies (**Figure 9.8A**), we might think that we see the suggestion of two peaks in the data, possibility representing the difference in flies with more and fewer partners. In Exercise 9.2, you will compute the means of these two groups. As we might have predicted, the mean lifespan of the single-mated flies is 10.7 days longer than the mean of the multi-mated group. Moreover, if we separate the groups and make new histograms, we can see this difference in the means more clearly (**Figure 9.8B**).

Exercise 9.2

Using the provided fruit fly lifespans, use R to compute the mean of the multi-mated and single-mated flies (http://qbio.statgen.ncsu.edu/data/Statistics/fruit_flies.dat). You can split the full dataset into these two groups with the commands:

- `fruitflies<-read.table("fruit_flies.dat")`
- `names(fruitflies)<- c("ID", "Partners", "Type", "LifeSpan", "Thorax", "Sleep")`
- `multimate<-subset(fruitflies, Partners>1, select=c(Partners, Type, LifeSpan, Thorax, Sleep))`
- `singlemate<-subset(fruitflies, Partners<2, select=c(Partners, Type, LifeSpan, Thorax, Sleep))`
- `hist(singlemate[[3]])`
- `hist(multimate[[3]])`
- `mean(singlemate[[3]])`
- `mean(multimate[[3]])`

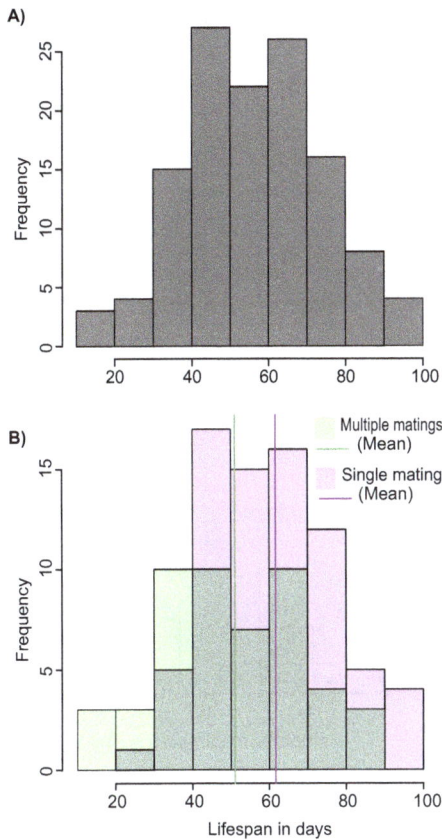

Figure 9.8 Lifespans of male fruit flies that have undergone different numbers of matings. (**A**) The individuals that mated once and more than once are illustrated together; notice the suggestion of a double peak. (**B**) Overlapping histograms for the single-mated (*purple*) and multi-mated flies (*green*).

However, we must remember that these data represent only a pair of samples from the entire population of flies with a single mating and the population with multiple matings. If we took another pair of samples, those new samples—and their means—would be different. **Figure 9.9** shows examples of this problem: I have drawn samples from a pair of populations A and B that have identical underlying distributions. However, because I only observed a few individuals from population A and a few from B, the two samples do *not* have identical means. Hence, the differences we see in Figure 9.9B and 9.9C are not in some sense "real"; they only result from sampling effects. As a result, the question arises as to whether or not the difference we see in the lifespans of our fruit flies could also be attributable to sampling effects.

Now we can see the value of the central limit theorem. Because the distribution of sample means is expected to be normal, we can (almost) use the normal distribution of Equation 9.40 to ask whether our two sample means $\bar{x}_1$ and $\bar{x}_2$ are different enough to draw the conclusion that means in the underlying two populations are also different.

There are, however, a couple of wrinkles. Because we have two samples, we cannot apply Equation 9.40 directly. Moreover, that equation is framed in terms of the population variance σ^2, when in practice we can only estimate that variance with the *sample* variance s^2. If we use the sample variance as an estimate of the population variance, particularly if n is small, that choice will introduce some error into our calculations [1].

There is another continuous probability distribution we can use to correct for imperfections in our estimate of σ^2. That distribution is known as *the Student's t distribution*, and it corrects for this issue by making n a parameter of the distribution. It is from this distribution that we get the ubiquitous *t-test*. (1). The t distribution is broader than a normal distribution for any particular σ^2, corresponding to the extra uncertainty in the estimation of σ^2. This extra uncertainty decreases as n increases. **Figure 9.10** illustrates how the t distribution approaches a normal distribution as n increases.

A)

B)

C)

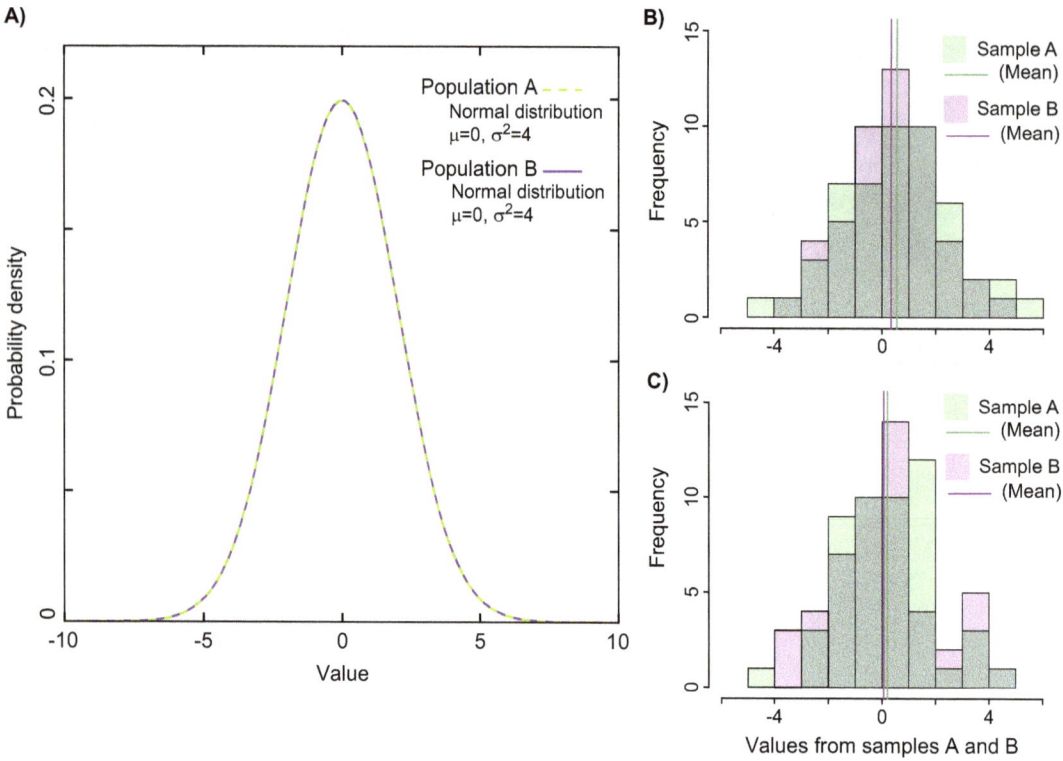

Figure 9.9 Samples drawn from identical underlying distributions can have differing sample means. (A) Populations "A" and "B" are both normally distributed with mean $\mu = 0$ and variance $\sigma^2 = 4$. **(B)** Samples drawn from A and B with the means illustrated. **(C)** Same as for **B** but with a different sample pair.

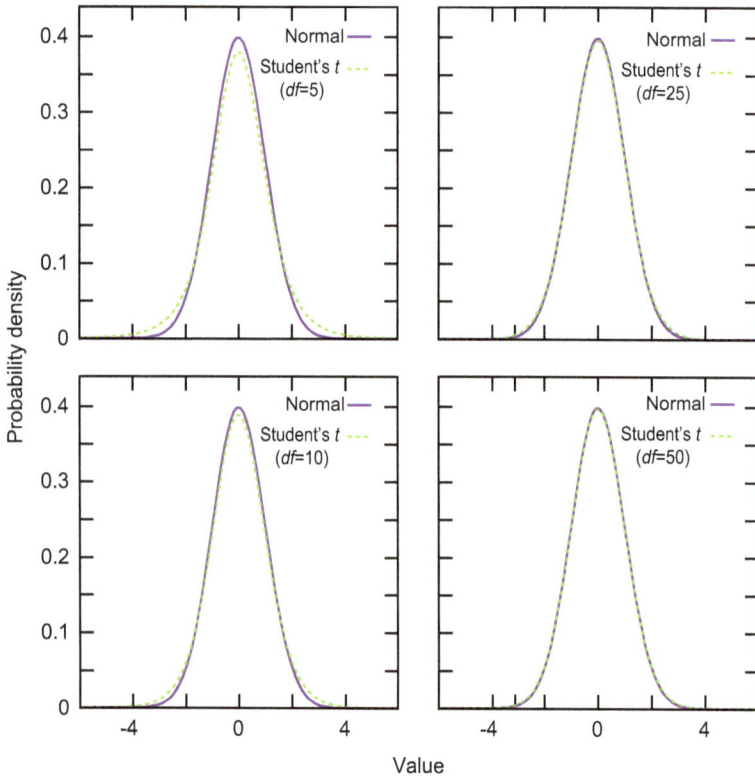

Figure 9.10 Comparison of the shape of a normal and a t distribution for different values of the degrees of freedom (df = sample size − 1) of the t distribution. By the time $n \approx 50$, the two are visually indistinguishable.

FRAMING AND TESTING A STATISTICAL HYPOTHESIS

It is very tempting to rush into the question of comparing our two sets of flies. However, it will be better to clarify first what the question we are asking is and what possible answers we might see.

We will frame the problem in terms of the population means. Hence, the question can be framed as: Are the population mean lifespans of the single-mated and multi-mated flies the same?

Taking this question, how can we use the central limit theorem? First, we will notice that we have framed our question in terms of a *null* hypothesis, namely that there is no difference between the mean lifespan of the two populations of flies. We can now use an elaboration of the t distribution to generate a new probability distribution T. We first define $d = |\overline{x_1} - \overline{x_2}|$: the absolute value of the difference between $\overline{x_1}$ and $\overline{x_2}$. The cumulative distribution $T(d, n)$ gives the probability of two means from samples of size n being *up to d* units apart, given that the two population means are the same. Recall that, for our actual data, we observed $d_r = 10.7$ days of difference in their mean lifespans. We use $T(d, n)$ to compute a statistic known as a P value to assess our null hypothesis:

$$P = T(d, n)\big|_{d_r}^{\infty} \qquad (9.41)$$

In other words, P gives the probability of observing a d value at least as big as the actually measured d_r. In Exercise 9.3, we will conduct such a t-test for the fly data, from which we will find that $P = 0.0007$.

Exercise 9.3

Using the provided fruit fly lifespans, use R to perform a two-sample t-test of the null hypothesis of equal population means for the single-mated and multi-mated flies. Use the R commands from Exercise 9.2 to load the data; the t-test can be run with the following command:

```
t.test(multimate[[3]], singlemate[[3]], var.
equal=TRUE, paired=FALSE)
```

Given that P values are ubiquitous in science, it will pay to spend a moment talking about their uses and meaning. First, recall that our P value was framed in terms of a null hypothesis. Here, that hypothesis was that these two populations of flies have the same mean lifespan. The P value thus expresses the chance that two samples of size n would have a difference in their *sample* means as large as or larger than d_r even though the two populations in fact did not differ in their means. Often, we will state that we *reject* this null hypothesis if the computed P value is relatively small (perhaps < 0.05).

Rejecting a null hypothesis is equivalent to claiming that it is so improbable that the null hypothesis generated our actual data, we are comfortable claiming that the null hypothesis is not correct. Notice, however, that because we are discussing probabilities, for a given d_r, there is some chance that two samples that follow the null hypothesis *could* generate a result that extreme. Hence, another way to define the P value is that it gives the probability that we are *falsely* rejecting the null hypothesis.

In those terms, how could we discuss our fly example? First, notice that $P = 0.0007 = 7 / 10,000$. In other words, if we took 10,000 samples of the size of our actual samples from two populations of flies that in fact had identical means, in 7 of those 10,000 samples the difference in the two sample means would be $d_s = 10.7$ or more days (e.g., as large or larger than the actual observed difference for Exercise 9.2). Hence, if we declare that mating reduces fruit fly lifespans, we have accepted a 7-in-10,000 chance that this claim is false and that our samples only show these differences due to chance effects. We are in fact quite comfortable with this very small chance of error and will claim that these data support the idea of an evolutionary trade-off between reproduction and lifespan.

REFLECTIONS AND PREVIEWS

In this chapter we first made the interesting observation that the means of samples drawn from a variety of probability distributions have a distribution that looks surprisingly bell-shaped. To understand why this might be, we took a long digression on the subject of random walks in discrete and continuous space. Using some rather complex mathematics, we showed that a continuous random walk generates final positions that follow a normal distribution. We then saw that we could apply this result to sample sums and means, a result known as the *central limit theorem*.

We then applied the central limit theorem to the problem of testing whether two sample means were significantly different from each other. This idea of *hypothesis testing* is central to statistics, and you have probably encountered it before. There are arguably two other larger conclusions we can take from this chapter. First, on the positive side, we have a powerful framework for thinking about and modeling random effects and including that uncertainty in developing models of biological systems. But second, even these rather simple systems involved some rather complicated mathematics to analyze. Would we be able, for instance, to add random contact effects to our SIR model of Chapter 2?

On the other hand, thinking back to Chapter 7, we can now understand a slight mystery there. I had mentioned there that cells can reduce expression noise in their protein levels by producing many mRNA molecules but only translating a few of them into proteins. Why it is that this approach reduces noise? In fact, because of the central limit theorem. By increasing our sample size of mRNA molecules, the variation about the mean mRNA level is reduced, meaning that the mRNA concentration from which the cell makes the translated proteins is more uniform than it would have been with lower mRNA levels. We will return to this question of expression noise one more time in Chapter 14, where again the central limit theorem will help us understand a strategy for reducing expression noise.

In the next chapter we will take another pause from biology to introduce ourselves to the basics of computer programming. With that experience in hand, Chapter 11 will show us a completely different approach to modeling systems with a random component, namely simulations that employ agent-based modeling.

APPENDIX 9A: PROPERTIES OF THE DERIVATIVES UNDER THE FOURIER TRANSFORMATION

Time derivatives of the heat equation under the Fourier transformation:

We can take the derivative with respect to t of the transformation of $c(x,t)$ as follows:

$$\frac{\partial}{\partial t}\left(\mathcal{F}\{c(x,t)\}\right) = \frac{\partial}{\partial t}\left(\int_{-\infty}^{\infty} e^{i\zeta x} \cdot c(x,t)dx\right) \tag{9.42}$$

By what is known as Leibniz's integral rule (19), we can rewrite this as:

$$\frac{\partial}{\partial t}\left(\mathcal{F}\{c(x,t)\}\right) = \int_{-\infty}^{\infty} \frac{\partial}{\partial t}\left(e^{i\zeta x} \cdot c(x,t)\right)dx \tag{9.43}$$

However, since $e^{i\zeta x}$ is a constant with respect to t, this expression becomes:

$$\frac{\partial}{\partial t}\left(\mathcal{F}\{c(x,t)\}\right) = \int_{-\infty}^{\infty} e^{i\zeta x} \cdot \frac{\partial}{\partial t}\left(c(x,t)\right)dx \tag{9.44}$$

or:

$$L\frac{\partial}{\partial t}\left(\mathcal{F}\{c(x,t)\}\right)=\int_{-\infty}^{\infty}e^{i\zeta x}\cdot\frac{\partial c}{\partial t}dx \tag{9.45}$$

From 9.45, we can see that the derivative with respect to t of Fourier transformation of $c(x,t)$ just the Fourier transformation of the derivative of $c(x,t)$ with respect to t:

$$\frac{\partial}{\partial t}\left(\mathcal{F}\{c(x,t)\}\right)=\mathcal{F}\left\{\frac{\partial c}{\partial t}\right\} \tag{9.46}$$

What about the transform of derivative of $c(x,t)$ with respect to x?

$$\frac{\partial}{\partial x}\left(\mathcal{F}\{c(x,t)\}\right)=\frac{\partial}{\partial x}\left(\int_{-\infty}^{\infty}e^{i\zeta x}\cdot c(x,t)dx\right) \tag{9.47}$$

By the definition of this transformation, it can be written as:

$$\frac{\partial}{\partial x}\left(\mathcal{F}\{c(x,t)\}\right)=\int_{-\infty}^{\infty}e^{i\zeta x}\cdot\frac{\partial c}{\partial x}dx \tag{9.48}$$

We now turn to an integration technique called "integration by parts" [20], which is taken from the product rule of differentiation, namely:

$$\frac{d}{dx}\left(f(x)\cdot g(x)\right)=f'(x)\cdot g(x)+f(x)\cdot g'(x) \tag{9.49}$$

From this fact, we could then recover the original functions f and g by:

$$f(x)\cdot g(x)=\int g(x)f'(x)dx+\int f(x)\cdot g'(x)dx \tag{9.50}$$

Looking at Equation 9.47, the right-hand side has the form of $\int g(x)f'(x)dx$, where $g(x)=e^{i\xi x}, g'(x)=i\xi e^{i\xi x}$ and f is $c(x,t)$. We next rewrite Equation 9.50 using those values for f and g:

$$c(x,t)\cdot e^{i\zeta x}\mid_{-\infty}^{\infty}=\int_{-\infty}^{\infty}e^{i\zeta x}\frac{\partial c}{\partial x}dx+\int_{-\infty}^{\infty}c(x,t)\cdot i\zeta e^{i\zeta x}dx \tag{9.51}$$

Meaning that, with a rearrangement, we can write the integral on the right of 9.47 as:

$$\int_{-\infty}^{\infty}e^{i\zeta x}\frac{\partial c}{\partial x}dx=c(x,t)\cdot e^{i\zeta x}\mid_{-\infty}^{\infty}-i\zeta\int_{-\infty}^{\infty}c(x,t)\cdot e^{i\zeta x}dx \tag{9.52}$$

Now, we know that $c(x,t)$ approaches 0 as x approaches negative or positive infinity because of the nature of the random walk, meaning that the first term on the right here goes to zero, leaving us with:

$$\int_{-\infty}^{\infty}e^{i\zeta x}\frac{\partial c}{\partial x}dx=-i\zeta\int_{-\infty}^{\infty}c(x,t)\cdot e^{i\zeta x}dx \tag{9.53}$$

or, substituting back for the original desired Fourier transformation:

$$\frac{\partial}{\partial x}\left(\mathcal{F}\{c(x,t)\}\right)=-i\zeta\int_{-\infty}^{\infty}c(x,t)\cdot e^{i\zeta x}dx \tag{9.54}$$

In other words, the Fourier transformation of $c(x,t)$ with respect to x is just $-i\xi$ times the Fourier transformation of $c(x,t)$ itself. We actually want the second derivative of $c(x,t)$ with respect to x, but that simply entails running this argument again to give:

$$\frac{\partial^2}{\partial x^2}\left(\mathcal{F}\{c(x,t)\}\right)=(-i\zeta)^2\int_{-\infty}^{\infty}c(x,t)\cdot e^{i\zeta x}dx \tag{9.55}$$

which, since $i^2=-1$, gives:

$$\frac{\partial^2}{\partial x^2}\left(\mathcal{F}\{c(x,t)\}\right)=-\zeta^2\int_{-\infty}^{\infty}c(x,t)\cdot e^{i\zeta x}dx \tag{9.56}$$

or:

$$\frac{\partial^2}{\partial x^2}\left(\mathcal{F}\{c(x,t)\}\right)=-\zeta^2\,\mathcal{F}\{c(x,t)\} \tag{9.57}$$

APPENDIX 9B: DIFFERENTIATING THE SOLUTION TO THE HEAT EQUATION YIELDS THE ORIGINAL DIFFERENTIAL EQUATION

Taking the derivative of Equation 9.28 with respect to t by the product rule gives us:

$$\frac{-1}{2\cdot t^{3/2}\sqrt{4D\pi}}\cdot e^{-x^2/4Dt}+\left(\frac{1}{\sqrt{4Dt\pi}}\cdot e^{-x^2/4Dt}\right)\cdot\frac{x^2}{4Dt^2} \tag{9.58}$$

which we can simplify to:

$$\left(\frac{-1}{2t}+\frac{x^2}{4Dt^2}\right)\cdot\frac{1}{\sqrt{4Dt\pi}}\cdot e^{-x^2/4Dt} \tag{9.59}$$

or:

$$\frac{\partial c}{\partial t}=\left(\frac{x^2-2Dt}{4Dt^2}\right)\cdot c(x,t) \tag{9.60}$$

If we similarly take the derivative of 9.28 with respect to x by the chain rule, we get:

$$\left(\frac{1}{\sqrt{4Dt\pi}}\cdot e^{-x^2/4Dt}\right)\cdot\frac{-2x}{4Dt} \tag{9.61}$$

If we apply the product rule to take the derivative again (i.e., the second derivative), we get:

$$\left(\left(\frac{1}{\sqrt{4Dt\pi}}\cdot e^{-x^2/4Dt}\right)\cdot\frac{-2x}{4Dt}\right)\cdot\frac{-2x}{4Dt}+\left(\frac{1}{\sqrt{4Dt\pi}}\cdot e^{-x^2/4Dt}\right)\cdot\frac{-2}{4Dt} \tag{9.62}$$

or:

$$\left(\frac{4x^2}{16D^2t^2}+\frac{-2}{4Dt}\right)\cdot\left(\frac{1}{\sqrt{4Dt\pi}}\cdot e^{-x^2/4Dt}\right) \tag{9.63}$$

or:

$$\frac{\partial^2 c}{\partial x^2}=\left(\frac{x^2-2Dt}{4D^2t^2}\right)\cdot c\left(x,t\right) \tag{9.64}$$

If we compare Equations 9.60 and 9.64, we see that if we multiply Equation 9.64 by D, we get Equation 9.60, in agreement with the original partial differential equation in Equation 9.9, meaning that the solution proposed satisfies our original differential equation.

REFERENCES

1. Illowsky B & Dean S (2018) *Introductory statistics* (OpenStax College, TX).
2. Galton F (1886) Regression towards mediocrity in hereditary stature. *The Journal of the Anthropological Institute of Great Britain and Ireland* 15:246–263.
3. Murray JD (2001) *Mathematical Biology, I: An introduction.* 3rd Edition (Springer, New York).
4. Feynman RP, Leighton RB, & Sands M (2011) *The Feynman lectures on physics, Vol. I: The new millennium edition: mainly mechanics, radiation, and heat* (Basic Books).
5. Logan JD (2013) *Applied mathematics* (John Wiley & Sons).
6. Bracewell RN (1989) The fourier transform. *Scientific American* 260(6):86–95.
7. Boyce WE & DiPrima RC (1992) *Elementary differential equations and boundary value problems* (John Wiley & Sons, New York), p. 680.
8. Hoel PG, Port SC, & Stone CJ (1971) *Introduction to probability: theory.*
9. Monaghan P, Charmantier A, Nussey DH, & Ricklefs RE (2008) The evolutionary ecology of senescence. *Functional Ecology* 22:371–378.
10. Kirkwood TB (2002) Evolution of ageing. *Mechanisms of Ageing and Development* 123(7):737–745.
11. Kirkwood TBL & Austad SN (2000) Why do we age? *Nature* 408(6809):233–238.
12. Kleiber M (1961) The fire of life. An introduction to animal energetics. *The fire of life. An introduction to animal energetics.*
13. Speakman JR (2005) Body size, energy metabolism and lifespan. *Journal of Experimental Biology* 208(Pt 9):1717–1730.
14. Ricklefs RE (1998) Evolutionary theories of aging: confirmation of a fundamental prediction, with implications for the genetic basis and evolution of life span. *The American Naturalist* 152(1):24–44.
15. Munshi-South J & Wilkinson GS (2010) Bats and birds: exceptional longevity despite high metabolic rates. *Ageing Research Reviews* 9(1):12–19.
16. Austad SN & Fischer KE (1991) Mammalian aging, metabolism, and ecology: evidence from the bats and marsupials. *Journal of Gerontology* 46(2):B47–B53.
17. Kim EB, Fang X, Fushan AA, Huang Z, Lobanov AV, Han L, Marino SM, Sun X, Turanov AA, & Yang P (2011) Genome sequencing reveals insights into physiology and longevity of the naked mole rat. *Nature* 479(7372):223–227.
18. Partridge L & Farquhar M (1981) Sexual activity reduces lifespan of male fruitflies. *Nature* 294(5841):580–582.
19. Flanders H (1973) Differentiation under the integral sign. *The American Mathematical Monthly* 80(6):615–627.
20. Larson RE & Hostetler RP (1986) *Calculus with analytic geometry.* 3rd Edition (D. C. Heath and Company, Lexington, MA), p. 1013.

Lather, Rinse, Repeat
A Gentle Introduction to Computer Programming

10

"The fact that all of this was happening in virtual space made no difference. Being virtually killed by virtual laser in virtual space is just as effective as the real thing, because you are as dead as you think you are."
—Douglas Adams, Mostly Harmless

WHY PROGRAMMING?

A bit of personal history. My family purchased its first computer when I was in fourth grade in the mid-1980s, while I bought my own first machine when I left for college in 1994 (Figure 10.1). We were not the first of the families in our neighborhood to get a computer, but they were not yet terribly common. It is probably no accident that I ended up with an interest in computers: Douglas Adams has written [1]:

I've come up with a set of rules that describe our reactions to technologies:

1. Anything that is in the world when you're born is normal and ordinary and is just a natural part of the way the world works.
2. Anything that's invented between when you're fifteen and thirty-five is new and exciting and revolutionary and you can probably get a career in it.
3. Anything invented after you're thirty-five is against the natural order of things.

As personal computers have gone from an exotic consumer good to part of the background noise of our schools and offices, I think that some of the "magic" of

Figure 10.1 The author's first computer, purchased in 1994. The machine had a 60-Mz processor, 16 megabytes of memory and a 500-megabyte hard disk. It had a modem but no internet connection: in fact, my college had only a single cable connecting the entire campus to the internet. Instead, we used floppy disks and CDs to move data around.

DOI: 10.1201/9781003687504-10

what they can allow has been lost. In this chapter we are going to pause and think a bit about the machines we have been using throughout the book and then learn just a bit about how to program them.

A History of Hardware

Asking when the first computer was built is a question that could have several different answers. **Figure 10.2** shows the Antikythera mechanism, a machine that used a series of bronze gears to represent the motions of astronomical bodies [2]. This device was made before Augustus became the first Roman emperor, perhaps between 100 and 60 BCE. It was recovered from an ancient shipwreck in the Greek isles in the 20th century. Although the machine itself is unique in its survival to the present, classical sources suggest that other similar machines were in use around this time. Much later, Charles Babbage designed, but never completed, a programmable mechanical computer in the 19th century. Simultaneously, Lady Ada Lovelace developed the mathematical theory that allowed that machine to be programed [3], in many ways prefiguring the modern delimitation of hardware and software.

Of course, a key advance in computing was the ability to use electricity rather than mechanics. Until surprisingly recently, part of the history of electronic computers was unknown to the general public. The reason for our ignorance is that the first electronic computers were developed secretly by British intelligence to decipher German radio codes during World War II. With the usual caution of intelligence services, the full details of this endeavor were only revealed in 2000 [4].

After the war, computer development proceeded rapidly, with a key innovation being designing hardware that could be easily programmed to perform different operations [3], leading to the modern situation where computer programs can be adapted to run on many different hardware platforms. This improvement was coupled to the use of transistors and integrated circuits, which eventually made it possible to economically construct the desktop computers we now use [3]. Because these computers were electrical in design, it was natural that they would represent numbers in binary format, where each binary digit of the number could be represented as a presence or absence *bit* (think of electrical current being on or off through a circuit). **Table 10.1** lists the numbers 0–31 in binary format to give the flavor of this idea. What is very interesting about these developments is that, for many applications, we need to understand very little about the underlying hardware to be able to use the computer for our purposes. So, how should we as scientists approach the computer?

A)

B)

Figure 10.2 Two pictures of the Antikythera mechanism found in an ancient shipwreck in the Greek isles (see text) [17, 18]. (Both images from Logg Tandy, on Wikimedia, published under CC BY-SA 4.0 license.)

TABLE 10.1 BINARY NUMBERS

Standard (base 10)	Binary (base 2)	B10	B2	B10	B2	B10	B2
0	00000	8	01000	16	10000	24	11000
1	00001	9	01001	17	10001	25	11001
2	00010	10	01010	18	10010	26	11010
3	00011	11	01011	19	10011	27	11011
4	00100	12	01100	20	10100	28	11100
5	00101	13	01101	21	10101	29	11101
6	00110	14	01110	22	10110	30	11110
7	00111	15	01111	23	10111	31	11111

Using Computers: Programming as Abstract Steps

Once computing hardware became available and general purpose, the topic of this chapter became possible: *computer programming*. What is meant by a computer program can encompass a great range of goals: photo editing, games, mathematical modeling, data storage and retrieval, random event simulation, and interactive text editing, to name a few. Rather than trying to cover this expansive range, for this book we will limit ourselves to one insight and use that as our paradigm for a computer program. This insight (you may notice a pattern here) is due again to Douglas Adams, who noticed that many of the early programs written for personal computers, such those for word processing and spreadsheets and even the first corporate web pages, sought to replicate preexisting things, namely typewriters, calculators, and brochures [1]. He then argued that instead of seeing computer programs as versions of those things, we can see the computer as a *modeling engine* that allows us to recreate things with which we are familiar from the wider world inside the computer.

This book, of course, is at its heart concerned with models of the living world. So far we have thought of those models primarily in mathematical terms (differential equations) and, to a lesser extent, in statistical terms. But of course, lurking in the corner of almost everything we have done was a computer. The reason for its presence was explained right off the bat, in Chapter 1. There we posed the inverse problem, namely when we knew a quantity's rate of change (derivative) and wanted to use that rate of change to calculate the value of the quantity Q itself. We saw that we could use a computer to approximate Q by computing its derivative Q' and then treating Q' as constant over short intervals. When we then summed the resulting interval changes, we produced an estimate of Q.

Strictly speaking, we do not *need* a computer for this inverse problem, even for problems where analytic solutions to the differential equations do not exist. We could manually compute the numerical solutions to the differential equations, as was done during the construction of the atomic bomb in the 1940s [5]. But for all save the simplest problems, analytic solutions are not forthcoming, and hand-computed numerical ones are too onerous [6]. On the other hand, even when analytic solutions do exist, nothing prevents us from approximating that solution with a computer, as we did for the falling ball in Chapter 1. Hence, computation offers us a window into science that only opened in the later part of the 20th century. In the years since then, computing has changed entirely how we relate models to external reality. Examples of new problems that could never have been approached without computational science include the construction of global climate models [7], simulations of the evolution of our solar system [8], large-scale particle physics simulations [9], and the sequencing of the human genome [10]. Here we will work on an almost trivial example: recreating our falling ball solution from Chapter 1 with a computer program.

Basics of Computer Programming for Modelers

What do we need to be able to tell a computer to do to use it for modeling? There are seven key operations we need from a computer to create models, which are listed

---

TABLE 10.2 SEVEN CLASSES OF OPERATIONS NEEDED FOR COMPUTATIONAL MODELING AND THEIR ASSOCIATED PYTHON SYNTAX

```
1. INPUT:
       a. input(): 
2. MATH:
       a. +: Add
       b. -: Subtract
       c. *: Multiply
       d. /: Divide
3. LOGIC:
       a. <: Returns true if left-hand is less than then the right
       b. ==: Returns true if left-hand is equal to the right
       c. >: Returns true if left-hand is greater than then the right
4. BRANCH:
       a. if()/else: Perform operation if enclosed argument is true,
          otherwise perform else (when present).
5. LOOP:
       a. while(): Perform operation so long as condition is true
       b. for(): Perform operation specified number of times.
6. OUTPUT:
       a. print(): Echo the enclosed text and variables to the screen
7. ASSIGNMENT:
       a. =: Assign value on the right to the variable on the left
```

---

in Table 10.2. We will use the falling ball example to see how these seven features are applied. To start with, we can extend our ball model by letting the initial height of the ball differ each time we run our modeling program. We can see that this decision is an example of INPUT. INPUT is critical because it allows the *user* of a program to change the model from run to run without that user needing to know how to program. (Imagine having to rewrite the code of your word processor every time you wanted a new font!) We will also INPUT the timestep size for the same reason.

Of course, the heart of our model will be MATH. In this case we only need the operations of multiplication and addition for the timestep computations. If we wanted to compute the actual position from the analytic solution, we might also need a MATH tool for calculating powers, namely:

$$p_y(t) = d - \frac{9.81 m/s}{2} \cdot t^2 \tag{10.1}$$

Our calculation itself will not need to BRANCH. However, we should use a BRANCH at the very beginning of the program to check that the user has given us a reasonable value for the initial position and step size. If not, we could terminate the program. In this chapter, that refinement to our code is omitted for clarity, but we will use several BRANCH statements in the next chapter.

We will LOOP our computation of the falling of the ball until the (predicted) position is 0 or smaller, which we will check with a LOGIC step. Finally, once we have finished our LOOP, we will OUTPUT the time the ball has taken to fall. Table 10.3 gives a *pseudocode* description of our program; an idea we will describe more carefully in a couple of paragraphs.

TABLE 10.3 PSEUDOCODE DESCRIBING HOW TO APPROXIMATE THE POSITION OF A FALLING BALL

```
• INPUT  → ballpos
• INPUT  → stepsize
• time=0
• LOOP (while ballpos>0)
       o ballpos = ballpos + ((-9.81*(time+0.5*stepsize))*stepsize
       o time = time + stepsize
• OUTPUT time
```

Words in blue are our basic language constructs from Table 10.2, while the four program variables are shown in green.

Variables

If you compare the preceding discussion to Table 10.3, you will immediately notice that we have forgotten something. We want to OUTPUT a time—but what is a time? In addition to our operations of Table 10.2, we need to add the concept of *variables* in our programs. Variables are an analogy to a concept you will recall from algebra in secondary school: x and y in $y = mx + b$. Now, in those classes your goal was most often to "de-variablize" your variables: to find a pair of numeric values for x and y that satisfied, for instance:

$$y = 4 \cdot x - 8$$
$$y = -3 \cdot x + 6$$

(10.2)

(As it happens, $x = 2$ and $y = 0$ satisfies both of these equations.)

In a computer program, variables function slightly differently. They allow us to mark numbers of which we know the existence but not the values while we are creating the program. Those variables will take on specific numeric values only when the program runs. In our simple case, we need only three variables in our program: our ball position, the step size, and the current value of time. However, because computers represent numbers using base 2, or binary, notation (see Table 10.1), we need to be more specific about our number choices and their representation. In particular, up to a certain limit, computers can perfectly represent integers in binary format. However, real numbers, being potentially infinite in their representation (think of π), will only be approximately represented as *floating point* numbers. For our program, all three variables are real values and will need to be stored as individual floating point variables.

As an aside, while the variables in *this* program are numbers, we will see in the next chapter that variables can also be of other kinds, such as letters or logical values (true/false). Interestingly, all those other types of value will nonetheless be stored as binary numbers inside the computer. There will then be a look-up table for converting them from the internal numeric format to letters or values that we see as users.

The concept of variables also requires us to create the last of our seven programming operations: ASSIGNMENT. Assignments allow us to change the value of a variable: in our example, we will perform a calculation that changes the value of *ballpos*.

Pseudocode

Perhaps surprisingly, we are now in a position to express our falling ball problem in a computer program–like form. This form is often called *pseudocode*, and Table 10.3 gives a solution to our problem in this format. Table 10.3 may appear so oversimplified as to be useless. However, in fact that simplification has some interesting benefits.

First, most skilled computer programmers can take Table 10.3 and quite easily convert it into running code in a language of their choice. It might appear that vague-sounding constructs like "INPUT" would be a problem, but in fact they represent a valuable flexibility. In the program we will write, INPUT will correspond to the user typing these values for us, but in a more complex program, INPUT could correspond to an input box in a website that sends values to our program. Hence, Table 10.3 captures the core of the model we want and leaves some of the details abstract to add flexibility.

The second advantage of this pseudocode format is the flexibility in programming language, a topic we will consider next. In my career, I have used seven different programming languages with some seriousness (Table 10.4). This number is neither impressive nor unusual compared to other people in this field. Thus, it can be helpful to represent computer programs in a way that is agnostic as to the actual programming language in which they will be written.

TABLE 10.4 A SUBJECTIVE TAKE ON SEVEN PROGRAMMING LANGUAGES

Language	Compiled/ Interpreted?	Difficulty	Scientific computing support	Computational efficiency
C	Compiled	High	Extensive	Very high
c++	Compiled	High	Extensive	High
Fortran	Compiled	High	Extensive	High
Pascal	Compiled	Moderate	Negligible	Unsure
Basic	Interpreted	Moderate	Negligible	Unsure
Perl	Interpreted	Moderate to easy	Extensive	Moderate
Python	Interpreted	Moderate to easy	Extensive	Moderate

PROGRAMMING LANGUAGES: HOW TO CHOOSE

For a beginning programmer, the great number and variety of programming languages can itself be a daunting concern. Most programmers have strong—and in my own case, probably unfounded—prejudices when it comes to the choice of language. For what follows, however, I am reasonably confident in my advice only because I will be recommending a language that I neither commonly use nor particularly like, but which I think is a good choice as a first language.

Compiled versus Interpreted Languages

The most elementary division of languages is probably that between compiled and interpreted languages [11], although to the beginner this difference is not very intuitive. Generally speaking, compiled languages use a second program, called a compiler, to convert their high-level programming constructs (see Table 10.2) into a set of lower-level machine instructions the computer will actually perform. Those instructions are then packaged into an *executable* program that can be run repeatedly. Almost all the computer programs you use on a daily basis have been compiled in this manner.

Because these machine instructions differ depending on the underlying hardware, each different type of computer will have a different compiler. However, if the code is correctly written, a program should compile and run essentially identically on all these different machines, even if those machines are as different as a supercomputer and a cellphone. Also, if the compiler is written correctly, the code should run as fast as is possible for that hardware, meaning that compiled languages generally yield relatively fast programs.

Looking at the code alone, interpreted languages will look very similar to some compiled languages, but they do not use a compiler. Instead, when the program is run, another program, called an interpreter, is called. This call is done *each time* the interpreted program is run. You can contrast this approach with that of the compiled programs, where the compiler converts the computer code to an executable program, which then can be run repeatedly. Hence, the interpreted programs pay the cost of interpretation at every run instead of just once during compilation. The interpreter also essentially executes the program line by line. Because no intermediate translation to machine instructions is made and because the interpreter is called at every run, programs written in interpreted languages will tend to run slower than compiled ones, with some individual exceptions [11]. However, interpreted language programs are easier to write and improve on than are programs written in a compiled language.

Python as a First Language: An Acceptable Choice

Table 10.4 gives my rather subjective take on the seven programming languages with which I have reasonable familiarity. While this list does not exhaust the set of programming languages in common use, with the possible exception of Java and some not-quite-languages such as the R environment [12], it does encompass most of the languages practicing scientists use to produce computer programs.

From this list, we can eliminate BASIC and Pascal immediately because they are not commonly used for scientific computing, and so the amount of prewritten software you could draw on for common tasks in scientific computing is minimal. As a result, using them requires too much work to reinvent things that already exist in other languages. Among the compiled languages, Fortran was once *the* language of scientific computing [3], and a huge catalog of tools written in it exists. These tools include the common *library* for linear algebra, called LAPACK [13]. (Here a library is simply a set of existing functions for tasks like matrix multiplication that the programmer can use without having to recreate them.) However, Fortran has a few deficiencies relative to c and c++. Moreover, any existing Fortran library can be called from a c or c++ program, so from a practical standpoint, c or c++ are better choices today for writing scientific codes. Between the two, c can be roughly thought of as a subset of c++, and so which of the two you might use is, to a degree, a matter of taste.

However, to a novice programmer, all the compiled languages, and particularly c, have the major disadvantages of being both difficult to learn and allowing new programmers to make coding errors that are both hard to find and hard to correct. One might think that the benefits of faster program execution for c would make it worth paying these costs, but a surprising fraction of the interesting science we would like to use computers to support does not require long-running programs. This book more or less illustrates this point, as you will notice that none of the exercises we have done so far has required long running times to yield their answers.

Hence, interpreted languages like Python and Perl can allow you to accomplish perhaps 80% or 90% of the computational tasks you need to—and even the remaining small percentage of tasks can often be accomplished by calling fast c or c++ programs from *within* your Perl or Python programs. For that reason, I believe the choice of a first language really comes down to picking between Perl and Python. That choice is not a completely obvious one: I happen to use Perl and find it more appropriate for the text manipulation and program integration I often need to perform. However, there are two major strikes against Perl. First, it uses some rather opaque syntax for much of its operation. As a result, when I taught Perl, I constantly found myself saying, "Don't ask why this works, it just does." Second and more important, Python enjoys a larger user community than does Perl, and this difference is arguably growing [11, 14, 15]. Hence, it is likely that the future will hold more development of useful routines for Python than for Perl. It is therefore probably a better choice as a first programming language.

CODING IN PYTHON

Table 10.2 also gives a limited set of the Python syntax for the seven key operations. With one minor and one major exception, these commands are sufficient for our falling ball example.

The minor exception is that we need to use parentheses to indicate the order of mathematical operations. We have already used parenthesis for precisely this function in Table 10.3 without mentioning it. You were likely taught in mathematics that different arithmetic operations have different precedence. Thus, multiplication has higher precedence than addition, meaning that, in the algebraic expression $a + b \times c$, first b and c are multiplied together and then a is added to the result [16]. It is an odd quirk of my mental makeup that I refuse to memorize these rules. Why is that? The reason is that I believe it is bad programming practice to write code that presumes the reader always remembers these rules. Instead, I use parentheses to specify the order in which I want the mathematical operations done.

The more serious omission from Table 10.2 is the concept of *functions*, which in computer programming is essentially an extension of the concept from mathematics. In mathematics, a function can have one or more numerical *arguments*, for example the variable x in $\sin(x)$. The function then returns exactly one numerical value, in this case the *sine* of the input variable x. In programming, the concept of a function is extended so that the number of arguments can range from zero to

TABLE 10.5 PYTHON PROGRAM FOR ESTIMATING THE IMPACT TIME OF THE FALLING BALL, GIVEN A USER'S INPUT OF THE INITIAL POSITION AND THE TIMESTEP SIZE TO USE IN THE APPROXIMATION

```python
#!/usr/bin/python3

print('Input initial height:')
ballpos=float(input())
print('Input time step size:')
timestep=float(input())

time=0.0

while(ballpos>0.0):
    ballpos = ballpos + (-9.81*(time+0.5*timestep))* timestep
    time = time + timestep

print('Time of impact: ', time)
```

The first line simply tells the computer's operating system that this is a Python program.

many, and those arguments need not be numeric. Likewise, the function may or may not return a single value, and if it does, that value need not be numeric. However, as in mathematics, the arguments to the function are given as a list within parentheses. With this knowledge, we can see that input() and print() are special functions in Python, with the former taking no arguments but returning one value and the latter taking potentially many values but returning nothing to the program, instead "typing" those values to the user's screen.

Table 10.5 presents the full code for the falling ball, completing Exercise 10.1. We use the new function *float*() to convert our input text values of the ball's initial position and the timestep size into the floating point variables we need to run the program. With those two slight changes, I think it is striking how similar this program is to the pseudocode of Table 10.3.

Exercise 10.1

Write a Python program to find the height of a falling ball given a user-specified initial height and a timestep size.

The program in Table 10.5 is not what we might call "production ready." One problem we already mentioned is that we have not checked that the user's inputs are sensible (What if the height is negative? Or given as "duck"?). Indeed, if you are writing programs for others, you need to give a good deal of thought to what an ignorant or malicious user might do and make sure your program does not crash the computer or manifest other unpleasant behaviors. Nonetheless, Table 10.5

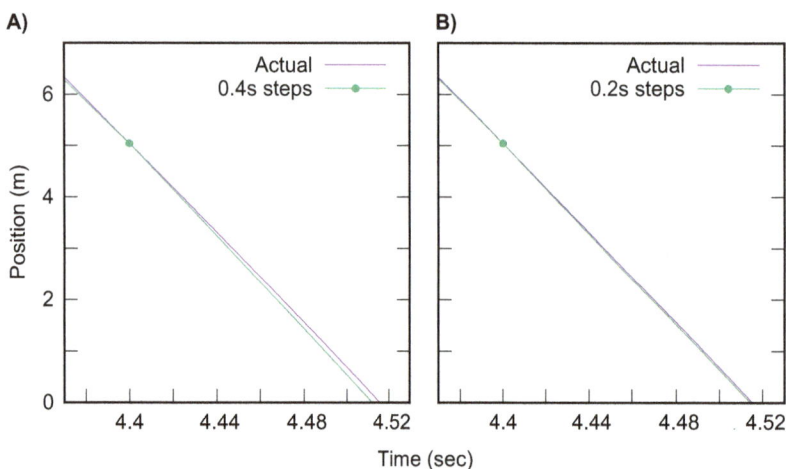

Figure 10.3 Comparison of the output of our Python program with a timestep of 0.4 sec (A) and 0.2 sec (B) to the true position of the ball, computed with Equation 10.1 from Chapter 1 (*purple*). The program's predictions are in green. Notice that the predicted impact differs a bit from the actual impact time in **A** but much less so in **B**.

does a good job of solving the problem from Chapter 1 (**Figure 10.3**). In particular, we can test the improvement in our predictions from differing step sizes without all the tedious copying and pasting that we used in the spreadsheet analysis from that chapter.

REFLECTIONS AND PREVIEWS

This chapter is the last of our interlude chapters with little biological content. Adding our new tool of computer programming to the mathematics and statistics of the previous chapters, we now have all the quantitative tools we need to explore biology. In the next chapter we will create a simulation that considers the evolution of genes in populations. We will also extend the concept of programming variables to include vectors and matrices, or lists and two-dimensional tables, if you prefer. In Chapter 12 we will consider even more complex ways of using computer programming, including both new *algorithms* for biological data and some comments on how to use languages like Python to automate running programs on large datasets. In both cases, you may be surprised how few additional programming tools are needed to accomplish these tasks. And, with our new focus on computing, in the last third of the book we will be in a position to enrich our analyses with much larger sets of biological data that we can use to parameterize our models and discover new biological principles.

REFERENCES

1. Adams D (2002) *The salmon of doubt: Hitchhiking the universe one last time* (Harmony).
2. Seiradakis J & Edmunds M (2018) Our current knowledge of the Antikythera Mechanism. *Nature Astronomy* 2(1):35–42.
3. O'Regan G & O'Regan (2008) *A brief history of computing* (Springer).
4. Durand-Richard M-J & Guillot P (2017) From Poznań to Bletchley Park: the history of cracking the ENIGMA machine. *CIIT lab workshop on history of cryptography* (Faculty of Electronic Engineering, Nis, Serbie), pp. 1–43.
5. Feynman RP (1985) *Surely you're joking, Mr. Feynman!* (Bantam Books, Toronto).
6. Dirac PAM (1929) Quantum mechanics of many-electron systems. *Proceedings of the Royal Society of London. Series A* 123:714–733.
7. Randall DA, Wood RA, Bony S, Colman R, Fichefet T, Fyfe J, Kattsov V, Pitman A, Shukla J, & Srinivasan J (2007) Climate models and their evaluation. *Climate change 2007: The physical science basis. Contribution of Working Group I to the Fourth Assessment Report of the IPCC (FAR)* (Cambridge University Press), pp. 589–662.
8. Nesvorný D (2018) Dynamical evolution of the early solar system. *Annual Review of Astronomy and Astrophysics* 56:137–174.
9. Harrison M, Ludlam T, & Ozaki S (2003) RHIC project overview. *Nuclear Instruments and Methods in Physics Research Section A: Accelerators, Spectrometers, Detectors and Associated Equipment* 499(2–3):235–244.
10. Venter JC, Adams MD, Myers EW, Li PW, Mural RJ et al., (2001) The sequence of the human genome. *Science* 291(5507):1304–1351.
11. Nanz S & Furia CA (2015) A comparative study of programming languages in rosetta code. *2015 IEEE/ACM 37th IEEE International Conference on Software Engineering* (IEEE), pp. 778–788.
12. R Development Core Team (2008) *R: A language and environment for statistical computing* (R Foundation for Statistical Computing).
13. Anderson E, Bai Z, Bischof C, Blackford S, Demmel J, Dongarra J, Du Croz J, Greenbaum A, Hammarling S, McKenney A, & Sorensen D (1999) *LAPACK users' guide* 3rd Edition (Society for Industrial and Applied Mathematics, Philadelphia, PA).
14. Bissyandé TF, Thung F, Lo D, Jiang L, & Réveillere L (2013) Popularity, interoperability, and impact of programming languages in 100,000 open source projects. *2013 IEEE 37th Annual Computer Software and Applications Conference* (IEEE), pp. 303–312.
15. Ezenwoye O (2018) What language?-The choice of an introductory programming language. *2018 IEEE Frontiers in Education Conference (FIE)* (IEEE), pp. 1–8.
16. Stroustrup B (2000) *The C++ programming language* (Addison-Wesley, Reading, MA) Special Edition Ed.
17. Tandy L (2025) Back of "Fragment A" of the Antikythera mechanism. https://commons.wikimedia.org/wiki/File:Antikythera_Fragment_A_(Back).webp
18. Tandy L (2025) Front of "Fragment A" of the Antikythera mechanism. https://commons.wikimedia.org/wiki/File:Antikythera_Fragment_A_(Front).webp

Agents of Change
Computational Models of Genetic Drift

"I do not know what I may appear to the world, but to myself I seem to have been only like a boy playing on the seashore, and diverting myself in now and then finding a smoother pebble or a prettier shell than ordinary, whilst the great ocean of truth lay all undiscovered before me."
—Isaac Newton

AGENT-BASED MODELING

Chapters 8 and 9 covered some mathematical and statistical methods for studying biological processes that involve randomness. But one of the things you might have noticed in these chapters is that the mathematics we used were fairly involved, even when the underlying biological processes were not terribly complex, like the lifespan of fruit flies. In fact, there are mathematical approaches that allow us to, for instance, add stochastic effects in SIR models [1]. The only difficulty with these approaches is that they are, shall we say, difficult. So you might wonder if there are ways that we could study complex phenomena with less elaborate mathematical machinery. Building on the programming skills we learned in the preceding chapter, we will answer that question now by creating a computational model of just such a system. We will then show how this computational model relates to an analytical model of the same biological process. In so doing, we will find strengths and weaknesses both in the computational and in the analytic approaches.

EXAMPLE: GENETICS OF A POPULATION IN TIME

As we discussed in Chapters 4 and 5, Darwin described his idea of evolutionary innovation by natural selection without knowing that inheritance had a particulate behavior [2, 3]. Indeed, his verbal model of blended inheritance was actually misleading in the construction of an evolutionary theory [4]: the particulate model of genetics resolved many of the difficulties with his theory. In this chapter we will explore another aspect of evolution that could not be considered without a particulate view: random variations in the genetics of a population. Once we have created such a model, we will revisit briefly the history of evolutionary thought through the light of that model. We will discuss how mathematicians and biologists have solved the very complicated mathematics of an evolving population and its genes and compare their results to our agent-based simulations. However, before writing our simulations, we will improve our intuition by examining some experimental results showing how the genes of a real population change in time.

DOI: 10.1201/9781003687504-11

Evolution in Real Time

As modern scientists, it is easy claim a keen understanding of the process of evolution because we have available to us the almost unlimited capacity to survey the genetics of life on this planet. That capacity derives from the cheap and ubiquitous DNA sequencing technologies that have appeared in the last two decades [5]. When, at the end of this chapter, we compare our simulations to mathematical models of evolution, we must remember that the creators of those mathematical models developed them *without access to a single DNA sequence*. We should therefore approach all of what follows with a dose of humility. Nonetheless, it is still breathtaking to me that with these new sequencing tools we can actually watch evolution happen before our eyes. In fact, the field of *experimental evolution* is one of the most exciting areas in biology right now [6]. We now have hundreds of examples of evolutionary trajectories where we can trace the genetics of how organisms have gained new traits through mutation and selection. The example I have picked for this chapter is quite an old one. My reasons for using it are its intrinsic importance and the warning it provides us about our hubris in attempting to tame the world around us.

The Evolution of HIV Resistance to the Drug AZT

Although the disease is in fact much older [7, 8], the first human infections by the human immunodeficiency virus (HIV) were noted in the early 1980s [9], a time I am just old enough to recall. The fear this epidemic caused was rooted in several factors; two of the most important were the lack of any available treatments and the fact that HIV was essentially uniformly fatal. Thus, I remember the excitement when the existing anticancer drug AZT (**Figure 11.1**) was found to effectively treat AIDS symptoms as well [9]. Unfortunately, AZT's effectiveness in patients was not permanent; eventually patients would start to become ill again. When researchers examined the virus particles present in patients with and without viral resistance to AZT, they found that the genomes of viruses from AZT-treated patients differed from those of newly infected patients. It became clear that the drug had induced evolution in the viral genomes, fixing mutations that made the viruses carrying them resistant to AZT [10]. We should not be surprised by this fact, given our discussion of resistance to cancer treatments in Chapter 4. But it is still striking that viral evolution in a single host can be tracked by sequencing viral samples from that individual over time [11]—evolution on fast-forward.

Figure 11.2 diagrams a laboratory evolution experiment where viruses resistant to AZT were placed in competition with wild-type viruses in a cell culture treated with AZT [12]. Exactly as Darwin would have predicted, the resistant viruses increase in frequency over the experiment when the drug is present (left side of Figure 11.2). From this experiment we can say that individual viruses possessing the mutation have higher *fitness* than do wild-type viruses. Fitness is a complex topic in evolutionary theory [13], but for the moment we can say that an allele or mutation confers a fitness *advantage* on its possessors if, on average, those possessors tend to have more offspring surviving to the next generation than do possessors of another allele (in this case the wild-type allele).

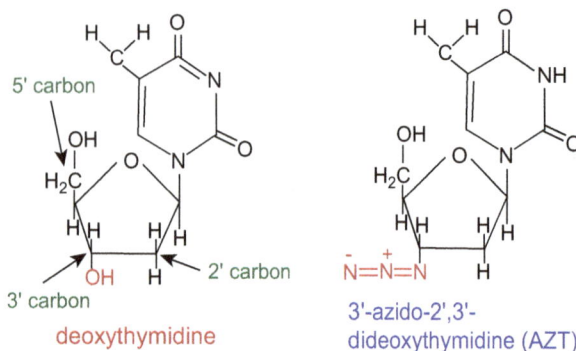

deoxythymidine

3'-azido-2',3'-
dideoxythymidine (AZT)

Figure 11.1 Comparison of the structure of the standard nucleoside deoxythymindine to that of the drug AZT. The key difference is at the 3' carbon: the usual OH group has been replaced in AZT with a nitrogenous group that does not allow the next DNA base to attach during DNA synthesis.

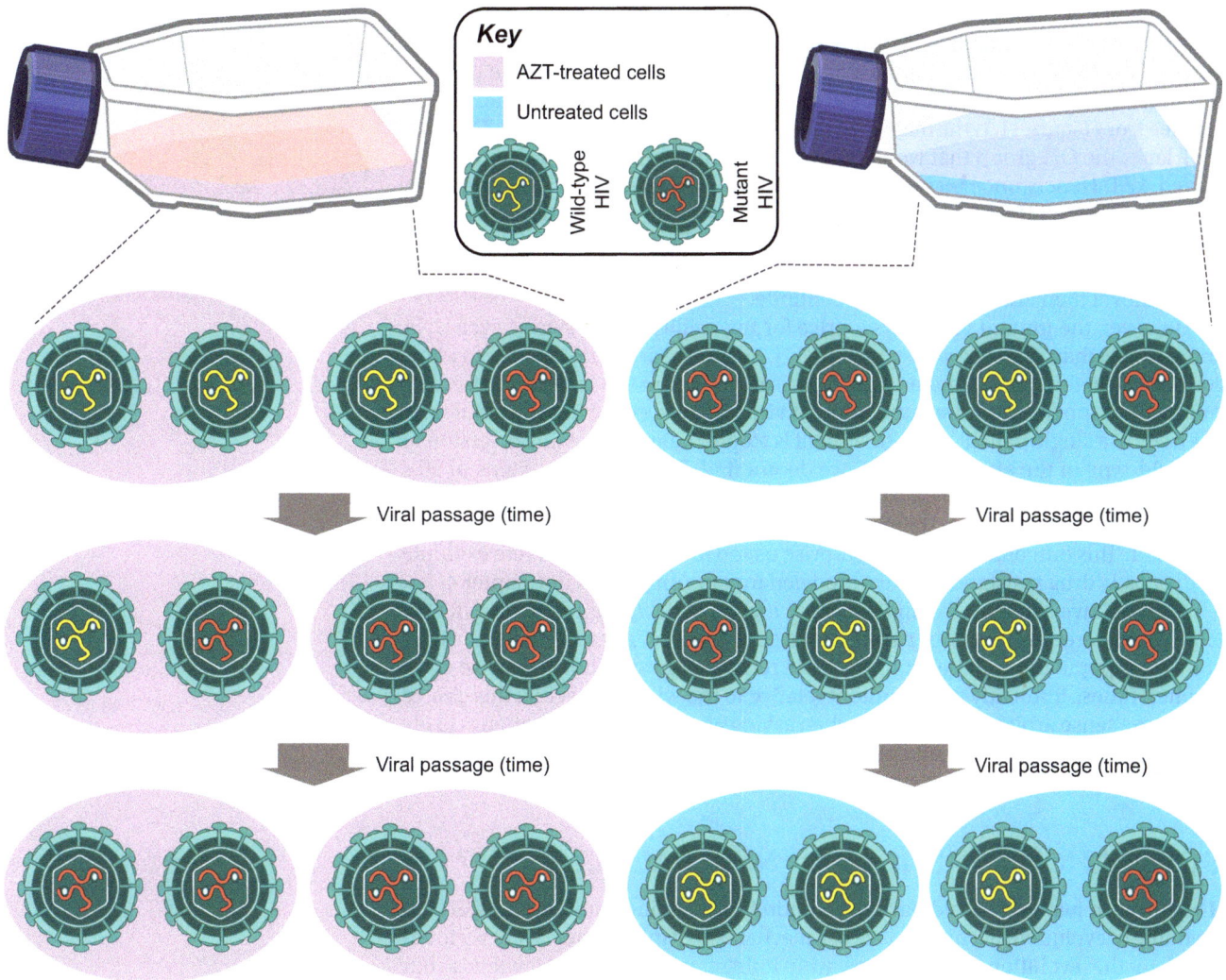

Figure 11.2 AZT and the HIV virus. Shown is a cartoon of competing wild-type HIV (*yellow* RNA) to mutant HIV (*red* RNA) with and without the presence of AZT in the cell culture. On the left, cells are grown in a flask treated with AZT. We start with cells infected with primarily wild-type HIV. As we allow time to pass in the experiment, we find that the mutant viruses are replacing the wild-type ones due to their ability to grow better in cell cultures treated with AZT. On the right, the same mutant and wild-type viruses are again grown in competition in the cell culture. However, in this case, no AZT is present. Although we start with a population that is primarily composed of AZT-resistant mutants, over time the wild-type viruses replace the mutants [42, 43]. These results are adapted from [12]. (Low-friction flask and retrovirus art from NIAID, with permission.)

What is striking, however, is to compare the left and right sides of Figure 11.2. On the right I show the outcome when the same mutant and wild-type viruses are again evolved in competition, but this time without the AZT. We find that, in the absence of the drug, the mutant viruses are actually at a fitness *disadvantage* relative to the wild-type viruses. In the cartoon version of evolution, this result would seem to make little sense: evolution is supposed to gradually build organisms of supreme fitness. But reality is not a cartoon: in actual living systems, an organism's fitness is always a feature or product of its environment. Allele combinations with high fitness in one environment can easily have low fitness if that environment changes [13].

The main reason I have chosen this example is that we know in great detail how the mutations confer resistance to AZT. We therefore can explain both the source of the fitness advantage for the mutants when AZT is present and of the disadvantage in its absence. AZT is what is known as a *nucleoside analog* [14], meaning that it is chemically similar enough to the nucleotide thymidine that a DNA polymerase from the virus can mistake it for a thymidine and incorporate it into a growing DNA chain.

Recall that a DNA polymerase is the enzyme responsible for synthesizing the new side of a DNA double helix from an existing DNA template [15]. However, since HIV is a retrovirus, its genome is actually RNA. It hence uses a similar enzyme called a "reverse transcriptase" (RT) to synthesize DNA from its genome. As you can see from Figure 11.1, the main difference between a thymidine and AZT is that AZT lacks the OH group that would allow RT to add the next base below a newly added AZT. Hence, once incorporated, AZT blocks the addition of the next DNA base, stopping synthesis and hence viral replication. The mutant RT achieves its resistance to AZT by being able to remove the blocking AZT molecules from the growing helix through a somewhat complex "proofreading" mechanism [14, 16]. This removal then allows synthesis to continue.

However, the mutant virus pays a price for this ability: its RT synthesizes DNA more slowly than does the wild-type one [17]. Since the virus must use RT to allow its lifecycle in the cell, the mutants are at a disadvantage in the absence of the drug and will disappear from the population of viruses. This extinction occurs *within the body* of a single patient when treatment is stopped. Because of this reversion to the wild-type in the absence of AZT, we do not find drug-resistant strains of HIV taking over from the wild-type strain across the world. Instead, individuals are infected with the wild-type stain and only evolve resistant viral strains if treated with AZT. This outcome should not surprise us too much: it is a concrete example of the costly drug resistance that we modeled for our tumor cells in Chapter 4.

Again, I have used this example because it gives a nicely complete overview of how two different versions, or *variants*, of a gene might change in frequency over time in a population, depending on their relative fitness. Notice that we have defined fitness implicitly as the mutant relative to the wild type. While this choice is in some sense arbitrary, it is useful when asking whether a new mutation is likely to spread through the population (an advantageous mutation) or be removed from the population (a disadvantageous mutation).

POPULATION GENETICS AND NEUTRAL ALLELES

The story we just told would have been entirely in keeping with Darwin's understanding of evolution, despite his lack of a mechanism of inheritance. Were we interested in the evolution of organisms with infinite numbers of individuals, our understanding would be almost complete at this point. Indeed, we developed a similar model in Chapter 4. (We could still ask how the difference in fitness changes the rate of allele frequency change.) But, of course, real populations are not infinite. It took much of the last century and the work of some of the best minds in statistics and genetics to formulate a series of statistical models of how evolution occurs in real, finite populations. That story starts with three names we saw in Chapter 5, J. B. S. Haldane, Sewall Wright, and R. A. Fisher [18], but it does not end there. In what follows we will describe some of their discoveries, but we will first see how far we can get on our own, using the simulation tools we started to create in the last chapter.

We first need to ask ourselves: Is our taxonomy of allele comparisons complete? In other words, if we limit ourselves to two versions of a gene, will one version always have a fitness advantage over the other? There has been a good deal of sometimes bitter ink spilled over this question by evolutionary biologists [19]. Fortunately, we can use two decades' worth of data from the study of multiple complete genome sequences to answer it.

From those genomes we know that the process by which the sequence of the genome (the *genotype*) maps into the form and function of the organism (the *phenotype*) is very complex [20]. Of particular interest to us, there appear to be places where changes in the genome sequence *do not* change that form and function. Two examples of such places are the *synonymous* codons and the introns [21].

Synonymous Differences

Figure 11.3 shows a diagram of the universal genetic code. The code is a conceptual representation of the chemical process of translation we modeled in Chapter 7. It shows how the 3-letter codons of DNA or RNA are translated to the 20-letter

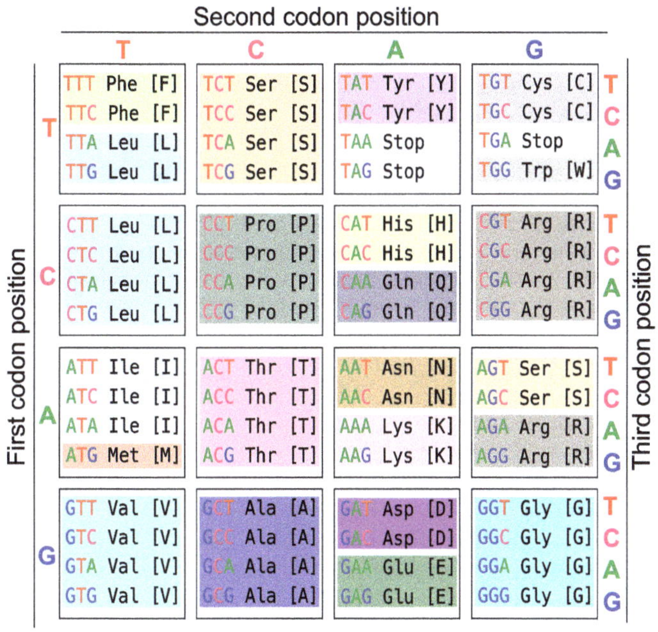

Figure 11.3 The universal genetic code most organisms use to translate the 4-letter alphabet of DNA and RNA into the 20-letter alphabet used in proteins [21]. Because there are 64 possible codons (61 of which code for amino acids) and only 20 possible amino acids, there are groups of codons that code for the same amino acid: these are illustrated here with the colored blocks.

amino acid alphabet. One thing we failed to emphasize in Chapter 7 is that, because there are 64 possible 3-letter combinations of DNA and only 20 possible amino acids, the genetic code is *redundant*. In other words, there are groups of different codons that all encode the same amino acid, a pattern shown as colored blocks in Figure 11.3. Given that proteins are the primary functional molecules of the cell [15], it is reasonable to think that two different alleles of the same gene that differ *only* by a single synonymous difference (e.g., an ATT in one and an ATC in the other, both encoding an isoleucine) should not differ a great deal in their fitness. (In unicellular organisms, the different codons do alter fitness a bit, but we will disregard this complication [22].). Hence, synonymous allele variants probably do not differ in fitness in multicellular plants and animals.

Introns

The discovery that many genes of eukaryotic organisms had *introns*, regions of the DNA that were transcribed into RNA but then spliced out before the resulting mRNA was translated [21], was a surprise to molecular biologists and geneticists [23]. As with the synonymous differences, however, it is hard to argue that genetic variants within an intron will have much prospect of altering form and function. Hence, they represent a second class of variants that should not alter fitness.

Lessons from the Genomic Era: Most DNA Base Changes Probably Do Not Produce a Selective Effect

The existence of synonymous variants and variants within introns clearly suggests that some genetic differences within a population are likely to be invisible to natural selection. However, the completion of the sequencing of the human genome revealed that these invisible variants were far more common than even those examples had hinted. In the human genome, protein-coding genes occupy less than 5% of the sequence; much of the remainder of the genome consists of various *repetitive* sequences [24]. As an example, **Figure 11.4** shows the same genomic region around the amylase genes that we considered in Figure 6.10. However, in this version of the figure, I have added markers for just three classes of repetitive elements, illustrating how much of the genome consists of such elements.

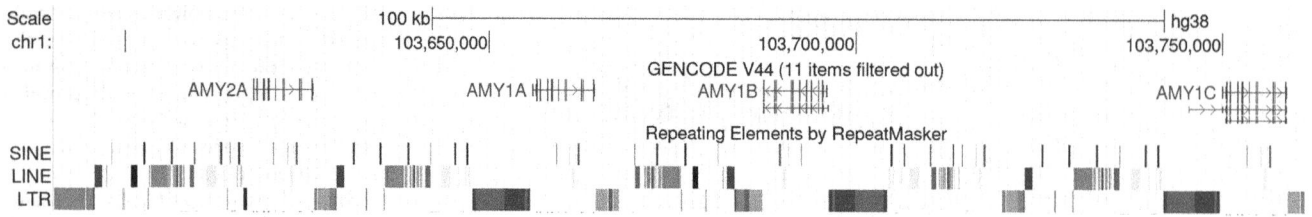

Figure 11.4 **A region of human chromosome 1 containing multiple copies of the amylase gene discussed in Chapter 6.** Now also included, however, are instances of just three classes of common repetitive regions (called SINEs, LINEs and LTRs) found throughout the human genome [24] in this chromosomal region [44]. (Generated with the UCSC Genome Browser.)

What Are Repetitive Elements?

When we say a sequence is repetitive, we mean that it is one of many very similar copies of the same sequence that are spread throughout the genome. These sequences are given a variety of names, including transposons and retrotransposons, both of which are DNA elements that can move and copy themselves within the genome. (Retrotransposons use an RNA intermediate as part of their copying process [25].) Pioneering work exposing the existence of such mobile DNA was performed by Barbara McClintock in the 1950s. Using maize [26], she proved that DNA elements were moving about in the genome, changing the form of the plants as they did. When we look at a genomic region such as that in Figure 11.4, we are reminded of the prescience of her work and of the key role of evolutionary processes other than Darwinian selection in shaping our genetics [27].

What are the origins of these mobile elements? At least some of them are the remnants of ancient *retroviruses*. Retroviruses are those viruses like HIV that can insert their genomes into that of the host [28]. Often these viruses merely use the host genome as a hiding place and emerge and replicate themselves after months or years. However, sometimes the insertion is passed on to the host's offspring [29]. If such a virus remains in the host genome over multiple generations, it can lose the ability to excise itself and create viral particles [30]. At this point, while still able to replicate and move within the host genome, it cannot leave, tying its evolutionary fate to that of the host.

Our purpose is not so much to consider the evolutionary effects of these elements and their movement, though this topic is a fascinating one [25]. Rather, the existence of large swathes of the genome composed of such elements or their decayed remnants (e.g., Figure 11.4) puts paid to the proposal that any change, at any DNA base in a human genome, has some measurable effect on fitness.

NEUTRAL GENETIC VARIATION

Instead, we can see that these three ideas: the degeneracy of the genetic code, the presence of introns, and the abundance of transposable elements in genomes, collectively indicate that we need consider a third class of variant, namely *neutral* genetic variants. Such variants have neither a positive nor a negative effect on the organism's fitness. The point is not that every variant in those three groups is evolutionarily neutral. Rather, I am arguing that the counterclaim—that every change in a genome sequence alters fitness in some way—is unsupportable [31].

How might the frequency of such neutral variants in a population behave over time? We can first acknowledge that, as we discussed in Chapter 4, were the population size infinite, their allele frequencies would not change in time. However, in a finite population, we can imagine that we might observe small changes in neutral allele frequencies as we go from generation to generation. There are at least three quasi-random processes that might drive these fluctuations:

1. Individuals are subject to random mortality that is unrelated to which allelic variant they carry.

2. Individuals may give birth to differing numbers of offspring, again in a manner unrelated to the variant carried.

3. In individuals *heterozygotic* for the variant (individuals that have a copy of each variant), meiosis will randomly select one of the two variants to pass on to each offspring.

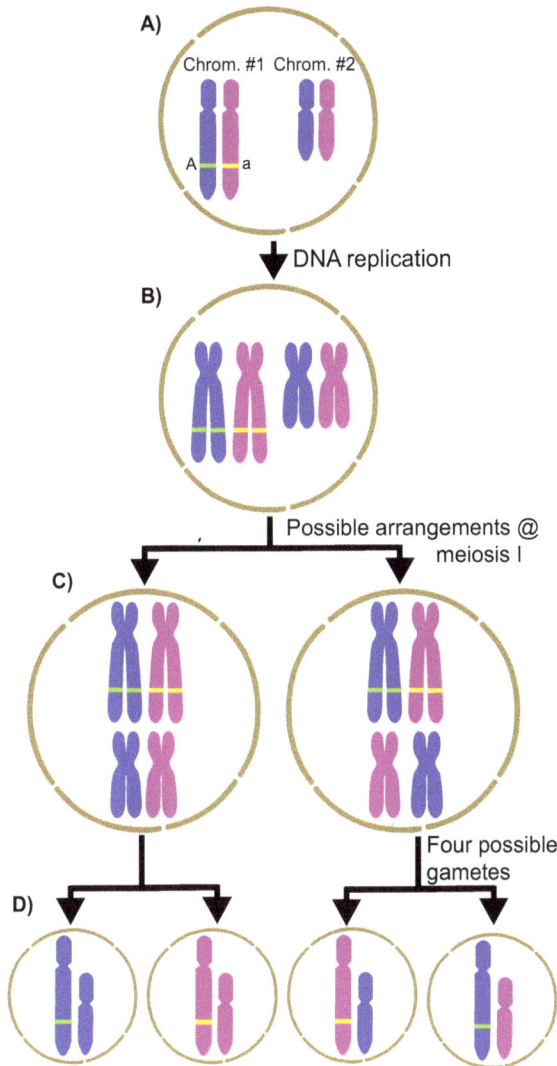

Figure 11.5 The contribution of random and independent assortment of chromosomes in meiosis to the fluctuations in variant frequencies in a population. Here chromosome 1 is host to a variant (**A** or **a**), and the depicted individual is *heterozygous* for this variant (**Aa**), meaning that it inherited an **A** from one parent and an **a** from the second parent. In the first step of meiosis (**B**), the DNA is replicated, and then the sister chromosomes pair up (**C**) in one of two possible configurations. If we disregard recombination, each of these two configurations can produce two possible gametes (i.e., an egg or sperm cell, **D**). Hence, 50% of the gametes will carry **A** and 50% will carry **a**. (Adapted from Pierce BA (2008) *Genetics: A Conceptual Approach*, Macmillan.)

This third point is worth exploring further. In **Figure 11.5**, we start from a variable position in the genome that has a wild-type allele **A** and a mutant allele **a**. (You can think of **A** as having a "T" at that position and **a** having a "C.") We can then consider how a *heterozygous* individual with one copy of each allele might behave (recall our similar discussion in Chapter 5). In a sexually reproducing population, each individual contributes only one of each of their two sister chromosomes to an offspring through the *gametes* (i.e., eggs or sperm) they produce [21]. As a result, a heterozygous individual can only pass on one of the two variants it carries to a given offspring. Moreover, if that individual reproduces again, the choice of which allele to pass on to the second offspring is statistically independent of that given to the first offspring. We can see, therefore, that a heterozygous individual that has two offspring has only a 50% chance of passing on both alleles, with a 25% chance of only passing on the **A** allele and a 25% chance of only passing on **a**.

AGENT-BASED SIMULATIONS OF ALLELE FREQUENCIES

Our argument above describes the situation from the perspective of a single individual. What are the consequences of these patterns at the level of an entire population? One obvious next step, having considered the patterns seen in meiosis, would be to use the ideas of random walks from Chapter 9 to understand how meiosis might contribute to changes in allele frequencies. And indeed, at the end of the chapter, we will see that this is exactly what statistical geneticists have done. However, as foreshadowed, the mathematics of doing so are complex. Can we use the computational tools from the last chapter to more easily approach this problem?

The approach we will use is termed an *agent-based model* [32], although our example is a very simple type of such a model. Our agents are the individuals of the population. In the spirit of simplification, we will only consider one piece of information about each individual: its two chromosomes and whether each carries an **A** or an **a**. Some other assumptions we will make for computational simplicity are:

1. The population is of constant size.

2. The population completely replaces itself at each generation. In other words, all the individuals of the current generation mate, produce offspring, and then die and do not directly contribute to the formation of the succeeding generations.

3. The agents/individuals are sexually reproducing hermaphrodites incapable of self-fertilization.

4. Alleles **A** and **a** are identical in their impact on the organism's fitness.

5. Each time the simulation runs, a new random population with the requested allele frequencies will be generated.

6. The initial population will be in Hardy-Weinberg equilibrium, which just means that there is no bias for or against heterozygotes in the population [21].

Exercise 11.1

Outline pseudocode for a simulation program that takes as input a population size and an initial frequency of the **A** allele and simulates the changes in the frequency of **A** and **a** from generation to generation.

TABLE 11.1 PSEUDOCODE DESCRIBING OUR SIMULATION OF THE CHANGES IN ALLELE FREQUENCIES IN TIME DUE TO RANDOM EFFECTS OF MEIOSIS AND REPRODUCTIVE SUCCESS.

```
• INPUT ➔ pop_size
• INPUT ➔ freqA
• numA= freqA*(2* pop_size)
• genN=Array(pop_size,2)
• genNp1=Array(pop_size,2)
• LOOP for(i=0; i<pop_size)
            o genN[i][0]="a"
            o genN[i][1]="a"
• cntA=0
• LOOP while cntA<numA
            o indiv= Random(0,pop_size)
            o allele=Random(0,1)
            o BRANCH if (genN[indiv][allele] is = "a")
                ■ genN[indiv][allele]="A"
                ■ cntA=cntA+1
• gencnt=0
• LOOP while ((numA>0) AND (numA<(2*pop_size)))
            o LOOP for(i=0; i<pop_size)
                ■ parent1= Random(0,pop_size)
                ■ allele1= Random(0,1)
                ■ parent2= Random(0,pop_size)
                ■ allele2= Random(0,1)
                ■ LOOP while (parent1 is = parent2)
                    • parent2= Random(0,pop_size)
                ■ genNp1[i][0]=genN[parent1][allele1]
                ■ genNp1[i][1]=genN[parent2][allele2]
            o genN=genNp1
            o numA=0
            o LOOP for(i=0; i<pop_size)
                ■ BRANCH if (genN[i][0] is = "A")
                    • numA= numA+1
                ■ BRANCH if (genN[i][0] is = "A")
                    • numA= numA+1
• gencnt= gencnt+1
• OUTPUT gencnt
```

"is =" is used to denote the test of equality, while "=" is used to denote assignment.

Exercise 11.2
Using your pseudocode or that in Figure 11.5, write a Python script for simulating the changes in allele frequency in a population. A few extra Python commands you might need include:

- `import random`
    - o Gives us access to random number generation
- `import sys`
    - o Allows us to read the population size and frequency of "A" from the program's command line arguments
- `population_size = int(sys.argv[1])`
- `Initial_Freq_A = float(sys.argv[2])`
    - o Read the population size and frequency of "A" from the command line arguments and convert them to numbers (`int()` and `float()`)
- `random.seed()`
    - o Start the random number generator
- `current_gen = [[0 for x in range(2)] for y in range(population_size)]`
    - o Create the 2-D array for the current generation
- `for individual in xrange(0, population_size):`
    - o Run a for-loop between 0 and population_size-1
- `individual = random.randint(0, population_size-1)`
    - o Generate a random integer between 0 and population_size-1

At various points, we will comment on the possibility of relaxing some of these assumptions. In Exercise 11.1, you will write pseudocode for our simulations—try not to peek at Table 11.1 until you have tried it on your own. The pseudocode in Table 11.1 assumes we have a method for generating random integers on an interval (the Random function) and a function for creating new two-dimensional tables to hold each generation (the Array) function. In Exercise 11.2, you will write your own simulation in Python or a language of your choice from the pseudocode of Table 11.1. Using this program, you can simulate the evolution of populations of different sizes, making sure to run 10–20 independent simulations at each size and observe the results.

DYNAMICS OF ALLELE FREQUENCY VARIATION AND FIXATION

The most important result your simulations will show is that, in finite populations, genetic variation is transitory. Even in the absence of selection, your simulations eventually end with either the **A** or the **a** allele being lost (**Figure 11.6**). (There is in fact a non-zero probability that the two alleles could "float" for an arbitrary length of time, but if your simulations are running for a very long time, the more likely explanation is a programming bug.) When one allele is lost, it is necessarily the case the other is now found in every member of the population. We therefore refer to the second allele as having been *fixed* in the population [33].

While we understand from examples like Figure 11.2 why a beneficial allele would become fixed, the reasons for the fixation of a neutral allele are clearly different. To understand this process, notice what happens in the population when the last copy of one of the alleles is lost. To bring that allele back would require a new mutation, but the chances of a new mutation in exactly that position in the genome for most species is so small as to be effectively zero on our timescale [34]. For all intents and purposes then, once one of the variants is lost, that position in the population's genomes will be monomorphic for the immediate future. One phrase used to describe such processes is a *random walk with absorbing*

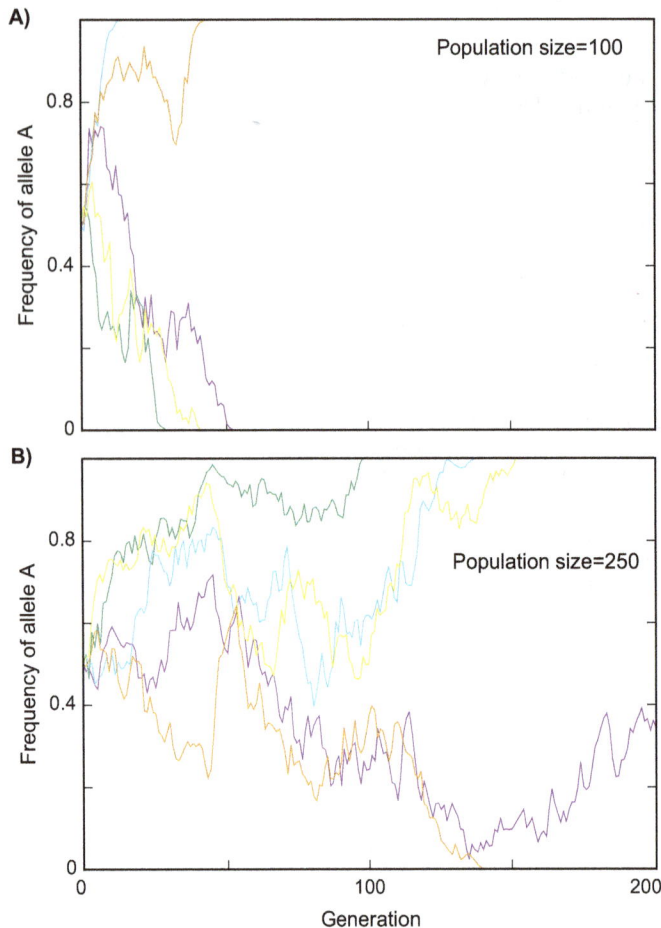

Figure 11.6 Results of five independent genetic drift simulations from Exercise 11.2 for a population size of (**A**) 100 and (**B**) 250, each with an initial frequency of 0.5.

boundaries. Metaphorically, a random walk between two cliffs will always end with the walker falling, even if that walker has no bias in the direction of the steps taken. Because we cannot recover genetic variant once lost, genetic drift has a similar undirected, yet effectively determined, outcome.

Time to Fixation

How long does it take to fix a neutral variant? Is this number independent of population size? If you run enough simulations of different population sizes, you will observe that the time it takes for an allele to fix is *not* independent of the population size: larger populations take longer to fix. The reason for this difference is that the generation-to-generation variation in allele frequencies is smaller for large populations (Figure 11.6). We can infer the relationship between population size and fixation time by writing a *second* computer program that runs our original script many times for different population sizes and averaging the result (Exercise 11.3). In fact, the fixation time is directly proportional to population size (**Figure 11.7**).

We actually already understand why this is from our work in Chapter 9. There we were considering the sample mean as a function of sample size. In larger samples (think populations), the difference between the sample and population means was smaller. In our simulations, the same logic applies: in a larger population, the sampled allele frequencies in the next generation will be closer to the frequencies in the current one than in a small population. Because they vary less from generation to generation, it takes longer for the frequency of **A** to reach 0 or 1. We will explore some of the implications of this fact in the next section.

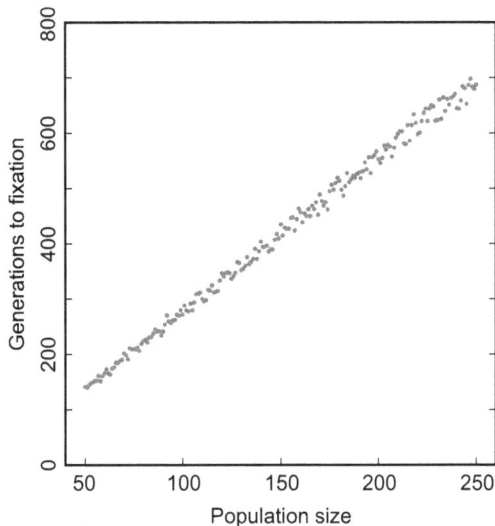

Figure 11.7 Time to fixation for different populations sizes, starting with an initially equal frequency of A and a alleles in these population. On x is the population size N; hence there are $2N$ alleles in the population. On y is the average number of generations to the fixation of either **A** or **a** over 1,000 runs of the simulation program from Exercise 11.2.

Exercise 11.3

Write a new Python script that takes as input a range of population sizes and a sample size S. For each population size, run the drift simulation code from Exercise 11.2 a total of S times. Then your code should average the number of generations to fixation and print these values as output.

Genetic Drift versus Natural Selection

Many of you probably have been longing to read the words "genetic drift" [33] rather than this cumbersome "variation in allele frequencies in time" for quite some time now. And indeed, in our simulations, we have (re)discovered genetic drift. An obvious question is how different the trends in Figure 11.6 would be if one of the two alleles had a selective advantage over the other. We can add natural selection to our simulations fairly easily. To do so, we allow the chances that an individual reproduces in the population to increase when its allele configuration induces higher fitness (see Exercise 11.4). In **Figure 11.8**, I compare the dynamics of populations experiencing drift at a particular variant position with those experiencing selection in favor of the **A** allele. There are two key differences we see in the population experiencing selection. First, the chances of **A** becoming fixed rather than **a** are much higher (3/3 populations rather than 1/6 for the populations experiencing drift). Obviously, these are only simulations, but this pattern of fixation chances is indeed expected in general [35]. The second difference is that the populations under selection see fixation of **A** much more quickly than is seen under drift. Thus, all three populations under selection fix in about 100 generations, while two of the six populations under drift have not fixed even after 300.

Exercise 11.4

Modify your script from Exercise 11.2 to allow the user to define different fitness values for the **aa**, **Aa**, and **AA** genotypes. An easy way to do this is to define an array of fitness values where the fitness of each individual is represented. Sum up the total population "fitness" T as the sum of the elements of this array. You can then sample from the parental generation by drawing a uniform floating point number between 0 and T. This procedure will allow parents of higher fitness to be selected for reproduction more often. The python command for generating a uniform random floating point number is `random.uniform (0, total_fitness)`.

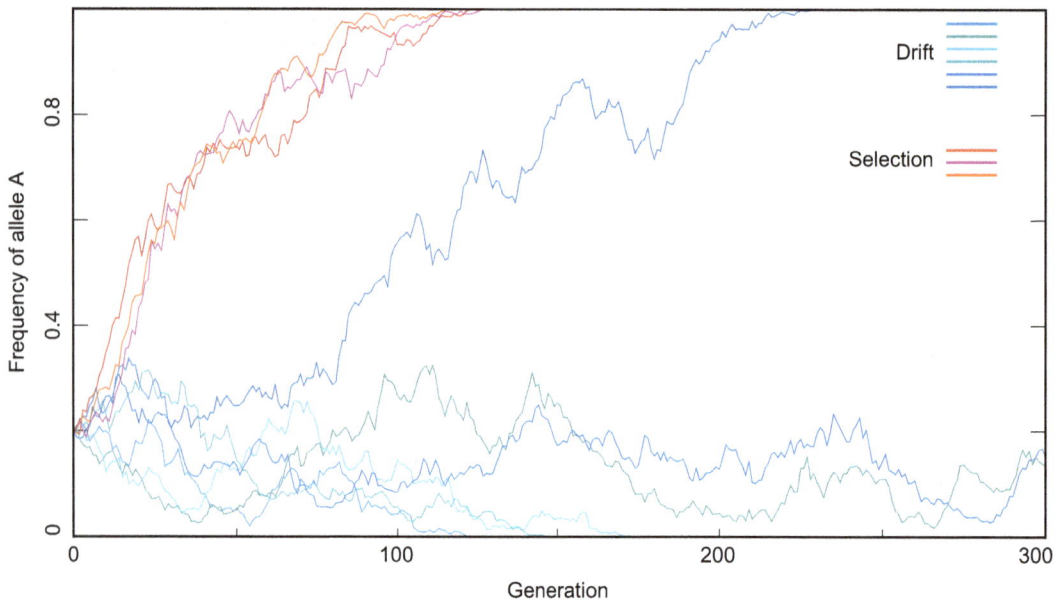

Figure 11.8 **Results of six independent simulations of a population of size 1,000 experiencing genetic drift (*blue*), compared to three populations, also of size 1,000, where allele A enjoys a dosage-dependent selective advantage, with genotype Aa having a 5% advantage and AA having a 10% fitness advantage over individuals that are homozygous for the a allele (*red*).** On the x-axis is the time in generations and on the y-axis is the frequency of **A** in the population. In all cases, the simulation starts with an initial frequency of **A** of 0.2.

generations. The reason for this difference is simply the favorability of the **A** allele in the creation of the next generation.

For completeness, I will note that our simulations are somewhat artificial in that we started from a frequency of **A** of 0.2. In a real population of size N, a new beneficial allele might originally arise by mutation at a frequency of $1/2N$. However, we have not performed our simulations in Figure 11.8 with these starting conditions because, had we done so, most of the populations experiencing drift would have lost the allele almost immediately, making it hard to generate intuition.

Even with that slight caveat, our agent-based model of evolution has given us several valuable insights:

- Genetic variation is transitory in finite populations.

- Small populations lose variation more quickly than do large ones.

- Alleles under selection change frequency much more quickly than do those experiencing drift.

We will now take these insights and see how they compare to what we can learn from mathematical models of evolution under drift and selection.

Statistical Approaches to Genetic Drift: The Wright–Fisher Model

As alluded to at the start of this chapter, the logic of random walks allows us to create statistical models of how allele frequencies drift in time. In fact, the most straightforward approach is a comparatively simple modification of the discrete random walk we considered in Chapter 9. The basic version of this *Wright–Fisher* model [33, 36, 37] assumes that a diploid population of N individuals carries two variants, allele **A** with frequency p and allele **a** with frequency $1-p$. Because these frequencies may change in time, we will refer to the frequency of **A** in generation n as p_n. We now would like some approach to computing p_{n+1}, namely the frequency of **A** in the next generation. We notice that p_{n+1} is obtained by randomly sampling the alleles in this generation $2N$ times, just as we did in our simulations. In a discrete population, there are a total of $2N+1$ possible values of p_{n+1}, ranging from 0 to 1.0. We can express the possible values of p_{n+1} using an index i and the expression:

$$p_{n+1}^i = \frac{i}{2N}, 0 \leq i \leq 2N \tag{11.1}$$

Here i is simply the number of **A** alleles in generation $n+1$. Thinking back to Chapter 8, we can reason that the probability of finding any one of these values of p_{n+1} could be computed using the binomial distribution. To see this fact, notice that the probability that we have i **A** alleles in generation $n+1$ is the probability of flipping an unfair coin with head probability p_n $2N$ times and getting exactly i heads. In terms of the binomial distribution, we would write this statement as:

$$p_{n+1}^i = \frac{2N!}{i! \cdot (2N-i)!} \cdot (p_n)^i \cdot (1-p_n)^{(2N-i)} \tag{11.2}$$

What is useful about Equation 11.2 is that we can use it to create a *transition probability matrix* T. This matrix gives the chance of going from any allele frequency p_n to any other frequency p_{n+1} in a single generation. To see how we might construct T, let j be the current number of **A** alleles in the population, such that $p_n = j/2N$. There are of course $2N+1$ possible values of j, and they define the $2N+1$ *rows* of T. Once we have selected a row representing the current generation, the probability of any of $0 \leq i \leq 2N$ values for p_{n+1} are given by element $T_{j,i}$: in other words, the ith column of the jth row of T. For compactness, we will also write $\frac{2N!}{i! \cdot (2N-i)!}$ as $\binom{2N}{i}$.
We then write T as:
Let $f_1 = 1/2N$ and $f_{2N-1} = (2N-1)/2N$.

$$T = \begin{pmatrix} 1 & 0 & \cdots & 0 \\ \binom{2N}{0}(1-f_1)^{2N} & \binom{2N}{1}(f_1)(1-f_1)^{2N-1} & \cdots & \binom{2N}{2N}(f_1)^{2N} \\ \cdots & \cdots & \cdots & \cdots \\ \binom{2N}{0}(1-f_{2N-1})^{2N} & \binom{2N}{1}(f_{2N-1})(1-f_{2N-1})^{2N-1} & \cdots & \binom{2N}{2N}(f_{2N-1})^{2N} \\ 0 & 0 & \cdots & 1 \end{pmatrix} \tag{11.3}$$

If we now define a vector P_n to contain the probabilities of each possible p_n (i.e., the current probability that allele **A** has a particular frequency in generation n), then the vector P_{n+1} containing the probability of each possible p_{n+1} in the next generation can simply be computed as the matrix-vector product:

$$P_{n+1} = T \cdot P_n \tag{11.4}$$

For P_0, we can let, for instance, $P_n^{i=N/2N} = 1$. This value of p_n simply tells us that our population starts with 50% **A** alleles and 50% **a** alleles. We then compute the probabilities of all possible allele frequencies out for as many generations as we wish by simply recurrently applying Equation 11.4 with P_{n+1} being the input vector to compute P_{n+2} and so forth. This computation for $2N = 16$ and eight generations is shown in **Figure 11.9A**. You can make the computation for yourself in Exercise 11.5.

An obvious question is how this Wright–Fishe model compares to the model we implicitly created by simulation. In **Figure 11.9B**, I show the average of the allele frequencies across 10,000 simulations from our script from Exercise 11.2. At least visually, the simulations and the Wright–Fishe model behave identically.

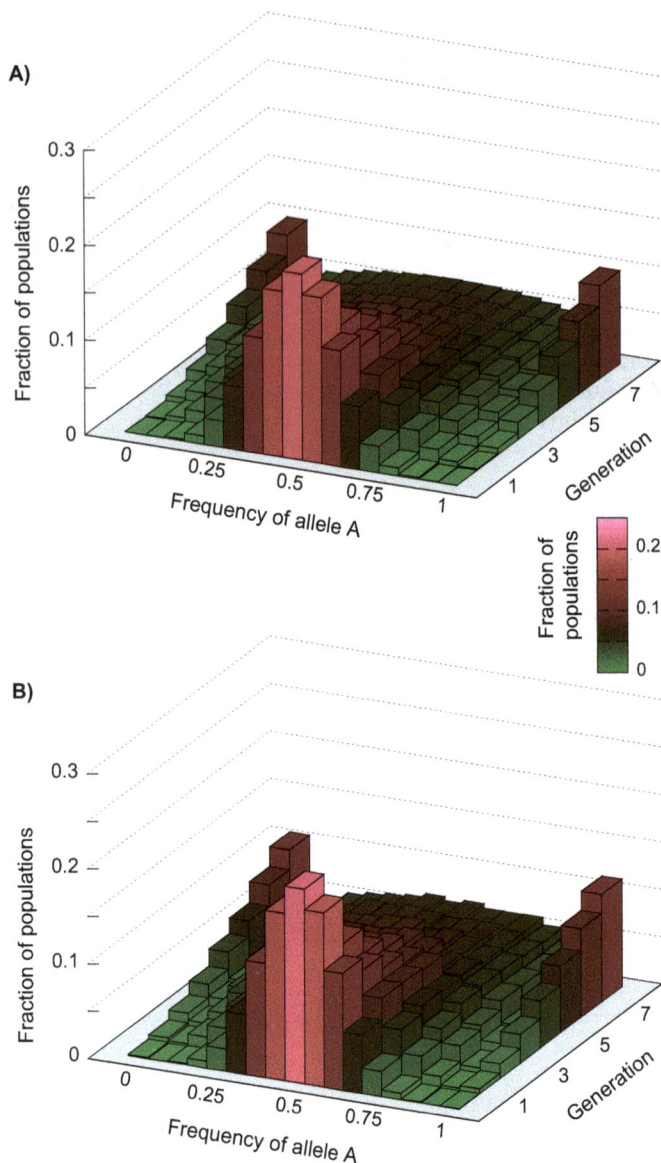

Figure 11.9 Comparing the Wright–Fishe model with our genetic drift simulations. (**A**) Predicted fraction of populations with a given frequency of **A** (x – axis) for eight generations of a population of size 2N = 16 under the Wright–Fishe model (generated with Equation 11.4). (**B**) Our simulation of a diploid population with random mating and self-incompatibility (2N = 16) for the same eight generations (results averaged from 10,000 simulations).

Exercise 11.5

Use R to compute the Wright-Fishe model of genetic drift for a diploid population of 50 individuals over 50 generations. R commands to use:

- `N=30`
- `TwoNp1=2*N+1`
- `transmatrix = matrix(data=1.0, nrow=2*N+1, ncol= 2*N+1)`
- `for(j in 1:TwoNp1) { for (i in 1:TwoNp1) {transmatrix[i,j]=choose(2*N,i-1)* (((j-1)/ (2*N))**(i-1)) * ((1.0-(j-1)/(2*N))**((2*N)-(i-1))))}}`
- `init <-double(TwoNp1)`
- `init[N+1]=1.0`
- `solnmatrix=matrix(data=0.0,nrow=2*N+1,ncol=41)`
- `solnmatrix[,1]=init`
- `for(k in 2:41) {solnmatrix[,k]=transmatrix%*%soln matrix[,k-1]}`
- `write.table(solnmatrix, "WF_N30_40gen.txt")`

MORE COMPLEX MODELS OF DRIFT AND SELECTION

The Wright–Fishe model is computationally slightly annoying, as it requires computing a series of matrix products. In 1955, Motoo Kimura formulated a model of the probability of allele frequencies changing in time, where both t and x (the allele frequency) were measured continuously [38]. If we express the probability of the allele frequency ϕ as a function of the frequency x and the elapsed time t, the differential equation for ϕ can be written as [33]:

$$\frac{\partial}{\partial t}\phi(x,t) = \frac{1}{4N}\frac{\partial^2}{\partial x^2}\left[x(1-x)\phi(x,t)\right]$$

(11.5)

This equation is extremely similar to Equation 9.9 that describes the continuous random walk (i.e., the heat equation) from Chapter 9. However, it has two extra terms in the second derivative with respect to x: x and $1-x$. These two terms imply that the variance/diffusion rate (D in Equation 9.9) is not constant over x: it is maximal when $x = 0.5$ and decreases in either direction from that point. We can understand this change as having the variance of the random walk decrease the farther we are from $x = 0.5$. Because of this nonconstant variance, we cannot use the Fourier transformation we used in Chapter 9 to solve Equation 11.5. Instead, the solutions are very difficult to write down [38], and we will not explore them directly. Instead, we can get a good conceptual understanding of the resulting model by returning to an equation we considered briefly in Chapters 3 and 4 [35]:

$$Q = \frac{N_e s}{N}\cdot\frac{1}{1-e^{-2N_e s}}$$

(11.6)

We can recall that Q gives the probability that a new mutant allele with fitness advantage ($s > 1$) or disadvantage ($s < 1$) eventually replaces the wild-type allele in a population of size N/N_e. For simplicity, let us consider a population where $N_e = N$, giving:

$$Q = \frac{s}{1-e^{-2N_e s}}$$

(11.7)

In **Figure 11.10**, I have plotted Equation 11.7 for some reasonable values of s and N_e. The main fact we should take home is that when s and N_e are large enough, we do reach a point where the deterministic equations of Chapter 4 are appropriate. However, there is a large domain of s and N_e where drift and selection are competing with each other, such that some beneficial alleles will not become fixed and some deleterious (i.e., harmful) ones will do so.

Recessive Lethals

To illustrate that the domain where we must consider both drift and selection is relevant, we need only return to the human genome. On average, most humans possess about one defective gene copy that, if both copies were equally defective, would be lethal [39]. In other words, they have a broken gene but survive because the copy of the gene they received from their other parent is working. These *recessive lethal* alleles are hard to find. The reason is that, if they prevent the successful development of the embryo, there are no individuals with both defective copies in the human population. However, there are at least two reasonably well-known examples of diseases caused by possessing two defective copies of an allele that manifest after birth: cystic fibrosis and Tay-Sachs disease [21, 40]. It might appear surprising that alleles causing such severe diseases can persist in the human population, given that natural selection ought to oppose such alleles.

The reason that recessive lethal alleles can persist, at least at very low frequencies, can be understood when computing the chances of observing the disease itself. If f is the allele frequency of the recessive lethal, the chance of producing an

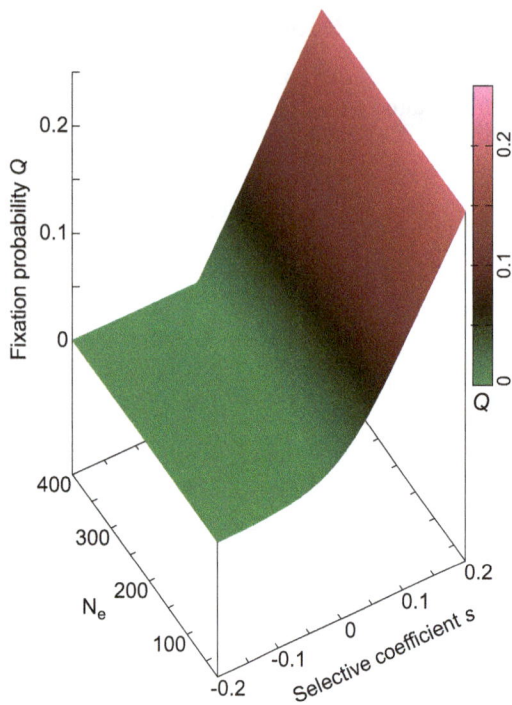

Figure 11.10 Equation 11.7 giving the chances Q ($z-axis$) of a new mutation replacing a wild-type allele for different values of its selective advantage/disadvantage s ($x-axis$) and the effective population size N_e ($y-axis$). (Adapted from Charlesworth B (2009) *Nature Review Genetics* 10:195–205.)

offspring having both nonfunctional copies is proportional to f^2. The reason for this rarity is obvious: it is necessary for two individuals who both possess this rare allele to meet and reproduce for such an offspring to be born. And the result of the rarity of embryos being formed with both defective copies is that the fitness cost of possessing one of these defective alleles is $s \approx 1 - f^2$. In order words, carrying a recessive lethal allele only reduces fitness by a tiny amount, so long as that allele remains rare. Looking back at Figure 11.10, we see that this value of s falls into the domain where drift dominates, explaining the continued existence of these low-frequency deleterious alleles in the human population. Of course, whenever the frequency of that allele starts to rise, the cost will rise as well, explaining why there are no *common* recessive lethal alleles circulated in the human population.

REFLECTIONS AND PREVIEWS

In this chapter we saw a new way to quantitatively model complex biological phenomena: *agent-based models*. We developed a simulation of how the frequency of two alleles changes in a population in time and compared our simulations for different population sizes. We found that genetic variation is unstable in finite populations even in the absence of natural selection acting on that variation. We also found that large populations change their allele frequencies more slowly than small ones do, reminiscent of the patterns seen with the central limit theorem in Chapter 9. We found that our simulations mirrored rather well the more formal predictions of Wright–Fishe models of *genetic drift*. We also revisited the drift-selection balance equation we had surveyed in Chapters 3 and 4. We found that we did indeed need to consider the random effects of drift when trying to understand evolution by natural selection, as we had predicted in those chapters.

Lest we think that drift is just a minor curiosity of genetic evolution, we will see in the next two chapters that not only must we consider the effects of drift when we compare the genes of different organisms, but that the existence of genetic drift allows us to realize one of Darwin's dreams: the construction of the *tree of life* (Chapter 13).

REFERENCES

1. Vynnycky E & White R (2010) *An introduction to infectious disease modelling* (OUP Oxford).
2. Darwin C (1859) *The origin of species by means of natural selection* (John Murry, London).
3. Gould SJ (1992) *Bully for brontosaurus* (W.W. Norton & Company).
4. Charlesworth B & Charlesworth D (2009) Darwin and genetics. *Genetics* 183(3):757–766.
5. Shendure J & Ji H (2008) Next-generation DNA sequencing. *Nature Biotechnology* 26(10):1135–1145.
6. Sandberg TE, Salazar MJ, Weng LL, Palsson BO, & Feist AM (2019) The emergence of adaptive laboratory evolution as an efficient tool for biological discovery and industrial biotechnology. *Metabolic Engineering* 56:1–16.
7. Zhu T, Korber BT, Nahmias AJ, Hooper E, Sharp PM, & Ho DD (1998) An African HIV-1 sequence from 1959 and implications for the origin of the epidemic. *Nature* 391(6667):594–597.
8. Korber B, Muldoon M, Theiler J, Gao F, Gupta R, Lapedes A, Hahn BH, Wolinsky S, & Bhattacharya T (2000) Timing the ancestor of the HIV-1 pandemic strains. *Science* 288:1789–1796.
9. Greene WC (2007) A history of AIDS: looking back to see ahead. *European Journal of Immunology* 37(S1):S94–S102.
10. Larder BA & Kemp SD (1989) Multiple mutations in HIV-1 reverse transcriptase confer high-level resistance to zidovudine (AZT). *Science* 246(4934):1155–1158.
11. Holmes EC, Zhang LQ, Simmonds P, Ludlam CA, & Brown A (1992) Convergent and divergent sequence evolution in the surface envelope glycoprotein of human immunodeficiency virus type 1 within a single infected patient. *Proceedings of the national Academy of Sciences* 89(11):4835–4839.
12. Harrigan PR, Bloor S, & Larder BA (1998) Relative replicative fitness of zidovudine-resistant human immunodeficiency virus type 1 isolates in vitro. *Journal of Virology* 72(5):3773–3778.
13. Futuyma DJ (1998) *Evolutionary biology*. 3rd Edition (Sinauer Associates, Inc, Sunderland, MA).
14. Arion D, Kaushik N, McCormick S, Borkow G, & Parniak MA (1998) Phenotypic mechanism of HIV-1 resistance to 3'-azido-3'-deoxythymidine (AZT): Increased polymerization processivity and enhanced sensitivity to pyrophosphate of the mutant viral reverse transcriptase. *Biochemistry* 37:15908–15917.
15. Alberts B, Johnson A, Lewis J, Raff M, Roberts K, & Walter P (2002) *Molecular biology of the cell*. 4th Edition (Garland Science, New York)
16. Meyer PR, Matsuura SE, Mian AM, So AG, & Scott WA (1999) A mechanism of AZT resistance: an increase in nucleotide-dependent primer unblocking by mutant HIV-1 reverse transcriptase. *Molecular cell* 4(1):35–43.
17. Kerr SG & Anderson KS (1997) Pre-steady-state kinetic characterization of wild type and 3'-azido-3'-deoxythymidine (AZT) resistant human immunodeficiency virus type 1 reverse transcriptase: implication of RNA directed DNA polymerization in the mechanism of AZT resistance. *Biochemistry* 36(46):14064–14070.
18. Crow JF (1987) Population genetics history: a personal view. *Annual Review of Genetics* 21(1):1–22.
19. Wagner A (2008) Neutralism and selectionism: a network-based reconciliation. *Nature Reviews Genetics* 9(12):965–974.
20. Pigliucci M (2010) Genotype-phenotype mapping and the end of the 'genes as blueprint' metaphor. *Philosophical Transactions of the Royal Society B: Biological Sciences* 365(1540):557–566.
21. Russell PJ (1994) *Fundamentals of Genetics* (Harper Collins, New York).
22. Shah P & Gilchrist MA (2011) Explaining complex codon usage patterns with selection for translational efficiency, mutation bias, and genetic drift. *Proceedings of the National Academy of Sciences of the United States of America* 108(25):10231–10236.
23. Jeffreys A & Flavell R (1977) A physical map of the DNA regions flanking the rabbit β-globin gene. *Cell* 12(2):429–439.
24. Lander ES, Linton LM, Birren B, Nusbaum C, Zody MC et al., (2001) Initial sequencing and analysis of the human genome. *Nature* 409(6822):860–921.
25. Feschotte C & Pritham EJ (2007) DNA transposons and the evolution of eukaryotic genomes. *Annu. Rev. Genet.* 41:331–368.
26. McClintock B (1950) The origin and behavior of mutable loci in maize. *Proceedings of the National Academy of Sciences* 36(6):344–355.
27. Dixon B (1983) McClintock's remarkable transposable elements—thirty years later. *Bio/Technology* 1(4):346–346.
28. Wells JN & Feschotte C (2020) A field guide to eukaryotic transposable elements. *Annual Review of Genetics* 54:539–561.
29. Coffin JM, Hughes SH, & Varmus HE (1997) Retroviruses.
30. Johnson WE (2019) Origins and evolutionary consequences of ancient endogenous retroviruses. *Nature Reviews Microbiology* 17(6):355–370.
31. Doolittle WF (2013) Is junk DNA bunk? A critique of ENCODE. *Proceedings of the National Academy of Sciences* 110(14):5294–5300.
32. Macal CM & North MJ (2005) Tutorial on agent-based modeling and simulation. *Proceedings of the Winter Simulation Conference, 2005* (IEEE), p. 14.
33. Hartl DL & Clark AG (1997) *Principles of population genetics*. 3rd Edition (Sinauer Associates, Sunderland MA).
34. Kumar S & Subramanian S (2002) Mutation rates in mammalian genomes. *Proceedings of the National Academy of Sciences* 99(2):803–808.
35. Charlesworth B (2009) Effective population size and patterns of molecular evolution and variation. *Nature Reviews Genetics* 10(3):195–205.
36. Wright S (1931) Evolution in Mendelian populations. *Genetics* 16:97–159.
37. Fisher RA (1930) *The genetical theory of natural selection* (Clarendon, Oxford).
38. Kimura M (1955) Solution of a process of random genetic drift with a continuous model. *Proceedings of the National Academy of Sciences* 41(3):144–150.
39. Narasimhan VM, Hunt KA, Mason D, Baker CL, Karczewski KJ et al., (2016) Health and population effects of rare gene knockouts in adult humans with related parents. *Science* 352(6284):474–477.
40. Turner TN, Douville C, Kim D, Stenson PD, Cooper DN, Chakravarti A, & Karchin R (2015) Proteins linked to autosomal dominant and autosomal recessive disorders harbor characteristic rare missense mutation distribution patterns. *Human Molecular Genetics* 24(21):5995–6002.
41. Mathews CK & Van Holde KE (1996) *Biochemistry*. 2nd Edition (The Benjamin/Cummings Publishing Company Inc., Menlo Park).
42. NIAID Visual and Medical Arts (12/13/2024) Low Friction Flask. in *NIAID BIOART Source*. https://bioart.niaid.nih.gov/bioart/303
43. NIAID Visual and Medical Arts (12/13/2024) Retro Virus. in *NIAID BIOART Source*. https://bioart.niaid.nih.gov/bioart/303
44. Kent WJ, Sugnet CW, Furey TS, Roskin KM, Pringle TH, Zahler AM, & Haussler D (2002) The human genome browser at UCSC. *Genome Research* 12(6):996–1006.
45. Pierce BA (2008) *Genetics: A conceptual approach* (Macmillan).

Ducks in a Row

Bioinformatics and Algorithmic Approaches to Biological Data

12

"On two occasions I have been asked, 'Pray, Mr. Babbage, if you put into the machine wrong figures, will the right answers come out?' I am not able rightly to apprehend the kind of confusion of ideas that could provoke such a question."
—Charles Babbage, inventor of an early mechanical computational device

COMPUTATION AND DATA

So far in this book, our focus has been on developing models of biological processes that we can represent using mathematics, statistics, or, in the preceding chapter, simulation. But if we think back to the metabolic modeling of Chapters 5–7, we notice that just because we can create biological models does not mean that it is possible to experimentally measure the parameters that would make those models predictive of the real biology. More generally, we might say that what has been missing from the book so far has been *data*. And now that we have started to think about genetic data in particular, we should recognize that there are rather a lot of those data around (**Figure 12.1**). Indeed, the sequencing of a single human

A)

B)

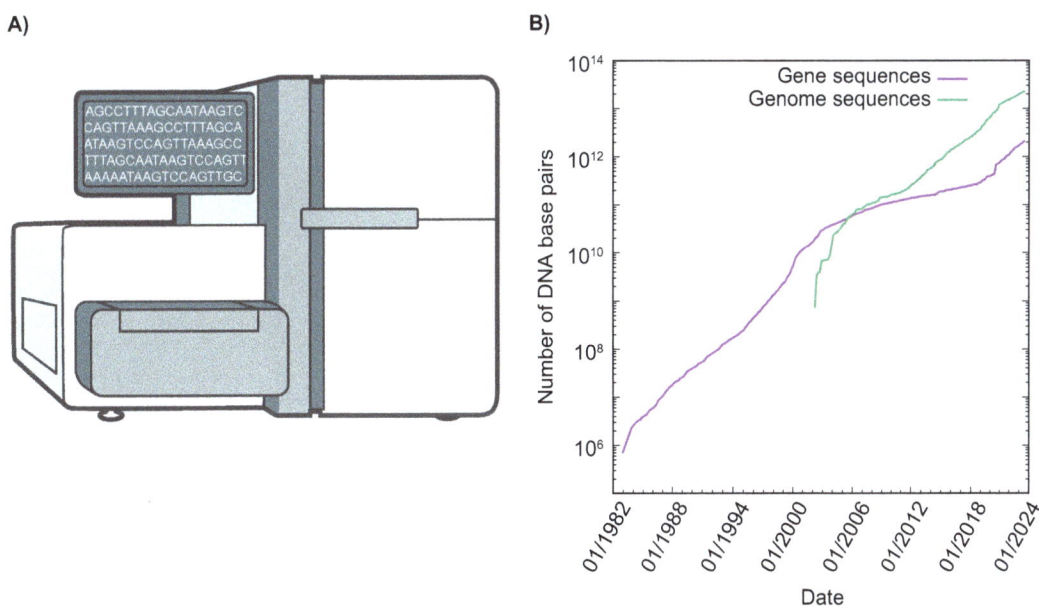

Figure 12.1 Growth in sequencing technologies and in the size of genetic databases. (**A**) New technologies for rapidly and cheaply sequencing DNA have resulted in enormous increases in the quantity of DNA sequence data available to researchers [4]. (**B**). The growth in the public data repository known as Genbank [8] over the past decades. Shown on the y − axis is the number of DNA bases from individual gene sequences (*purple*) and from genome sequences (*green*). Notice that the scale of the y − axis in **B** is logarithmic, meaning that the number of bases sequenced has increased by a factor of 3 million in the past 40 years. (Image in A reproduced with permission from NIAID.)

DOI: 10.1201/9781003687504-12

genome on modern sequencing machines costs less than $1,000 [9]. How do we incorporate such data into our modeling efforts? In this chapter, we will build on the work of the previous chapter and perform computations that describe how our observed gene sequences evolved. To do so, we will first need to create a rough qualitative model of that genetic evolution.

EVOLUTION OF DNA SEQUENCES THROUGH GENETIC DRIFT

In Chapter 11, we considered how a single position in a DNA sequence changed in time. We found that, for a finite population, any initially variable position in that population will eventually become fixed for one of the two variants (or alleles) because the other is lost due to genetic drift. We also saw that the time it took for this loss to occur was, on average, proportional to the population size.

Let us now consider how this process might look if we zoomed out both in time and in space. Rather than considering events over a timescale of hundreds of generations, we will consider them over a timescale of millions of years. And instead of considering a single position in the genome, we will consider a longer sequence of n basepairs. At this scale, we can effectively ignore the time it takes for fixation to occur and simply consider the fixation events themselves. These events will occur relatively rarely and at differing positions along the sequence. But before we consider those processes, we need to pause, because it will already have occurred to you that something is missing from this picture. That missing piece is, of course, the origins of the variants in the first place.

Mutation, Drift, and Fixation

In our simulations in Chapter 11, we always started with a population where there was preexisting variation, namely an **A** allele and an **a** allele, both of which were at a frequency of 50%. Of course, the obvious question is then where those variant alleles came from. The answer to that question is, bluntly stated, mutations [12]. However, that apparently obvious statement seems less satisfying if we consider it a moment longer. For the purposes of argument, let us consider a mutation that takes place in a single individual. We will assume that the base pair position where this mutation happens in one where everyone else in the population is currently identical. That assumption is unproblematic: on average, any two human genomes differ by only one base pair in a thousand [13], so much of the human genome is identical across every living person.

Mutations can occur by several different mechanisms. Two examples are chemical damage to the DNA that can change one base pair to another one and errors in DNA replication that can both change base pairs as well as add or remove those base pairs [14, 15]. When such a mutation occurs at a previously invariant site in a diploid organism, we now have one allele that is present exactly once in the population and another, *wild-type* allele that is present in every other individual. If we calculate the frequency of these two alleles, we find that the mutant allele has frequency $1/2N$ because its carrier has one copy of it and one wild-type allele. Meanwhile, the remaining $N-1$ individuals in the population each have two wild-type copies. Necessarily, then, the wild-type allele has a frequency of $1.0 - 1/2N$.

While we started our simulations in Chapter 11 with two alleles of equal frequency, there was no requirement that this be the case. In **Figure 12.2**, I used our simulation program from the previous chapter to run 20 simulations of a population of size $N = 20$. I set the starting allele frequency of **A** to $1/2N = 0.025$, which corresponds to a new mutation. (As an aside, I generated this figure by writing a new program that called our simulation program 20 times and stored the output of those runs so that the plot could be made.)

As one would expect, most of these mutations are quickly lost through drift, often in a single generation. However, in two simulations, the new **A** allele actually replaces the original **a**. Of course, in real populations that are larger, the chances of

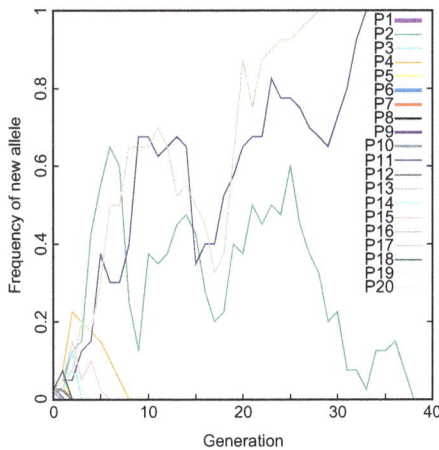

Figure 12.2 Allele frequency simulations using the program from Chapter 11 but assuming a new mutation has appeared in a small, simulated population of $N = 20$ individuals. Hence, that mutation has an initial frequency of 1 in $2N$ or $1/40$. Seven simulations have the new allele lost after a single generation and six have it lost after two generations. For the remaining seven simulations, five eventually lose the new mutation and two see it replace the original one.

the new allele replacing the original are correspondingly lower. However, in such a larger population, the number of new mutations increases in exact proportion to the decrease in the chances of a single mutation fixing. As a result, we expect that, over long periods of time, there should be a slow series of changes in the DNA sequences of a population of organisms as genetic drift *fixes* a (tiny) fraction of the new mutations [16].

It is worth pausing to better understand the dynamics of mutation fixation in a population, not so much for this chapter but because working it out will be quite useful in the next one. Let us assume that mutations are happening in the population at a rate of µ per genome per generation. Just for reference, in humans, and considering only DNA base pair changes, this mutation rate appears to be very roughly 50 or so new mutations per egg or sperm cell produced [17].

We argued a moment ago that a fraction $1/2N$ of those new mutations will fix by genetic drift. So how many mutations should fix in the population each generation? To compute this, we need to know the total number of new mutations appearing in that population, which is, simply enough, $2N\mu$ (the mutation rate per individual times the number of individuals' alleles in the population). Obviously, any new mutation in the population will take some time to become fixed, but since the mutations are occurring continuously, we need not worry about that. As result, we can write that the number of new mutations fixing in the population per generation is:

$$\frac{1}{2N} \cdot 2N\mu = \mu \tag{12.1}$$

There are two striking facts about this result. First, the fixation rate is a simple function of the mutation rate. Second, the rate is *independent* of population size. Why is that? Well, in a large population, many mutations are occurring, but under our rules of genetic drift, only a few are becoming fixed. On the other hand, a small population has fewer mutations, but each has a proportionally higher chance of fixation. Since these two effects cancel each other (under our relatively simply model here), we find that genetic drift drives populations to fix new neutral alleles at a frequency equal to the mutation rate, regardless of the population's size.

Expanding Mutation: INDELS

While our discussion above has implicitly focused on single base pair changes, small gains or losses of DNA bases can also follow this pattern. We call these gains and losses INDELS, for insertions and deletions. This name is intended to remind us that if we are comparing sequences from two modern organisms, we cannot easily tell if one of them has lost bases or the other has gained them.

Implications of Evolution under Drift with Gains and Losses

In Exercise 12.1, we will write a simple simulation program that takes two initially identical DNA sequences and evolves them for a set number of mutational steps. We can consider these simulations to be a very simplified version of how two populations would behave after they stop mating with each other. In other words, we are considering how two DNA sequences might diverge after a speciation event [12].

Exercise 12.1

Create a Python program that simulates the evolution of two initially identical DNA sequences of user-specified length n as they evolve separately under genetic drift after a split between the populations that carry them. You may assume that "gaps" (loss or gain of a base pair in one lineage) happen one-third as often as do base changes. Your simulation should run for a user-specified number of substitution events r.

Figure 12.3A, B shows two realizations of these simulations for a moderate number of mutations. In the images, about one-third of the simulated mutations are single base-pair gains or losses and the remainder are base changes.

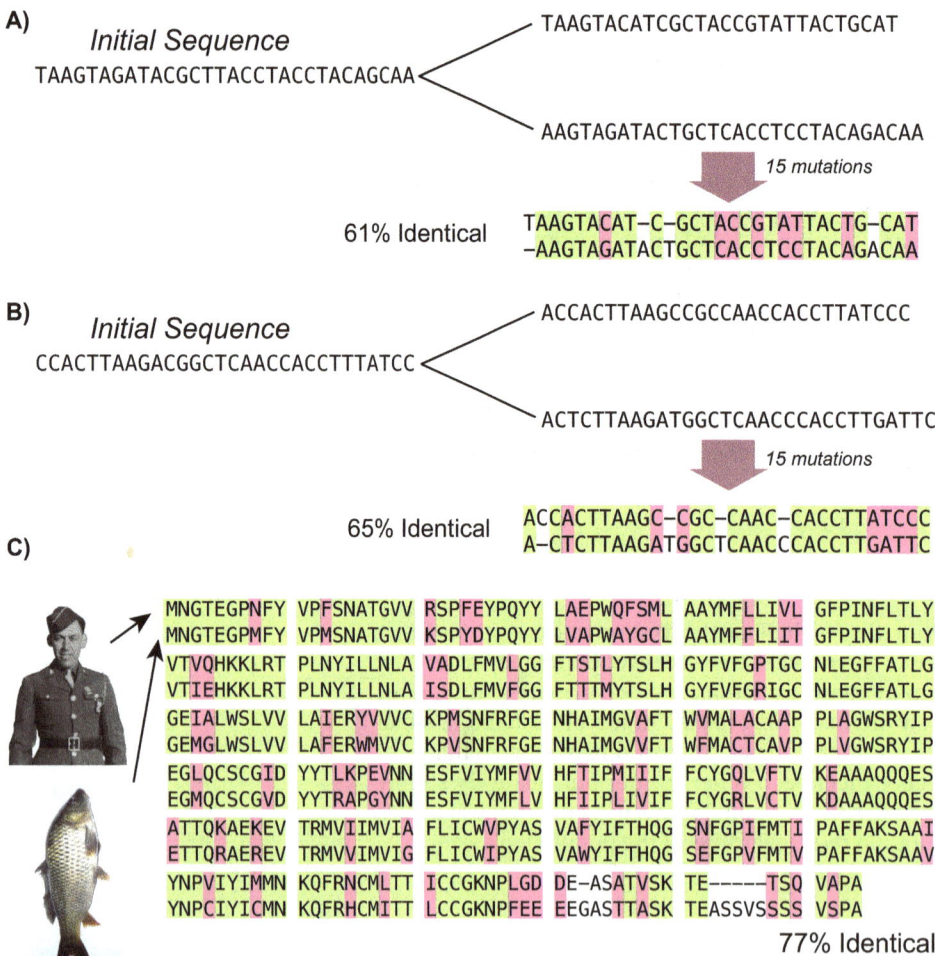

Figure 12.3 Sequence evolution when both base changes and insertions and deletions are occurring. (A) A simulation of mutation through base changes and insertions and deletions for 15 random mutations on sequences of length 30 base pairs. See Exercise 12.1. **(B)** A second simulation with the same parameters as **A**. **(C)** A comparison of two rhodopsin genes, one from humans (photo of the author's maternal grandfather) and one from carp [5]. These sequences are proteins because it is easier to visualize the functional differences in this format. Image courtesy of George Chernilevsky, on Wikimedia, published under CC BY-SA 3.0 license.

As you can see from these examples, comparing the resulting two sequences is quite a bit more difficult when the mutational process includes INDELS. Because of the INDELS, the two final sequences no longer need be the same length, nor does the first base of the first sequence need to be *homologous* to the first base of the second. (We will consider this concept of homology in just a moment.)

Just to illustrate that this problem exists for real biological sequences as well, **Figure 12.3C** shows the sequence of the rhodopsin protein from humans [18] compared to that from common carp [19]. Rhodopsin, as you may recall, is the pigment in your eyes used in dim light [14]. Carp have a version of this same gene that has essentially the same function [19]. Despite this shared function, the two proteins are not identical: they are not even the same length. As a result, we need to introduce what we will now call *gap characters* to allow us to accurately compare the two sequences.

ALIGNMENT AND HOMOLOGY

With Figure 12.3, we have rather gotten ahead of ourselves. The idea that we need to properly line up the two sequences is intuitive, but whether the way we lined them up in the figure itself is sensible remains to be seen.

First, we will need some terminology. We will call this lining up of a pair of strings so that each character in one corresponds to exactly one character in the other an *alignment*. The concept of an alignment extends to many fields besides biology. For example, you might want to compare files containing two versions of an essay you wrote to see where the changes in the second draft are relative to the first. If you did not account for the possibility of words having been added or deleted, you might conclude that two rather similar essays were quite different (e.g., Figure 12.3C).

If we restrict ourselves to biological sequences taken from different individuals or species, we can make a rather interesting observation about alignments. The columns of the alignment, at least those that are not gaps, should correspond to positions in the sequence that both individuals inherited from their ancient common ancestor. This fact is most obvious for the columns that are identical in the two sequences: the two "A"s in the second position of the alignment in Figure 12.3A correspond to the "A" in the second position of the original sequence. Meanwhile, notice that the "T" in the first position of the original sequence was lost in the second descendant sequence.

Figure 12.3B gives a slightly more difficult problem: the "A"s in the first position of the alignment are probably an alignment error. They correspond to a position where the first sequence experienced the insertion of that "A" and the second experienced a mutation of its first "C" to an "A." Hence, we have already found a case where alignment approaches are imperfect. However, on average, even the columns in an alignment that differ (pink in Figure 12.3A, B) reflect an ancestral shared base that has changed by fixation in one or both sequences.

Homology Defined

The idea of considering structures in modern organisms and inferring that their ancestor also possessed that structure has been given a name by evolutionary biologists: *homology*. Homology describes a structure jointly possessed by two or more organisms that was also possessed by their common ancestor [12]. Homology is most easily understood with examples: in **Figure 12.4A, B**, the homologous arm bones of a dinosaur and a tiger are shown. Claiming that these radius and ulna bones are homologous is simply to say that this dinosaur and tiger share a common ancestor, and that this ancestor also possessed both a radius and an ulna. An important corollary about homology is that it is a statement about ancestry rather than function. This dinosaur used its forelimbs differently from how tigers use them now, yet these bones are homologous because they were both inherited from a mutual common ancestor. To reinforce this idea, **Figure 12.4C** shows a bird feather: while the dinosaur in Figure 12.4A was

A)

B)

C)

D)

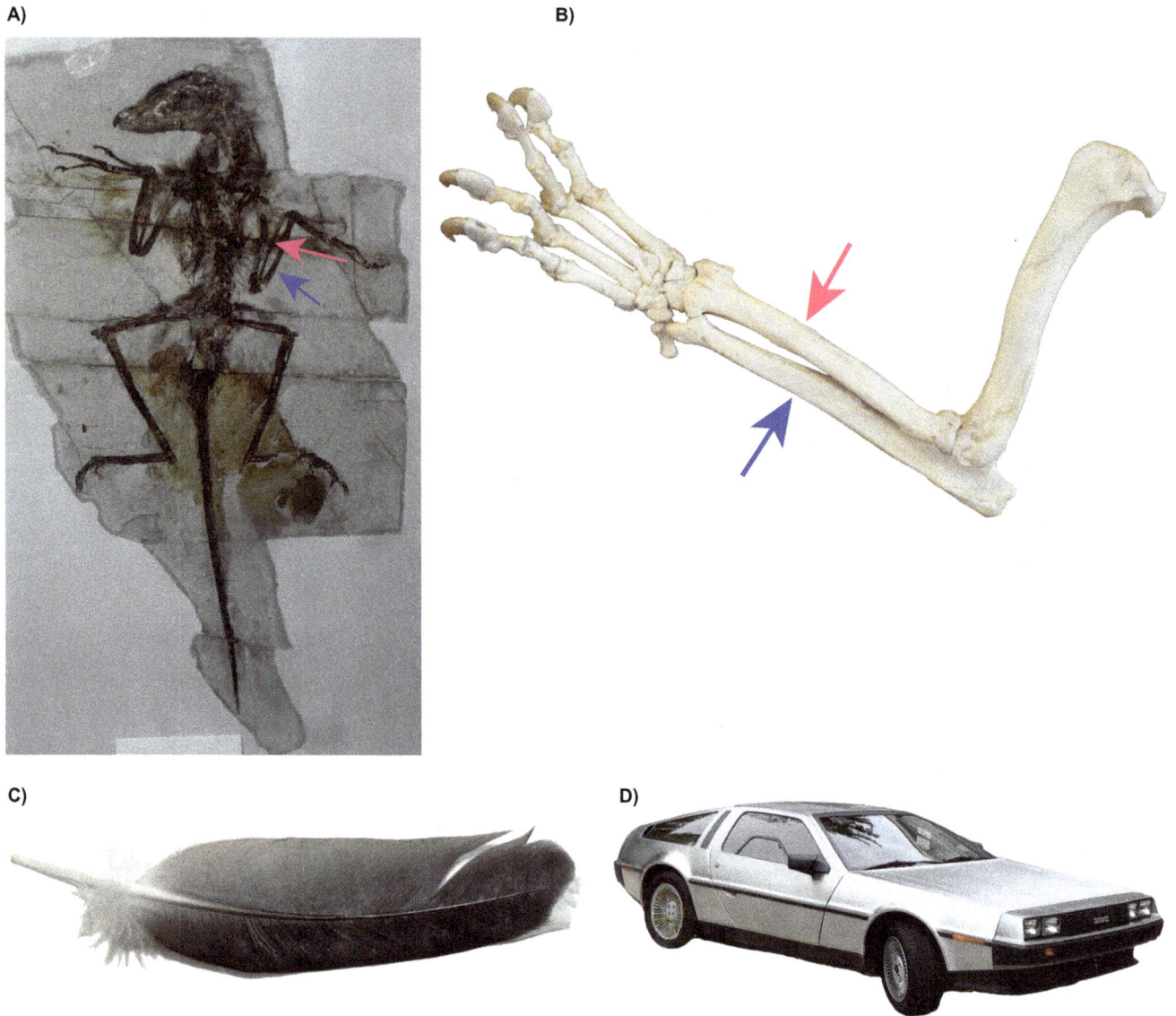

Figure 12.4 Evolution and homology. (**A**) A fossil of a feathered but non-flying dinosaur from China [3]. Its radius and ulna bones are indicated with pink and blue arrows, respectively. (**B**) The bones of a tiger's forelimb, with the *homologous* radius and ulna also indicated with pink and blue arrows. (**C**) A modern bird feather, which is homologous to the feathers from the dinosaur in **A** despite not sharing the same function (the dinosaur in **A** could not fly) (See Exercise 12.2). (**D**) The only (imaginary) scientific instrument that can *confirm* if two structures are homologous: a time machine that would allow us to examine the common ancestor of the individuals that carry that structure and confirm that it too possessed it (flux capacitor not yet installed). (Image in A courtesy of Jonathan Chen, on Wikimedia, published under CC BY-SA 4.0 license.)

feathered, it did not fly. Hence, its feathers were homologous to those of flying birds but did not share the function of flying. Indeed, it has been proposed that feathers probably first evolved for insulation (similar to hair in mammals) and were only later co-opted for flight in the bird lineage [20].

This concept of homology extends very naturally to gene sequences, meaning that when we compute alignments later in the chapter, we are, at least approximately, making the claim that the computational procedure of aligning a pair of sequences produces columns that are homologous. Of course, homology in this context is essentially always an inference or a hypothesis, since it makes a claim about the genes of an extinct organism: **Figure 12.4D** shows the type of instrument we would need to, for instance, test the hypothesis that two sequences, one from humans and one from carp, descend from a common ancestral organism that lived hundreds of millions of years ago. At the same time, it is statistically almost

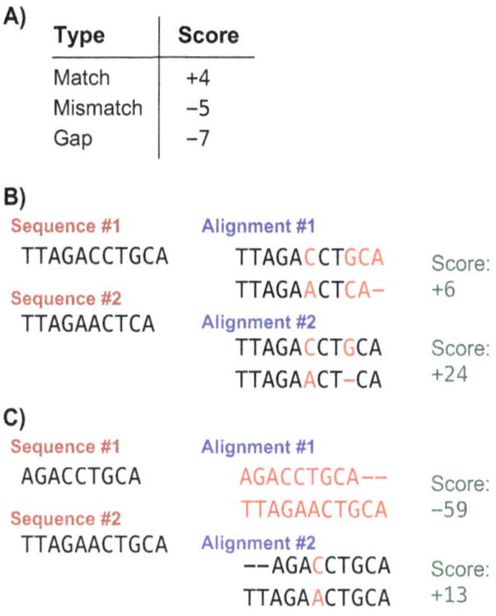

A)

Type	Score
Match	+4
Mismatch	−5
Gap	−7

B)

Sequence #1
TTAGACCTGCA

Sequence #2
TTAGAACTCA

Alignment #1
TTAGACCTGCA
TTAGAACTCA−   Score: +6

Alignment #2
TTAGACCTGCA
TTAGAACT−CA   Score: +24

C)

Sequence #1
AGACCTGCA

Sequence #2
TTAGAACTGCA

Alignment #1
AGACCTGCA−−
TTAGAACTGCA   Score: −59

Alignment #2
−−AGACCTGCA
TTAGAACTGCA   Score: +13

Figure 12.5 **Scoring sequence alignments.** (**A**) A numerical scheme to compute the score of a proposed alignment. (**B**) An example of two sequences and two potential alignments that are visually different and also differ in score. (**C**) A second pair of sequences where the "obvious" alignment is very poor but an alignment that begins with gaps is better, both visually and by way of the score.

unimaginable that two sequences as similar as those in Figure 12.3C would arise independently. As a result, while evolutionary biologists spend a fair amount of time improving their approaches to aligning sequences, the premise that similar biological sequences are homologous is uncontroversial.

What Is an Alignment? What Is a "Good" Alignment?

In some sense we now know what we want to do: line up a pair of sequences so that the homologous positions are in columns. How might we accomplish this feat? One starting point would seem to be finding a means to distinguish a good alignment from a poor one. As shown in **Figure 12.5**, it is rather easy to come up with a naïve alignment of two sequences that looks very poor and yet to have a potentially very good alignment of those two sequences "hidden" nearby. If we assign numerical rewards to matches (identical bases) and costs to mismatches and gaps (Figure 12.5A), we have a means for picking between alignments. Simply speaking, we will sum up the rewards and penalties and prefer alignments with higher total scores (Figure 12.5B&C). Assuming that the scoring protocol we have created is sensible, picking an alignment with a higher score should improve the chances that the columns correspond to homologous positions.

Indeed, we can go beyond comparing alignments using these scores and instead ask ourselves if there is one alignment that has the highest possible score for a pair of sequences and a given scoring protocol. And it is here that we depart, for a time, from the approaches we have used so far in this book and think a bit more about the computers we have been using throughout it.

ALGORITHMS AND OPTIMALITY

Having written computer programs a few times, we can now think about how to write a program that would align two sequences to yield a high alignment score. Unlike the programs we have written so far, we are going to impose some conditions. First, we want a program that guarantees that the alignment we find has the highest possible score. Second, we would like the program that finds that alignment to run as quickly as possible.

It is probably not too difficult to solve one or the other of these two problems on its own. If we were willing to compute the score of a very large number of alignments and simply retain the best one, we could be reasonably certain we had found the best alignment. On the other hand, we could write a fairly simplistic program that, say, aligns the first two bases if they match, otherwise inserting a gap if the first base in the second sequence matches the second in the first, and applying that rule of always looking forward one base beyond the current one to the rest of the two sequences. However, in such a scheme, we would not be even reasonably certain that we had found the best scoring alignment.

We should note before we continue that the question of writing a program that runs fast is a complex one, and we have not yet provided a very good definition of what we mean by fast. We will do so in the following sections, once we have an example from which to work.

Our First Algorithm: Dynamic Programming

We are now going to discover a solution to the alignment problem that meets both our criteria, but we are going to do so in a slightly oblique fashion. We will first define our two sequences s_1 and s_2 as having lengths m and n, respectively. We desire an alignment of length a, where $a \geq m, n$ and with a score x that is as large as or larger than any other possible alignment.

To achieve this goal, we are going to do something that looks very much like cheating. We will *assume* that we have already found three such optimal alignments. The first is the optimal alignment for the two substrings of s_1 and s_2 of length $m-1$ and $n-1$. We will also assume we have found the two optimal alignments of the substring of s_1 of length $m-1$ against all of s_2 and of all of s_1 against the first $n-1$ bases of s_2 (**Figure 12.6A**). In other words, if we assume we have the optimal alignments of every base in s_1 but the last one against every base in s_2 except the last one, as well as the optimal alignments of each complete sequence against all but one base of the other, can we work out the optimal alignment of all of s_1 against all of s_2?

Let the score of these threes subalignments be $x_{m-1,n-1}, x_{m-1,n}$ and $x_{m,n-1}$ (**Figure 12.6B**). Since only one base remains in each of the two sequences, we have only three options to finish the alignment. They are: (1) aligning the two bases against each other, (2) aligning the last base of s_2 against a gap, or (3) aligning the last base of s_1 against a gap. In each case, we need to know the score of the optimal partial alignment corresponding to the proposed complete alignment. For case 1, we are using bases from both sequences, so the score required is $x_{m-1,n-1}$, corresponding to *not* having used either of those bases already. In case 2, we need the score $x_{m,n-1}$, corresponding to having used all of s_1 but not all of s_2. Case 3 is analogous to case 2, using $x_{m-1,n}$.

With this information in place, the problem is actually very simple. We compute the cost (or reward) of aligning the last two bases to each other and add that to $x_{m-1,n-1}$. We similarly add a gap penalty to $x_{m,n-1}$ and to $x_{m-1,n}$. Of the three options, we then choose one that gives the highest score (see Figure 12.6). This highest score is the best possible alignment of all of s_1 against all of s_2.

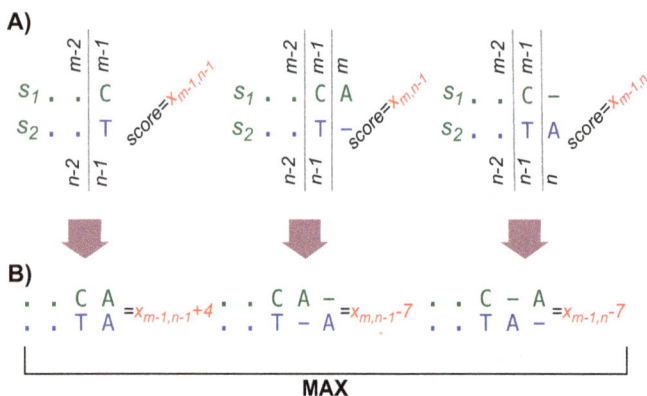

Figure 12.6 **Finding optimal sequence alignments through induction.** (**A**) Three optimal subalignments of s_1 and s_2. The first considers only the first $m-1$ and $n-1$ bases while the second two consider all of s_1 but only $n-1$ bases of s_2 or all of s_2 and only $m-1$ bases of s_1, respectively. (**B**) Using the scores of these three optimal subalignments, we can compute three possible full alignments of all of s_1 against all of s_2 by way of the three scores just calculated. Given these three possibilities, the optimal alignment is the one with the highest score.

Induction

Now, of course, we started by cheating, namely assuming that we already had three optimal subalignments. One might therefore think that not too much has been achieved. However, the scheme of Figure 12.6 is in fact a general one. It applies to *any* two substrings of s_1 and s_2. To see this, we can replace m and n in the score functions above with i and j. Here, we will let i denote the position in s_1 and j that in s_2. If we know the three subalignments giving the scores $x_{i-1,j-1}$, $x_{i,j-1}$, and $x_{i-1,j}$, we can compute $x_{i,j}$ using the scheme of Figure 12.6.

There are only three exceptions to the scheme above. They are the three special scores $x_{0,0}$, $x_{i,0}$, and $x_{0,j}$, where $i = \{1..m\}$ and $j = \{1..n\}$. In other words, if we knew the values of these three special case scores, we could compute the optimal subalignments for all possible values of $1 \leq i \leq m$ and $1 \leq j \leq n$.

So, what are these special cases? The simplest is $x_{0,0}$:

$$x_{0,0} = 0 \tag{12.2}$$

The alignment of an empty substring against an empty substring is trivial and has no score. If we define our gap penalty as g, we find that $x_{i,0}$ and $x_{0,j}$ are also straightforward:

$$x_{i,0} = i \cdot g$$
$$x_{0,j} = j \cdot g \tag{12.3}$$

In other words, if we want to align a part of s_1 against no part of s_2, the only way to do this is with a series of gaps, meaning that we incur a penalty equal to the gap penalty times the length of the substring we are aligning. The argument is of course analogous for using part of s_2 and none of s_1.

A Matrix Schematic for Computing and Representing Optimal Alignments

With these three *base cases* in hand, our optimal alignment problem is solved: we know how to make the three trivial alignments, and once we have them, we can optimally extend them one base at a time in either one or both sequences. It is now convenient to represent this computation in the form of a rectangular matrix of size $m+1$ by $n+1$. That matrix shows how we sequentially compute partial optimal alignments until we reach the end of both sequences. At that point, we have the guaranteed optimal alignment of the full sequences. Each cell of this matrix is computed with the follow equation:

$$\max \begin{bmatrix} x_{i-1,j-1} + c \\ x_{i,j-1} + g \\ x_{i-1,j} + g \end{bmatrix} \tag{12.4}$$

which corresponds to using ether the diagonal score at $i-1$, $j-1$ or the score immediately above or immediately to the left. Here we take c to be:

$$c = \begin{cases} 5 & \text{If } s_1^i = s_2^j \\ -4 & \text{Otherwise} \end{cases} \tag{12.5}$$

In other words, we use the match penalty for the cell if the two bases are the same and the mismatch penalty otherwise.

Figure 12.7 shows the result of completing such a matrix for an example problem. With the first row and column filled according to our base cases, we then use Equation 12.4 to compute the remaining cells (as shown in Figures 12.6 and 12.7), always using the optimal subalignments to the left, on the diagonal, and above to find the optimal subalignment in that cell. For convenience, we can use a system

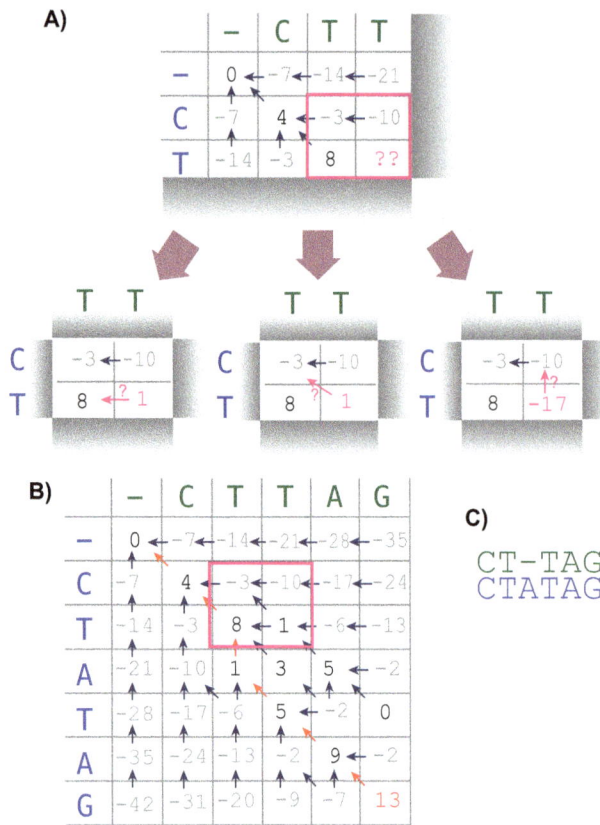

A)

B)

C)

CT-TAG
CTATAG

Figure 12.7 A sequence alignment algorithm. (A) Computing an optimal sub-alignment given three known neighboring optimal subalignments. Given those three values in the magenta box, there are three ways we might generate the optimal subalignment for the remaining cell: aligning against a left gap, aligning the two bases involved, and aligning against an upper gap. We save the choice(s) with the highest score. (B) We can fill out an entire table in this fashion, once we have initialized the first row and column of the table to a series of increasingly long gaps. For clarity, scores less than 0 are shown in gray. (C) Following the red arrows from the bottom right cell in B yields the optimal alignment, which you can verify has a score of 13. In Exercise 12.4, you can compute a local alignment of these two sequences.

of arrows in each cell to store not only the optimal score $x_{i,j}$ but also whichever of the three previous subalignments we used to achieve that subalignment.

Recovering the Optimal Alignment

We now know that the score in the bottom right of the matrix in Figure 12.7B is the score of the optimal alignment of s_1 and s_2. How do we obtain the alignment itself? If we have stored the choices/arrows that gave us those scores, as we did in Figure 12.7B, the problem is (almost) trivial: we simply use the arrows to decide if the last position of the alignment is a left gap, an upper gap, or the alignment of the two bases in question. Once we have deduced the last column of the alignment, we follow the arrow in question to the next cell and repeat the process. In Exercise 12.3, you can write a Python script to perform pairwise alignment.

There is one minor wrinkle to this approach. Notice that in the example of Figure 12.7A, both a left gap and an alignment of the two bases produce the same score, namely 1. No such divergences occur in our traceback, but if they did, we would have not one but two or more optimal alignments, all with the same score. This problem seems more serious than it is: most common alignment packages simply make an arbitrary choice in such cases. For instance, they might always go left if possible at such divergences, and if not to go diagonally and only if neither is possible go up. Such a choice will give us one of the optimal alignments, which is usually sufficient.

Algorithms Defined

The scheme we have just laid out has a technical name: an *algorithm*. For our purposes, an algorithm is a defined set of computational steps that, when followed, give a result of known quality [21]. Hence, our sequence alignment algorithm tells us that if we compute the cells of the matrix in Figure 12.7B, we have computed the optimal alignment of whatever two sequences we gave as input. The algorithm we have just described is known as *global* alignment and is canonically due to Needleman and Wunsch [10].

An advantage of this concept of a defined series of steps is that we can generally provide mathematical descriptions not only of the algorithm's output but of the time it will need to give that output. This idea of *computational complexity* turns out to be quite important in many areas of computational biology, so we will address it next.

DIGRESSION: COMPUTATIONAL COMPLEXITY

Several times in this chapter we have referred to the question of how long an alignment program might run, but we have not explored this idea closely. One reason for this vagueness is that the problem is more complex than it initially appears. Intuitively, we could measure the running time of the problem with a stopwatch: how many seconds are required for it to finish. There are two difficulties with this approach, one obvious and one less so. The obvious difficulty is that it will require different running times for differently sized problems: we would certainly expect longer sequences to require more time to align. The second difficulty is that different computers will also require different amounts of time to run the same program. In fact, over the last few decades, computers have become faster, a phenomenon commonly known as *Moore's law* (**Figure 12.8**). If one expects computers to become faster in the future, one might be able to forecast whether a problem that is intractable using existing machines may become feasible with future ones or remain computationally unachievable indefinitely.

The approach computer scientists use to deal with both of these problems is to focus on how the running time of the algorithm grows with increasing problem size. We will first make the approximation that s_1 and s_2 are about the same length, namely n. Let us now assume that the computation of a single cell of the matrix in Figure 12.7B requires eight computational steps: three additions to compute the three possible scores from $x_{i-1,j-1}$, $x_{i,j-1}$ and $x_{i-1,j}$, the comparison of each of those against the other two to find the maximum, and the storage of the new score and the directional information about that score. (We will see momentarily that we need not be overly concerned whether there are truly 8 steps or 9 or 15.) Hence, we can compute a single matrix entry in a constant number of steps. How many steps will it take to compute the full matrix? Since the matrix has n rows and n columns, we need $8n^2$ steps.

Let's now consider what happens as n grows large for two algorithms with differing *scaling* behavior. For clarity, let us first compare an algorithm where running time is proportional to a large constant multiplied by the problem size, $R_1 \sim 1000n$, to a second problem where it is proportional to a small constant multiplied by the problem size squared: $R_2 \sim 2n^2$. For small values of n, we see that $R_1 > R_2$. But once n becomes even reasonably large (here $n > 500$), $R_2 > R_1$. The general conclusion we

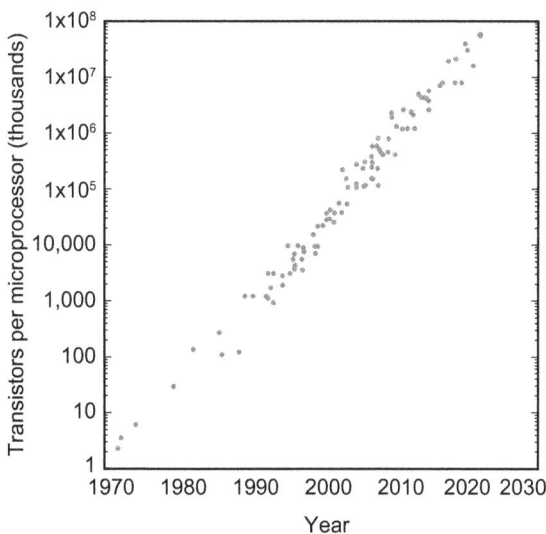

Figure 12.8 Increases in the number of transistors per microprocessor over the past 50 years of computing history. Note the logarithmic scale on y: the effective doubling of the number of transistors per processor every 18 months is known as *Moore's law* after one of the founders of Intel Computing [6]. These particular data are taken from Karl Rupp's curated dataset of microprocessor performance [7]. While not directly translatable to computing speed, these increases have driven increased computing performance.

TABLE 12.1 EXAMPLES OF ALGORITHMIC COMPLEXITY

Operation	Worst-case complexity	Ref.
Adding two numbers	$O(1)$	
Searching a sorted list for a value	$O(\log(n))$	[1]
Finding the maximum value in a list	$O(n)$	[1]
Sorting a list	$O(n \cdot \log(n))$	[1]
Global sequence alignment	$O(n^2)$	[10]
Find all parental chromosome assignments in a pedigree	$O(4^n)$	[11]

can draw from this example is that we need not concern ourselves with the numerical values in front of the proportionality constants of the running time. For our example here we do not need to know exactly how many steps the computation of each cell requires, so long as it is constant and not problem size dependent.

Instead, we can express the algorithm's behavior as just a general function of n. We could also consider the best-case, average-case, and worst-case running times. However, here we will just worry about the worst-case running time: for global alignment, all three cases are in fact the same. Conventionally, we express the worst-case running time as $O(n^2)$: the algorithm's running time grows as the square of the input size, meaning a problem of twice the input size will require four times the compute time to run.

In Table 12.1, I give some examples of computational problems and the complexity of the best-known algorithm for that problem ("best known" in the sense of being the optimal approach found so far, not in the sense of being most commonly cited). It is not important to worry about which particular algorithms I have listed, nor why their complexity is what it is. In fact, the main message to take away from Table 12.1 is the last line.

The problem in that last line is probably unfamiliar to you. It stems from the need to trace a genetic disease that has been transmitted through the pedigree of a family. Our goal in this case could be to locate the causative gene for the disease by getting genetic data from locations across the genome from every member of the family. Unfortunately, while we can measure the state of the genome at each of these positions, we do not know if a particular allele at such a location was inherited from that individual's grandfather or grandmother. As a result, if we need to compute the probability of the disease gene being at any one of our locations in the genome, we need to carry this uncertainty in inheritance through that calculation [11]. This uncertainty increases every time we add a new person to the pedigree, so that each new person that we add quadruples the running time of the analysis. Many years ago, some friends of mine and I worked on some software for this problem. We found that, with one of the biggest supercomputers available at the time, we could relatively easily analyze a pedigree of 21 individuals, but moving to 23 individuals would have impossible on that computer [22].

As we move toward the end of this book, we will move beyond building models of biological systems and start to try to apply those models to actual experimental data. In Chapters 5 and 6 we saw that there are mathematical models of the cell for which we cannot obtain the experimental data necessary to apply those models, even though the computations themselves were straightforward. In the next chapter we will study a problem where the data are straightforward to obtain and yet the problem itself cannot be solved even with all the computers in the world connected together.

SEQUENCE ALIGNMENT IN PRACTICE

Affine Gaps

Returning to the question of alignment, we could also make some refinements to the scheme we have used so far. First, in our simulations in Figure 12.3 we assumed that deletions and insertions always occurred a single base at a time. However, in real sequences, a single mutational event can remove or add several bases at once [14]. As a result, our scoring system of penalizing a gap of length l with a penalty of $l \cdot g$ is arguably too extreme. A better approach might be to use *affine* gap penalties [23]. These penalties state that a gap of length l incurs a penalty of $g_o + l \cdot g_e$. Here g_o is called the gap *opening* penalty (often on the order of 10) and g_e is the gap *extension* penalty (often on the order of 2). Affine gaps penalize longer gaps more than they do shorter ones, but not quite so extremely as the scheme in Figure 12.6. They therefore can prefer an alignment with one long gap over one with two shorter ones of the same total length. Although more difficult to write as a computer program, the computational complexity of sequence alignment with affine gaps is the same as that of the algorithms we have described [23].

Local Alignment

Another problem we may encounter is to have sequences s_1 and s_2 that contain homologous substrings but where s_1 and s_2 are not homologous over their entire length. For example, we may have a mRNA molecule from one species and be looking for the corresponding gene in a known genomic region of a related species. The fast evolution of intergenic DNA and the presence of introns [14] would mean that the alignment of all of s_1 against all of s_2 would not be meaningful. However, it turns out that there is a relatively simple modification of the Needleman and Wunsch algorithm that allows us to make *local* alignments. This algorithm is due to Smith and Waterman [2]. As shown in **Figure 12.9**, its structure is very similar to that for global alignment. The three differences are in the base cases, the selection of the optimal subalignments, and where we start the traceback of the alignment. For the base cases, we have the following scores:

$$x_{i,0} = 0$$
$$x_{0,j} = 0$$
(12.6)

In other words, we do not need to allow for the possibility of initial gaps, since we are seeking optimal substring alignments. Because of that goal, it is acceptable to discard any part of the leading sequence in either s_1 or s_2 without incurring the corresponding gap penalty.

We also change our cell computation slightly:

$$\max \begin{bmatrix} x_{i-1,j-1} + c \\ x_{i,j-1} + g \\ x_{i-1,j} + g \\ 0 \end{bmatrix}$$
(12.7)

where again:

$$c = \begin{cases} 5 & \text{If } s_1^i = s_2^j \\ -4 & \text{Otherwise} \end{cases}$$
(12.8)

As you can see, we have added a fourth case to the "Max" statement. This addition of a 0 implies that as soon as the local alignment score drops below 0, we no longer need to consider it. As shown in Figure 12.9, whenever the score drops in this manner, we also omit storing the directional information because whatever local alignment we recover should not include cells after the score has dropped to 0.

The final change to the algorithm is also shown in Figure 12.9: rather than starting the traceback at the lower right cell, we start it at the maximum score found in the matrix, wherever that might be. Once these three changes have been made, we have a $O(n^2)$ algorithm for finding the best alignment of some part of s_1 against some part of s_2.

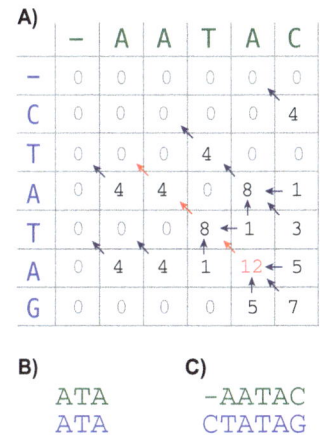

Figure 12.9 Local sequence alignment. (**A**) An algorithm for finding the best *local* alignment of two sequences [2]. We change the algorithm for global alignment in how we initialize the base cases and the optimal subalignment step. (**B**) The optimal local alignment. (**C**) The optimal global alignment.

Aligning Protein Sequences

Although we have discussed alignment in terms of DNA sequences, it is perfectly straightforward to extend it to protein sequences. Indeed, because the protein alphabet contains 20 letters rather than just 4, alignment is arguably more robust using protein sequences. The only significant change we need to make to our algorithm is in the scoring functions. If you refer back to the genetic code in Chapter 11, you will see that the codons for some amino acid pairs are closer than others, in the sense that fewer DNA base changes are needed to convert one of those codons to the other. Moreover, some pairs of amino acid residues are more chemically dissimilar from each other than are others [24]. As a result, the simple match/mismatch scheme of Figure 12.6 is insufficient for scoring protein alignments. Instead, there are amino acid scoring matrices that have been estimated from data; the most common of these are the BLOSUM series [25]. These matrixes are just symmetric 20 x 20 tables giving the match and mismatch scores for all possible pairs of amino acid residues. They provide the values for c in Equation 12.4, with the rest of the alignment algorithm remaining unchanged.

Multiple Alignment

A final extension to alignment algorithms are approaches for *multiple alignments*. Conceptually, we could imagine extending the paradigm of Figure 12.7 to three, four, or five sequences, using tables with the corresponding dimension. The running time of such a computation would be $O(n^s)$, where s is the number of sequences. In practice, such approaches are not used; instead, there are a number of approximate methods for aligning groups of more than two sequences. Generally, these approaches work by first computing all possible pairwise global alignments using the algorithm above and then using the alignment of the closest pair of sequences as a base and progressively adding other sequences [26]. In the next chapter we will use one such tool, called *T-Coffee* [27], to produce a sequence alignment that we can then employ to infer an evolutionary tree.

REFLECTIONS AND PREVIEWS

In this chapter we took a new approach to applying quantitative methods to biological data: the use of an *algorithm* to process or analyze those data. Algorithms are often a first analysis step when your biological data are large or complex. The algorithm will guarantee that the data are processed optimally according to some rule and can then be applied to further analyses.

Computation is integral to modern biology, and sequence alignment is just one of the tools needed to make sense of the large datasets researchers are producing. We also might think about the problem of *assembling* a genome sequence from a number of short DNA reads from a sequencer [28] or of *mapping* sequence reads taken from a cell's messenger RNAs (mRNAs) back to its genome [29]. If we think that the ultimate goal of the scientific process is to gain biological knowledge, there two lessons we can take from this chapter. First, raw data will often need to be fit or processed through a biological model or algorithm to answer the underlying scientific question. Second, this processing step need not be quick: indeed there will be biological problems of very great importance for which an exact computational solution does not exist.

In the next chapter we will take up a case in which an exact solution is not necessarily obtainable. It is nonetheless a very interesting and important area of computational biology because it combines statistics, algorithms (including sequence alignment), and models. That topic is the inference of evolutionary trees from DNA sequence data.

Exercise 12.2

Complete the included worksheet on different views of homology and the difference between homology and analogy.

Exercise 12.3

Develop a Python program for aligning two DNA sequences provided by the user as as two file names, each listed as a command line argument. You may assume the fixed scoring scheme given in Figure 12.6.

Exercise 12.4

Compute the best local alignment of the two sequences in Figure 12.7 and verify that in this case the optimal local and global alignments are identical.

REFERENCES

1. Xu Z & Zhang J (2021) *Computational thinking: A perspective on computer science* (Springer).
2. Smith TF & Waterman MS (1981) Identification of common molecular subsequences. *Journal of Molecular Biology* 147(1): 195–197.
3. Ji Q, Norell MA, Gao K-Q, Ji S-A, & Ren D (2001) The distribution of integumentary structures in a feathered dinosaur. *Nature* 410:1084–1088.
4. NIAID Visual and Medical Arts (8/29/2024) Next generation sequencing machine. in *NIAID BIOART Source*. https://bioart.niaid.nih.gov/bioart/386
5. Chernilevsky G (2008) Common carp or European carp (Cyprinus carpio). https://commons.wikimedia.org/wiki/File:Cyprinus_carpio_2008_G1_(cropped).jpg
6. Shalf J (2020) The future of computing beyond Moore's Law. *Philosophical Transactions of the Royal Society A* 378(2166): 20190061.
7. Rupp K (2022) Microprocessor trend data. https://github.com/karlrupp/microprocessor-trend-data
8. Sayers EW, Cavanaugh M, Clark K, Pruitt KD, Schoch CL, Sherry ST, & Karsch-Mizrachi I (2022) GenBank. *Nucleic Acids Research* 50(D1):D161–D164.
9. Gibbs RA (2020) The human genome project changed everything. *Nature Reviews Genetics* 21(10):575–576.
10. Needleman SB & Wunsch CD (1970) A general method applicable to the search for similarities in the amino acid sequence of two proteins. *Journal of Molecular Biology* 48:443–453.
11. Lander ES & Green P (1987) Construction of multilocus genetic linkage maps in humans. *Proceedings of the National Academy of Sciences, USA* 84:2363–2367.
12. Futuyma DJ (1998) *Evolutionary Biology: 3rd Edition* (Sinauer Associates, Inc, Sunderland, MA).
13. The International HapMap Consortium (2003) The international HapMap project. *Nature* 426(6968):789–796.
14. Lodish H, Baltimore D, Berk A, Zipursky SL, & Matsudaira P (1995) *Molecular cell biology*. 3rd Edition (Scientifc American Books, New York), p. 1344.
15. Russell PJ (1994) *Fundamentals of genetics* (Harper Collins, New York).
16. Li W-H (1997) *Molecular evolution* (Sinauer Associates, Sunderland, MA).
17. Ségurel L, Wyman MJ, & Przeworski M (2014) Determinants of mutation rate variation in the human germline. *Annual Review of Genomics and Human Genetics* 15:47–70.
18. Lander ES, Linton LM, Birren B, Nusbaum C, Zody MC et al., (2001) Initial sequencing and analysis of the human genome. *Nature* 409(6822):860–921.
19. Tsai H-J, Shih S-R, Kuo C-M, & Li L-K (1994) Molecular cloning of the common carp (Cyprinus carpio) rhodopsin cDNA. *Comparative Biochemistry and Physiology Part B: Comparative Biochemistry* 109(1):81–88.
20. Bakker RT (1986) *The dinosaur heresies* (Zebra Books, New York).
21. Campbell AM & Heyer LJ (2003) *Discovering genomics, proteomics, and bioinformatics* (Benjamin Cummings, San Francisco, CA).
22. Conant GC, Plimpton SJ, Old W, Wagner A, Fain PR, Pacheco TR, & Heffelfinger G (2003) Parallel Genehunter: Implementation of a linkage analysis package for distributed-memory architectures. *Journal of Parallel and Distributed Computing* 63:674–682.
23. Altschul SF & Erickson BW (1986) Optimal sequence alignment using affine gap costs. *Bulletin of Mathematical Biology* 48:603–616.
24. Mathews CK & Van Holde KE (1996) *Biochemistry*. 2nd Edition (The Benjamin/Cummings Publishing Company Inc., Menlo Park).
25. Henikoff S & Henikoff JG (1992) Amino-acid substitution matrices from protein blocks. *Proceedings of the National Academy of Sciences, USA* 89(22):10915–10919.
26. Thompson JD, Higgins DG, & Gibson TJ (1994) CLUSTAL W: Improving the sensitivity of progressive multiple sequence alignment through sequence weighting, positions-specific gap penalties and weight matrix choice. *Nucleic Acids Research* 22:4673–4680.
27. Notredame C, Higgins DG, & Heringa J (2000) T-Coffee: A novel method for fast and accurate multiple sequence alignment. *Journal of Molecular Biology* 302(1):205–217.
28. Nagarajan N & Pop M (2013) Sequence assembly demystified. *Nat Rev Genet* 14(3):157–167.
29. Langmead B & Salzberg SL (2012) Fast gapped-read alignment with Bowtie 2. *Nature Methods* 9(4):357–359.

Life on a Tree

Phylogenetics

13

"Thus, on the theory of descent with modification, the main facts with respect to the mutual affinities of the extinct forms of life to each other and to living forms, seem to me explained in a satisfactory manner. And they are wholly inexplicable on any other view."
—Charles Darwin, On the Origin of Species [1]

EVOLUTION AS A TREE

Darwin's name has become so synonymous in the modern mind with the concept of natural selection that the full scope of the argument he was making in *On the Origin of Species* might be overlooked (**Figure 13.1**). In particular, in the chapter epigraph, he describes something very important and particular about the species that live (and have lived) on this planet. They have many similarities with each other that are not explicable solely from their similar ecological niches (**Figure 13.2**). For instance, pandas have many anatomical similarities to other bears, despite being herbivores, unlike other bears [2]. At a molecular level,

A)

B)

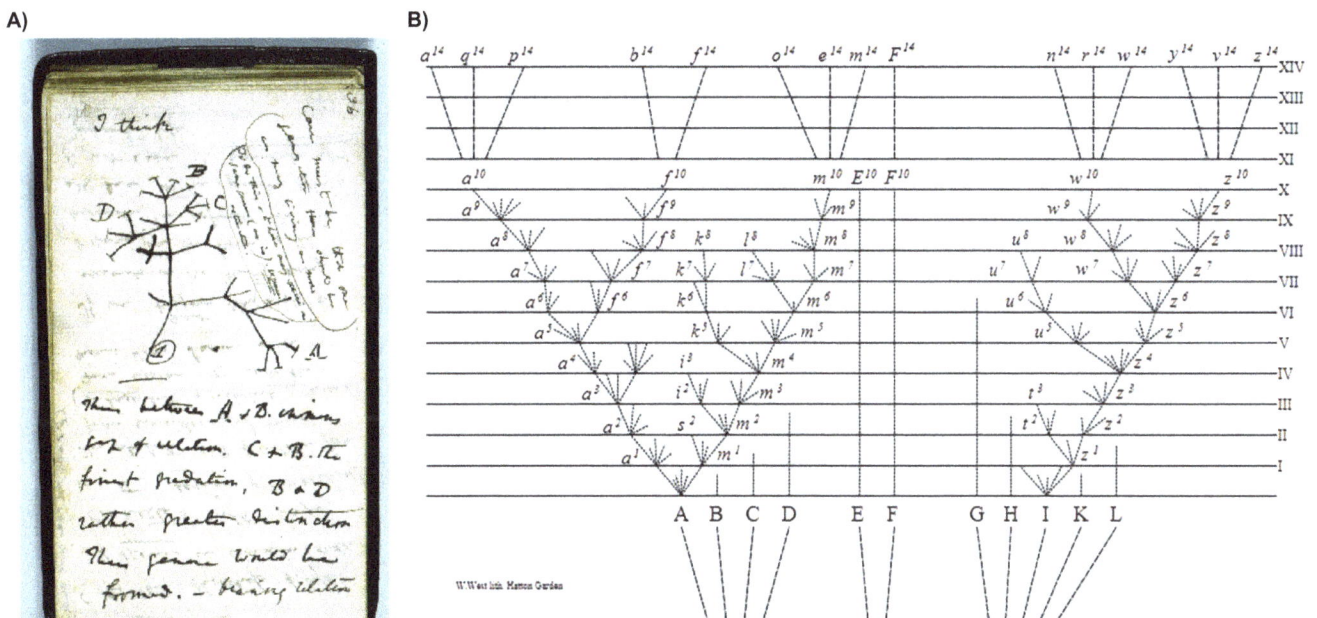

Figure 13.1 Charles Darwin's depictions of his view of the evolutionary relationships between the various species of life on Earth. (**A**) A page from one of Darwin's notebooks written shortly after his return from his trip on the *Beagle*. It depicts the idea of different species being related to each other through a tree-like structure [41]. (**B**). The only figure in Darwin's *On the Origin of Species*. It expands on the idea in **A** and includes the concepts that (1) most species that have ever existed are extinct, (2) there are gaps in the fossil record, (3) there are "living fossils" alive today (form *F*), and (4) even species with no apparent affinity for each other might still derive from a common ancestor (lowest lineages in capitals) [1].

DOI: 10.1201/9781003687504-13

A)

B)

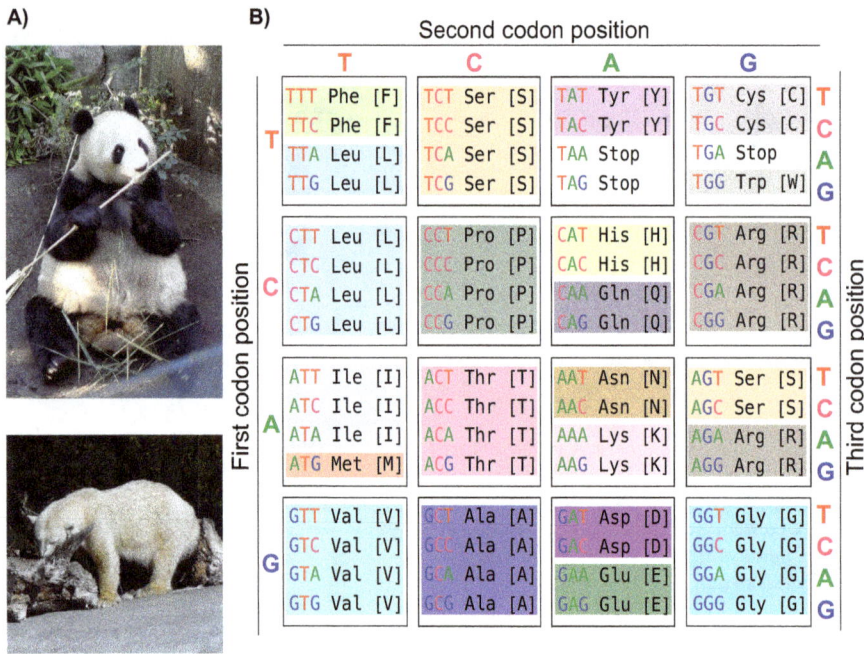

Figure 13.2 **Similarities between organisms that do not seem to be the result of tuning to a common environment.** (A) Polar bears and pandas are anatomically very similar despite their different ecological niches. (B) Nearly all life on Earth uses the same genetic code.

humans and chimpanzees differ in their genomes at approximately 1% of bases; the vast majority of those differences are in parts of the genome that do not code for proteins and are hence expected to have little or no effect on the form or behavior of the organism carrying them [3]. Even if we were to claim that humans and chimpanzees have similar ecological niches (and it is not clear that they do), the similarities in (most of) the *noncoding* regions of their genomes are not explicable under the hypothesis of genetic optimization for the same function.

Common Ancestry and the Genetic Code

Perhaps an even more striking example of a shared evolutionary feature that cannot be due to a shared selective advantage is the genetic code itself. This code, which we first encountered in Chapter 7, is nearly, but not entirely, universal [4]. Hence, while there are a number of known alternative codes, all of them are simply minor derivatives of the "universal" code shown in **Figure 13.2B** [5]. One hypothesis for this common code might be that its universality is due to some optimal property it possesses, notwithstanding the known alternatives. The optimality might take a few forms, but one of the most obvious is robustness to mutation. Such robustness could be accomplished by having single base-pair changes in the code result in amino acids that are chemically similar. If the code had this form, mutations would often create new proteins that were chemically similar to the originals. And, in fact, the universal genetic code is organized in this fashion [6, 7]. However, it is not unique in this property: there are other possible genetic codes that are even better at minimizing the costs of single-base pair mutations [8]. And trivially, we could make a genetic code with identical error minimization properties to the universal code by simply applying the base-pairing rules to it: the DNA complement of our code should work identically, yet it has never been seen on Earth. We need to find another explanation for the universality of the genetic code.

What Darwin realized, as he explains in the chapter epigraph, is that these many, many cases of similarity are explicable only if all of the different forms of life on the planet have evolved from a single and universal common ancestor. In this view, the shared genetic code is a relic of the common ancestry of life: it was chosen early in the history of life and has been kept ever since.

Representing Common Ancestry

Searching for a visual representation for this idea of the evolutionary relationships between the various species, Darwin used a tree (see Figure 13.1), drawing on a long history of representing familial relationships in this form [9]. It is surprising how useful and durable this visual metaphor has turned out to be. Thus, while it is true that a branching tree does not perfectly represent the story of the evolution of species, it works very well. The main reason for its utility as a representation is that, over evolutionary time, populations and species primarily *split*. In other words, an interbreeding population of individuals becomes divided into two populations that interbreed among themselves but where interbreeding *between* the two populations has become rare or nonexistent [10]. By and large, once such interbreeding has stopped, it is unlikely to start again, leading to the branching behavior of Figure 13.1. Now, it is not difficult to think of exceptions and qualifications to this claim. But rather than dwelling on them, let us explore the usefulness of this branching representation and where it leads us as quantitative biologists.

The Dream: The Tree of Life

Figure 13.3 is an incomplete first attempt to produce the object Darwin had in mind: a complete representation of the relationships between all the different forms of life on Earth. I call it incomplete not just because it lacks the vast majority of the at least 10 million species that are believed to live on our planet [11, 12]. More seriously, the species it does depict are biased toward plants and animals, omitting the huge diversity of prokaryotic organisms. Those organisms in fact can be divided into bacteria and archeons, groups as different from each other as either is from us [11, 13]. What is striking about Figure 13.3B is that, for all of this incompleteness, we still understand our human species to be one very small twig on an enormous tree. The remainder of this chapter will consider the mathematical, statistical, and computational problems associated with realizing the goal of fully recovering the tree of life.

THINKING ABOUT TREES

Before we delve into the problem of inferring evolutionary trees, we should spend a moment considering them purely as representational objects because they have some tricky properties. We can think of the most important of these properties as the *transitivity* of statements about relatedness. You may remember from high school that we describe addition as a transitive operation because $A + B = B + A$. In the same way, if we claim that the closest living relative of species A is species B, it is also true that the closest living relative of B is A. From the point of view of a tree diagram, transitivity is equivalent to saying that *rotation* around the nodes of the tree does not change the tree itself. Figure 13.4 illustrates this property and its results. The figure depicts four tree diagrams; however, these four diagrams represent only two distinct sets of evolutionary relationships. In that figure, panels 13.4A, 13.4B and 13.4C are in fact the same evolutionary tree: they differ by the rotations indicated. Figure 13.4D depicts a different set of evolutionary relationships, and no number of rotations can produce Panel D from any of the other three trees. Figure 13.4 also provides some terminology regarding these trees that we will return to later. In particular, we will have a good deal to say about the *branches* of the trees; we will also need to know that the *tips* of the tree represent the (usually) extant species that we would like to know the evolutionary relationships among. Fortunately, we will commonly have DNA sequences from those species that we will be able to use for this inference problem.

Another, more subtle problem the transitivity in trees introduces is an unconscious bias in where certain taxa are drawn. Figure 13.5 provides an example of this problem, where humans are drawn to the extreme left of the tree, producing the unconscious perception that they are more evolved than other organisms in

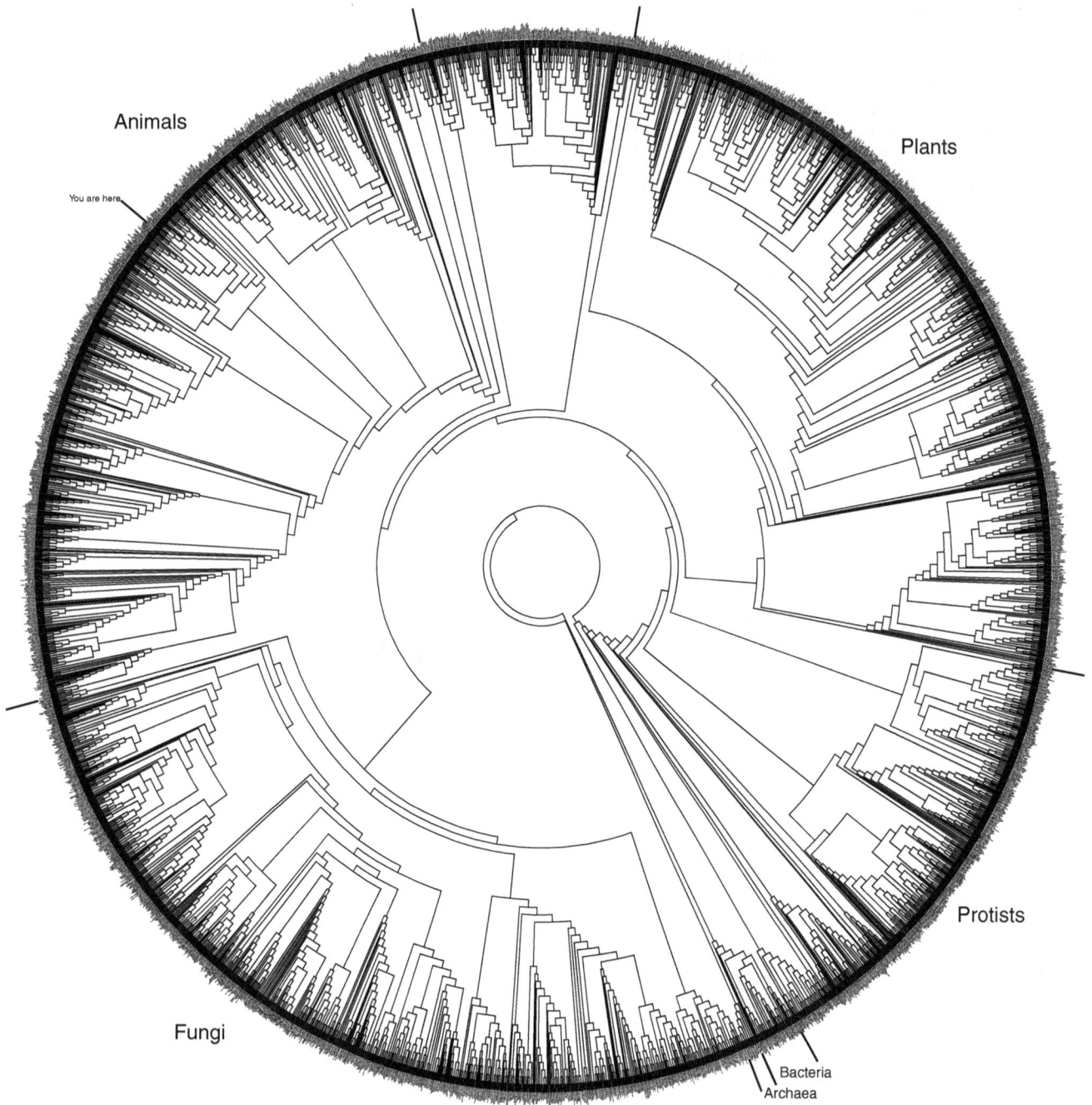

Animals

Plants

You are here

Fungi

Protists

Bacteria
Archaea

Figure 13.3 Humanity's place. The human species is a small twig on an enormous tree of life, a small part of which is depicted here [42]. (Special thanks to David Hillis for creating and sharing this illustration http://www.zo.utexas.edu/faculty/antisense/downloadfilestol.html).

the tree [14]. Of course, the rotations in Figure 13.5B demonstrate the fallacious nature of that presumption, since in this different representation of the same tree topology there is nothing special about the placement of humans. The problem is not limited to humans: biologists have worked extensively in "model" organisms such as *Drosophila melanogaster, Arabidopsis thaliana, Saccharomyces cerevisiae,* and *Caenorhabditis elegans.* In my own research on such species, I have often found myself drawing trees with those species depicted at the left of their respective trees, for no reason other than instinct. The only moral I can offer is that if your understanding of a biological problem changes upon tree rotation, then that understanding is wrong.

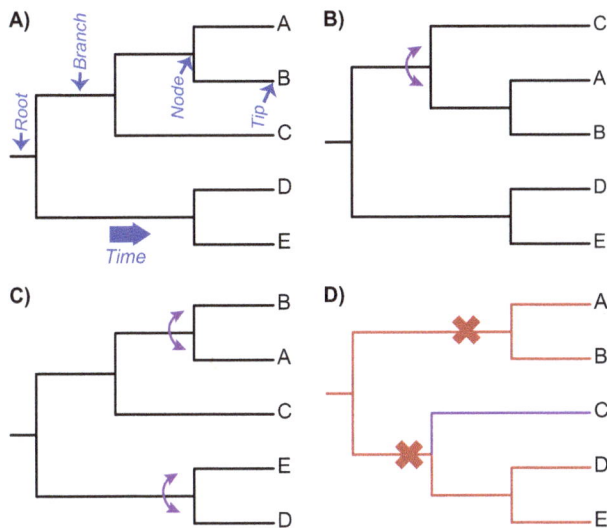

Figure 13.4 Complexities and terminology for the representation of evolutionary relationships as a tree. (**A**) A reference tree topology. Time on this tree increases to the right. It is composed of *branches* spread from the *root* (origin) of the tree. The *tip* taxa (**A–E**) are alive today, while *nodes* represent speciation events splitting extant species. (**B–C**) Two alternative depictions of the topology in **A** that differ by one **B** or two **C** rotations. (**D**) An alternative set of evolutionary relationships for these five taxa that differs in the placement of "C." Notice that even though the species are listed in the same order down the page in both **A** and **D**, the two topologies are not the same because no rotations of **A** can produce **D**.

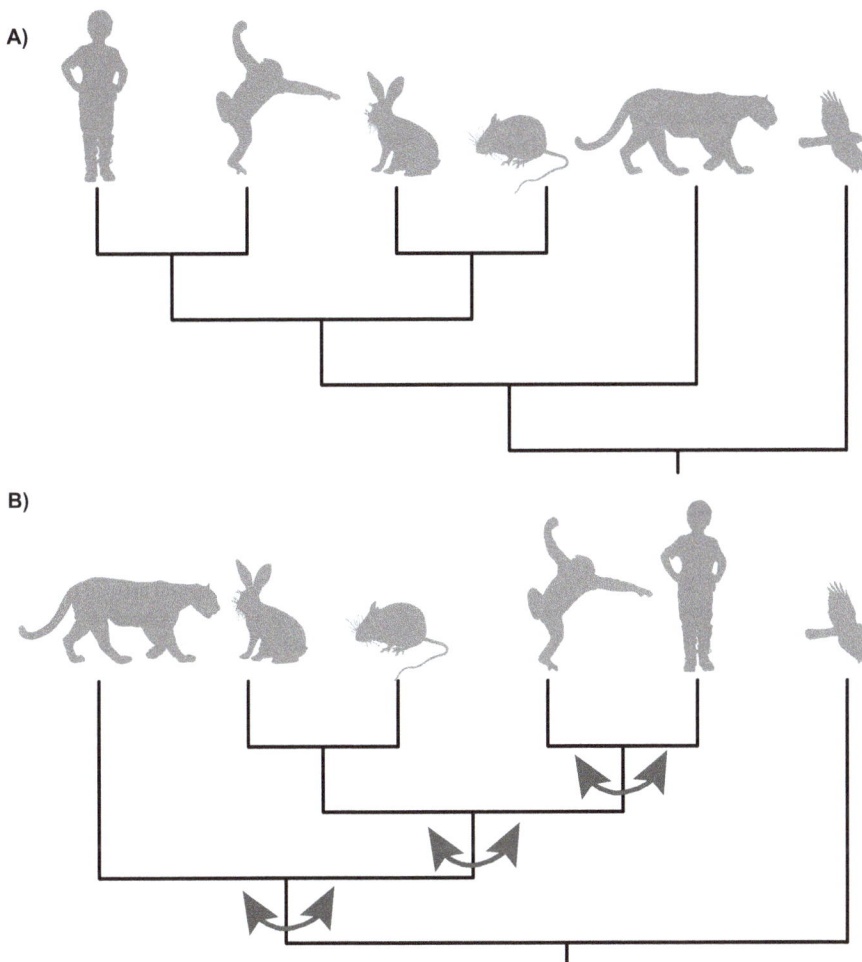

Figure 13.5 Choices in the depiction of the same evolutionary tree. (**A**) Often humans are shown in a "privileged" location on the right or the left of the diagram. That illustration scheme can promote the fallacy that mountain lions are more closely related to birds than to humans. (**B**) In this panel I show how three rotations of the tree in **A** produces a visualization that emphasizes that all mammals are equally closely related to birds. Silhouette references [43–48]. (Art courtesy of NIAID.)

Evidence for Reconstructing Evolutionary Trees

That the structures of organisms might give evidence of the relationships between them was, curiously enough, appreciated even before the idea that such relationships might derive from evolutionary history was accepted [15]. For obvious reasons, in the century that followed Darwin, there was a great deal of work on using animals' body structures (especially skeletal elements) to infer the evolutionary relationships between the various extant and extinct species [10]. However, in this book, we are going to consider instead a problem that appeared once it became possible to obtain first protein [16] and later DNA [17] sequences from living (and occasionally extinct [18]) organisms. *Can we use the changes in DNA sequences over time to infer the evolutionary relationships of their possessors?*

Having chosen to use DNA sequences in our evolutionary tree search, we can see how the rules of evolution by genetic drift from Chapter 11 and the work on sequence alignment from Chapter 12 will come together. We will start by trying to understand the meaning of a single column of a sequence alignment from an evolutionary perspective. Under the hypothesis that our alignment algorithm has been performed successfully, we will believe that all the DNA bases in that column are homologous to each other. We further know that the common ancestor of all the sequences in that alignment possessed a DNA base pair at that alignment position. However, we do not (usually) know which particular nucleotide was present at that position in the ancestor. We can nonetheless argue that any differences between the sequences within that column represent *substitution* events that have occurred since that common ancestor. For simplicity, we can assume that those substitutions happen through the fixation of new mutations by genetic drift, just as we simulated in Chapter 11.

First, notice that, just like our simulations, these substitutions are occurring in populations. **Figure 13.6** gives a hypothetical example of the evolution of one alignment column from a single ancestral population that eventually splits into three daughter species. In the real world, we will generally not know the states of the ancestral populations (shown in gray), but having them is useful for this illustration. In the initial ancestral population, any substitutions that occur by drift (or selection) will be inherited by all the descendent populations or species. In the example in Figure 13.6A, species 2 has a "T" because that was the last substitution that occurred prior to the split of species 1–3, and species 2 experienced no further substitutions. On the other hand, both species 1 and 3 *do* experience substitutions due to drift, which start as mutations and then drift to fixation. Recall that these substitution events will occur in such a population on a timescale of perhaps hundreds of generations.

Speciation is a hugely complicated topic in evolutionary biology [10], and we will not even try to delve into its many complexities. Instead, we will describe it as the result of a single ancestral population becoming subdivided into two populations that no longer exchange genetic material. A useful conceptual example would be the division of the ancestral range of the population into two regions that

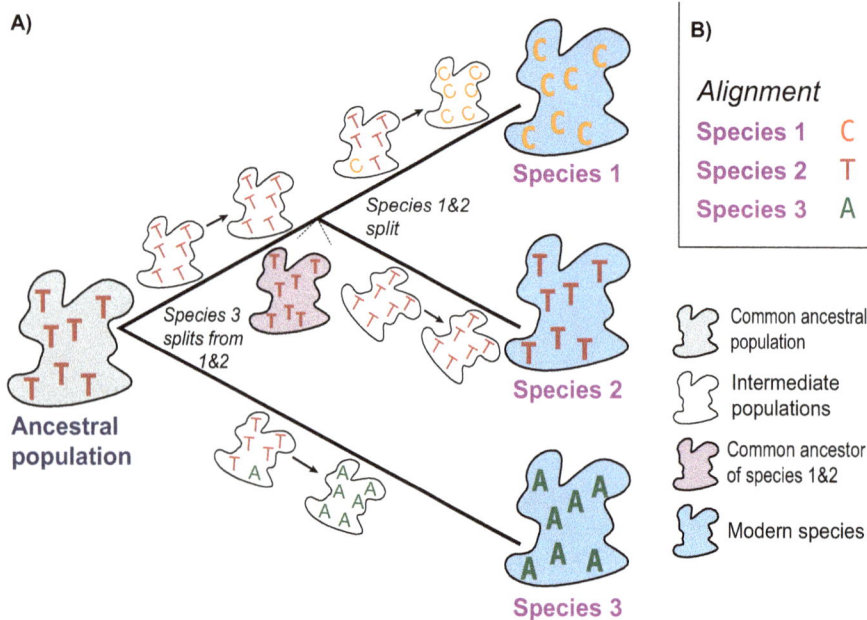

Figure 13.6 Conceptual view of the evolution of a single position from a sequence alignment in three species. (**A**) Although we cannot observe it, the position started as a *T* in the common ancestral population (gray). Once species 3 split from the common ancestor of species 1 and 2, the position undergoes a substitution converting the *T* to an *A* in that lineage. As illustrated, this substitution was a multistep process, with first a mutation that then drifted to fixation by the rules we saw in Chapter 11 (populations shown without shading). In the lineage that is the common ancestor of species 1 and 2, no change occurs, and the *T* is retained (pink). The internal node in the tree illustrates this intermediate population with its retained base *T*. Once species 1 and 2 split, species 2 preserves the *T*, while species 1 undergoes a substitution to a *C* that occurs in the same manner as discussed for species 3. (**B**) The sequence alignment we would observe in the three extant species from the evolutionary process in **A**.

are not connected, as might occur if, for instance, a mountain range arose in its center. Notice that the timescales on which these types of events are occurring will be much longer than the hundreds of generations required for genetic drift to fix a new mutation.

Once such a division has occurred, each population will continue to experience genetic drift just as the ancestor did. However, once a pair of species split, any new changes that occur are *independent* and *specific* to only one of those lineages. This independence is why the change from a "T" to a "C" is limited to only species 1. It is the combination of shared and independent changes that lets us use DNA sequences to infer trees. Even though we saw in Chapter 11 that the process of the fixation of a new allele requires time, here we will follow Chapter 12 in assuming that this fixation time is rather short relative to the speciation times we are interested in. Doing so allows us to treat substitutions as point events in the models that follow.

Data and Trees: Unrooted Trees

How might we make inferences of such trees? Looking at Figure 13.6, we can see that if we were to consider different evolutionary relationships between these three species, the set of possible shared and independent base changes implied by those different trees would also differ. We could therefore ask whether any particular tree is a better fit for the sequences we have than are the other trees. In a moment, we will describe some approaches for asking how well each evolutionary tree fits a given set of DNA sequences, allowing us to conduct a search for the best tree.

Unfortunately, to explore a concrete example of this idea, we need to make a slightly perplexing change to our trees, which we will now start to call *phylogenetic trees*. In Figure 13.6 we have represented the common ancestor of all the species in the tree as the *root* of the tree and assumed we knew the sequence that was present at that common ancestor. Of course, for real analyses, we do not know this ancestral sequence. Moreover, because we are using DNA bases as our data, we must assume there is no directionality in the pattern of sequence changes. Another way of saying this is that if we were to watch these sequences evolve for a very long time, we would not observe that any of the bases are accumulating or disappearing in that time. Instead, we will assume that:

$$P(X \rightarrow Y) = P(Y \rightarrow X)$$
$$\text{for all } X \text{ and } Y$$

(13.1)

In other words, we assume that over time an A substitutes for a T as often as a T substitutes for an A.

The combination of these two factors means that the methods for phylogenetic tree inference we are about to discuss will return to us *unrooted* phylogenetic trees (Figure 13.7B) rather than the rooted trees of Figure 13.7A. Essentially, an unrooted tree is just a rooted tree from which we remove the root and then unbend the two branches this root connects (Figure 13.7C). While unrooted trees feel much less familiar than do rooted ones, you should not let this detail trouble you too much. We will discuss how to convert those unrooted trees to rooted ones in just a bit. First, however, we show some examples of inferring trees from DNA sequence alignments (Figure 13.8).

MAXIMUM PARSIMONY AND TREE INFERENCE

Our first approach to the evolutionary tree inference problem will be what is known as the *maximum parsimony* inference method. As described by Joseph Camin and Robert Sokal in 1965 [19], given all the possible unrooted trees relating n species, we will prefer the topology that requires the fewest inferred nucleotide substitutions to explain our DNA sequence alignment. Figure 13.8 illustrates an example of this analysis for a simple case of four species and six aligned DNA bases.

In that figure, we can treat one of the sequences as *ancestral* for the purposes of the computation. We then find the set of inferred sequences at each of the internal nodes (shown in gray) that allow the remaining three observed sequences to evolve from that assumed ancestor in the minimum number of inferred substitutions (the red arrows). (Notice that the assumption of one ancestral sequence is purely a convenience: if you pick a different sequence and use the same approach, you will get the same scores; Exercise 13.1.) If we repeat this computation for all the possible unrooted trees, we will generally find that one of the trees requires fewer substitutions than do the others. We can then say that the tree that requires the fewest changes explains the data in the most *parsimonious* way possible and is hence called the *maximum parsimony tree*.

Exercise 13.1

Confirm that treating species #2 as the ancestral sequence for Figure 13.7 generates the same set of parsimony scores for the three possible trees.

Looking at Figure 13.8, you will probably wonder how to construct an algorithm like those in the last chapter to take a particular tree and set of aligned bases and compute the scores given. While the algorithm for doing so is not particularly difficult [20], we are not going to discuss it. There are two reasons for this omission. First, while the maximum parsimony approach superficially appears to be an ideal solution to our evolutionary tree inference problem, it actually has some failings. Second, later in this chapter we will learn another computational approach to tree inference, namely *maximum likelihood*. When we dissect that approach, we will find, somewhat surprisingly, that maximum parsimony can be viewed as a special case of it. As a result, the algorithm we develop for maximum likelihood inference will also allow maximum parsimony inference, allowing us to forgo the parsimony-specific algorithm.

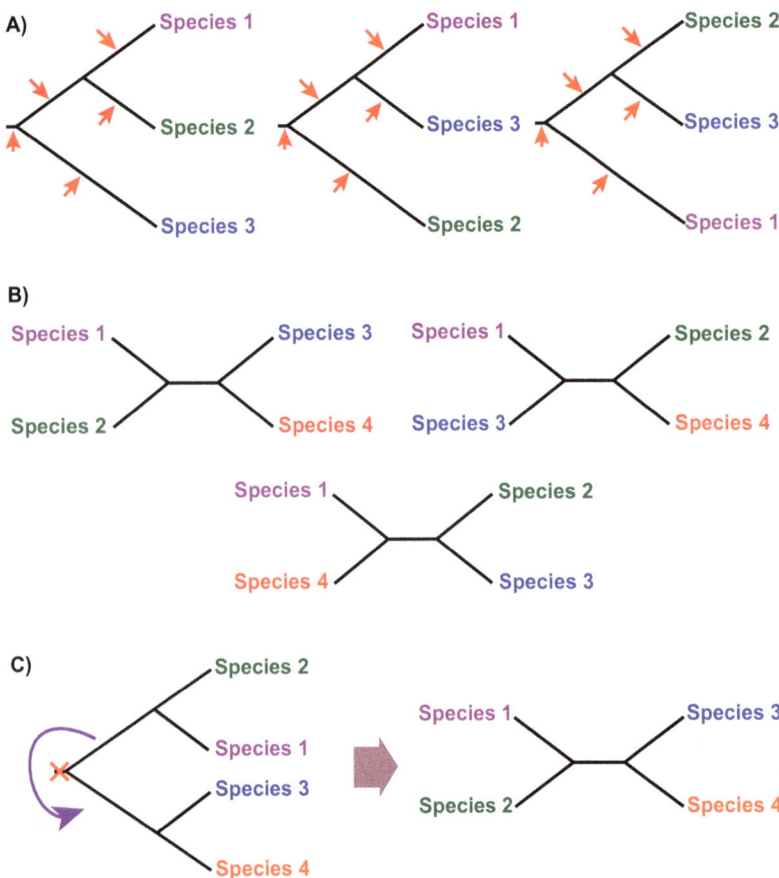

Figure 13.7 Possible evolutionary trees. (A) Following the example of Figure 13.6, we can propose three different sets of evolutionary relationships among three species. Notice that any other apparent relationships one could think of are simply rotations of the three *topologies* depicted here (see Figures 13.4 and 13.5). (B) When considering DNA sequence evolution, we must consider *unrooted* trees (see text). For the four species depicted here, there are only three possible unrooted topologies: again, other apparent topologies are simply rotations of these three. (C) Converting a rooted tree to an unrooted tree. The root branch is removed and the two branches it intersects are "straightened" into a single one.

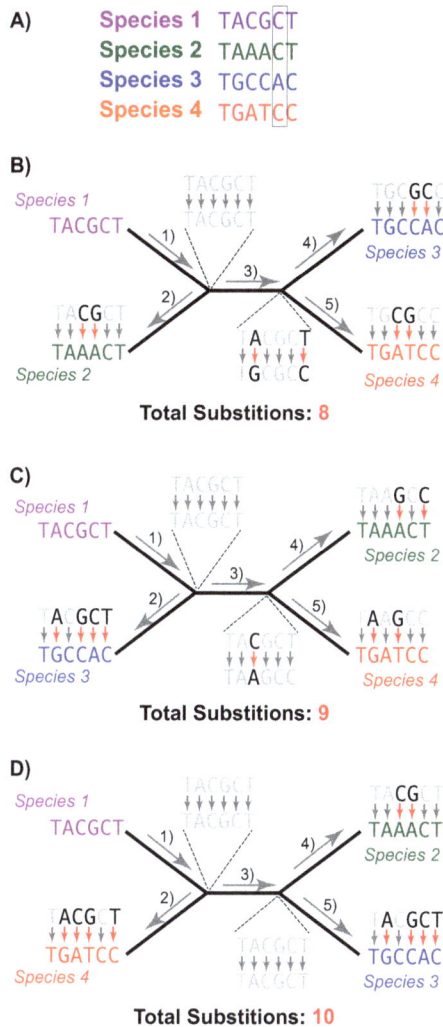

A)

Species 1 TACGC̲T̲
Species 2 TAAAC̲T̲
Species 3 TGCCA̲C̲
Species 4 TGATC̲C̲

B)

Total Substitions: 8

C)

Total Substitions: 9

D)

Total Substitions: 10

Figure 13.8 **Maximum parsimony approaches to tree inference. (A)** A sequence alignment for six DNA bases and four species. (**B**) We can treat species 1 as the ancestral state without a loss of generality because we assume the substitution process is time reversible. Hence, the inferred sequence at the first internal node can always be assumed to be the sequence of species 1 (arrow #1). Thereafter we propose a sequence at the second internal node (after arrow 3) that minimizes the changes we need to place after arrows #4 and #5. We then make any needed changes for #2 as well and count all changes. We then apply the argument from **B** to the other two possible trees (**C** and **D**).

What are the problems with parsimony? One of them is hinted at in Figure 13.8. If we look at the second-to-last column in the alignment (boxed), we see that three of the species share the same base and one of them has a different base. Under the parsimony algorithm, this column adds the same number of substitutions (namely one) to all three trees and hence is not *informative* regarding the tree under this approach. In fact, in a standard alignment, most of the columns will not be parsimony informative, meaning that this algorithm tends to ignore most of the sequence data it is given, which is at least disconcerting.

Another problem with parsimony is that several researchers have identified pathological cases where DNA sequences known to have evolved on one tree will cause parsimony to report a different and incorrect tree even when those data are unlimited [21, 22]. Hence, our search for an approach to finding trees from sequences should probably continue. However, before we go too much further, it is worth considering the computational scale of the problem we have set ourselves up for.

THE SET OF POSSIBLE TREES

How many different evolutionary trees are there for n species? In Figure 13.7A we saw that, for three species, there are only three possible phylogenetic trees (phylogenies). Once again, we should refer to Figure 13.4 to remind ourselves that any other apparent trees are simply rotations of these three.

However, while the problem looks very manageable for this small number of species, it quickly gets rather out of hand. To see why, we need to consider two numbers. The first is the number of branches for a tree with n species. Consulting Figure 13.7 and returning to rooted trees for a moment, we can convince ourselves that, in addition to the n branches corresponding to the n extant species, there are $n-1$ internal branches. There are therefore a total of $2n-1$ branches in a tree of n taxa [23]. If we look at Figure 13.7A, we can see that, were we to add a new species to one of these trees, we could do so at any one of those $2n-1$ branches. Hence, if there are three possible trees for three species, and we can add a fourth species at any one of $2n-1=5$ branches for each tree, there will be 15 (3×5) trees for 4 species. We can generalize this idea to write that the number of possible trees $T(n)$ for n species is given by:

$$T(n) = \prod_{k=2}^{n} (2k-1) \tag{13.2}$$

This product is what is known as a "double factorial": in other words, a product of the form of $n \cdot (n-2) \cdot (n-4) \ldots \cdot 1$. In this case, we know that $2n-1$ is always odd, and as result, Equation 13.2 can reduced to [23]:

$$T(n) = \frac{(2n-1)!}{2^{n-1} \cdot (n-1)!} \tag{13.3}$$

Exploring a few sample values for Equation 13.3 can be informative. For 8 species there are slightly more than 2 million possible trees; for 12 species there are more than 300 trillion trees; and for 20 species there are 3.2×10^{23}—half a mole of trees!

What about unrooted trees? From the perspective of topology, the only difference between a rooted tree and an unrooted one is that the rooted tree has an additional "pseudo" species at that root branch. Hence, if we substitute $n-1$ for n in Equation 13.3, we find the number of unrooted trees $T^u(n)$ is:

$$T^u(n) = \frac{(2n-3)!}{2^{n-2} \cdot (n-2)!} \tag{13.4}$$

Put more simply, there are as many possible rooted trees for $n-1$ species as there are unrooted trees for n species [23]. In Exercise 13.2, you can compute the number of unrooted trees for 10 taxa.

The implications of these numbers are very serious. The maximum parsimony approach just discussed assumes that we will compute the minimum number of required substitutions to explain the sequence alignment on every possible unrooted tree. The running time of the maximum parsimony computation for a single tree is simply $O(n)$—that is, linear in the number of bases in the alignment [20]. However, because we need to run that linear-time computation on every possible tree, the running time of the overall algorithm is proportional to Equation 13.4. In practice, therefore, any reasonably large phylogenetic inference problem will need to use computational tricks to avoid having to consider all the possible tree topologies [20, 24–27]. These tricks can be quite clever but are beyond our scope here. We should remember, however, that such methods cannot guarantee that the best tree (under whatever approach is being used) has been found.

EVOLUTION MEETS STATISTICS: MAXIMUM LIKELIHOOD PHYLOGENETICS

Maximum parsimony seems like a such a self-evident way to find the optimal evolutionary tree that you might be wondering how we could improve on it. Rather than try to answer that question right away, we are going to approach it obliquely by doing something we have done many times in this book. We will build a model. In this case we will model the process of base-pair substitutions in a DNA sequence over evolutionary time. To do so, we will start with the insights we generated from our population models in Chapter 11 and then extend our time horizon.

Recall that in Chapter 11 we saw that any currently variable DNA position in a population will tend to lose that variation at a rate inversely proportional to the population size. In Chapter 12 we extended that idea to the concept of the fate of a new mutation, concluding that almost all mutations are lost through genetic drift. However, on average, a fraction of $1/2N$ of them will become fixed.

A Model of DNA Evolution

For this new model we are going to make the assumption that we are interested in processes that are happening on rather long timescales relative to the population-level events of Chapter 11. Is this assumption reasonable? Not entirely, but if we consider, say, the divergence between humans and chimpanzees, it looks plausible. Our two species are believed to have split from each other approximately 6 million years ago [28]. The population size for both species is around 10,000 and generation times are (generously) 20 years. From these two values we can see that the population-level fixations of new mutations by genetic drift are indeed happening rapidly relative to the long divergence between the two species.

So, given this assumption, we will treat the substitution of one DNA base for another in time as being discrete events. We expect such events to be very rare: we have already developed a model for a process such as this one: the model of a cancer cell dying from the radiation gun of Chapter 8. In fact, if all we were interested in were the *change* of a DNA base, that probability would follow Equation 8.20:

$$P(t) = 1 - e^{-\lambda t} \tag{13.5}$$

Of course, time is now measured in millions of years, but otherwise the behavior is similar.

However, we want to know more than just if a change has happened. For a given time t, we would like to know what the *probability* of a given DNA letter being at that position is. We can see that we will need four equations to answer this question, one for each of the bases. We can write the rate of change of those four probabilities with the following four differential equations:

$$
\begin{aligned}
\frac{dP_A}{dt} &= -\mu \cdot P_A + \frac{1}{3}\mu \cdot P_C + \frac{1}{3}\mu \cdot P_G + \frac{1}{3}\mu \cdot P_T \\
\frac{dP_C}{dt} &= \frac{1}{3}\mu \cdot P_A - \mu \cdot P_C + \frac{1}{3}\mu \cdot P_G + \frac{1}{3}\mu \cdot P_T \\
\frac{dP_G}{dt} &= \frac{1}{3}\mu \cdot P_A + \frac{1}{3}\mu \cdot P_C - \mu \cdot P_G + \frac{1}{3}\mu \cdot P_T \\
\frac{dP_T}{dt} &= \frac{1}{3}\mu \cdot P_A + \mu \frac{1}{3} \cdot P_C + \frac{1}{3}\mu \cdot P_G - \mu \cdot P_T
\end{aligned}
\tag{13.6}
$$

While Equation 13.6 may look a trifle intimidating, it is not that different than the SIR and predator–prey models we considered at the beginning of the book. We can start with the parameter μ that appears in each equation. That value is the mutation rate and comes directly from Equation 12.1 of the previous chapter. Remember that we are assuming that we are considering neutral fixations, which are occurring at a rate μ in the population, independent of its size. In other words, although I have referred to μ as the substitution rate, in a population experiencing drift, that substitution rate is equal to the mutation rate μ, as we saw in Chapters 11 and 12.

Next, notice that if we sum up all four equations, we get 0, meaning that probability is conserved in these equations. In other words, at any given time, there must be *one* of the four bases present at that position.

Given these two facts, we can focus on dP_A/dt and try to understand what the equation is telling us. $P_A(t)$ is increasing in three ways, namely through mutations to the base A from the bases C, G, and T, each one occurring at a rate proportional to $\frac{1}{3}\mu$, since one out of three of the mutations/substitutions from those bases will give us an "A". On the other hand, we are losing probability from $P_A(t)$ through substitutions to the other three bases at rate μ, assuming the base was an "A" to start with.

Solutions to the Transition Probabilities

Unlike most of the systems of differential equations we have considered, Equation 13.6 actually has an analytic solution. Reaching that solution is slightly involved, so I have left that derivation to the Appendix. However, our work on mutations in Chapter 7 and the fact that all of the terms in Equation 13.6 are linear with respect to P_A, P_C, P_G, and P_T suggests that the solutions to Equation 13.6 will be some sort of exponential functions. Before we can examine those solutions, we need to recall that, as we have seen before, our solutions will only become definite if we apply initial conditions to them.

For this problem, the initial conditions are in fact rather intuitive: we need to know the probabilities for seeing each of the four bases at time t, given that we know which base was present at time $t = 0$. For the example of a base that was initially an A, we can express such probabilities with the notation $P_{A \to C}(t)$. We would read $P_{A \to C}(t)$ as the probability of an A transitioning to a C by time t. We can then express the solutions of Equation 13.6 as a 4×4 matrix:

$$
\begin{array}{c}
\phantom{P_{A \to X}} \quad\quad X=A \quad\quad\quad X=C \quad\quad\quad X=G \quad\quad\quad X=T \\
\begin{array}{c}
P_{A \to X} \\[6pt]
P_{C \to X} \\[6pt]
P_{G \to X} \\[6pt]
P_{T \to X}
\end{array}
\left[
\begin{array}{cccc}
\frac{1}{4}+\frac{3}{4}e^{-\mu t} & \frac{1}{4}-\frac{1}{4}e^{-\mu t} & \frac{1}{4}-\frac{1}{4}e^{-\mu t} & \frac{1}{4}-\frac{1}{4}e^{-\mu t} \\[6pt]
\frac{1}{4}-\frac{1}{4}e^{-\mu t} & \frac{1}{4}+\frac{3}{4}e^{-\mu t} & \frac{1}{4}-\frac{1}{4}e^{-\mu t} & \frac{1}{4}-\frac{1}{4}e^{-\mu t} \\[6pt]
\frac{1}{4}-\frac{1}{4}e^{-\mu t} & \frac{1}{4}-\frac{1}{4}e^{-\mu t} & \frac{1}{4}+\frac{3}{4}e^{-\mu t} & \frac{1}{4}-\frac{1}{4}e^{-\mu t} \\[6pt]
\frac{1}{4}-\frac{1}{4}e^{-\mu t} & \frac{1}{4}-\frac{1}{4}e^{-\mu t} & \frac{1}{4}-\frac{1}{4}e^{-\mu t} & \frac{1}{4}+\frac{3}{4}e^{-\mu t}
\end{array}
\right]
\end{array}
\tag{13.7}
$$

Equation 13.7 is a transition probability matrix. It is quite similar to the transition probability matrix we saw for the Fisher–Wright model in Chapter 11, except that time is now measured continuously rather than in discrete units of generations. In Exercise 13.3, you can take the derivative of 13.7 and confirm that you indeed recover 13.6.

The entries of this new transition probability matrix give the probability of starting with the DNA base on the left (row) and end at the base in a particular column. It is instructive to consider the case of $\mu t = 0$ for these equations. There, we find that $P_{X \to X}$ is 1.0 in all cases, and $P_{X \to Y}$ is 0 when $X \neq Y$. In other words, if there has been no time for a substitution to occur, the probability of maintaining the same base is 1.0. This particular model of transition probabilities is known as the Jukes–Cantor model, after its proposers [29].

Using the Jukes–Cantor Model

In Figure 13.9, I show an example of computing the probability of the transition from three bases to three other bases. In the example, we allow the combination of the divergence time t and the mutation rate μ to be $\mu t = 0.3$ units (we will explore the meaning of these *branch lengths* later in the chapter). You will notice that the chance of the base having a T at both the beginning and end of the branch is about 80.6%, while the chance of changing from that T to any other base is about 6.5%. Since $0.065 \times 3 = 0.195$, or 19.5%, we can see that the total probability of the four events $\left(P_{T \to T}, P_{T \to A}, P_{T \to C}, \text{and } P_{T \to G} \right)$ is 1.0, just as it should be.

Multiple Hits

One of the reasons maximum parsimony shows some of the problems it does is that it always "explains" the data in as few substitutions as possible. This approach seems highly sensible at first. However, what if the true evolutionary process saw an A change to a G and then back to an A? That sequence of events might seem so unlikely as to be safe to ignore. However, there are a couple of problems with that argument. First, we will be analyzing long sequence alignments, so we will need to account for even rare events. The second reason is hinted at in the equations

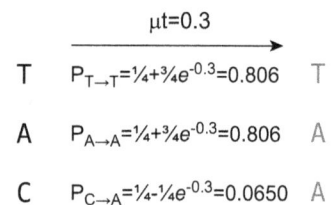

$$\mu t = 0.3$$

T $P_{T \to T} = \frac{1}{4} + \frac{3}{4}e^{-0.3} = 0.806$ T

A $P_{A \to A} = \frac{1}{4} + \frac{3}{4}e^{-0.3} = 0.806$ A

C $P_{C \to A} = \frac{1}{4} - \frac{1}{4}e^{-0.3} = 0.0650$ A

Figure 13.9 Computing the probability of one sequence evolving (gray) from another (black) using the equations comprising Equation 13.7. We assume that the combination of mutation rate and time separating the two sequences is $\mu t = 0.3$.

comprising Equation 13.7. As we let μt get very large, we see that the four equations (e.g., again, $P_{T \to T}, P_{T \to A}, P_{T \to C}$, and $P_{T \to G}$) all approach an equal ¼. Now, were we only thinking about single substitutions, this result seems a trifle too neat. On the other hand, we should consider what it means to have μt be very large. That condition would imply that every position in the sequence had undergone so many substitutions that it had effectively forgotten which base it started as. Under those circumstances, it makes sense that the chance of observing any base would be equal at each position.

Biologists refer to this problem of many substitutions having occurred as the *multiple hits* problem. As implied by the results above, our transition probability matrix accounts for this problem for us. In other words, the $P_{A \to C}$ in Equation 13.7 gives *not* the probability of starting with an A and seeing a substitution to a C, but rather the probability of observing a C at time μt, given that the starting base was an A. Wrapped up in that probability is the probability of all possible substitution paths (even very long ones) that could result in that observed C at time μt.

A mentor of mine, Paul Lewis, has explained this idea with the metaphor of a parking lot [30]. Imagine that you take a photo of a particular space in a parking lot, which happens to be occupied by a blue car, and then leave for lunch. When you return, you photograph a red car in that same position. (For argument's sake, we assume that there are only four car colors in this lot and that the cars are indistinguishable save for their colors.) The idea of multiple hits is simply to say that you do not know if, while you were away, the blue car stayed for a while and then left, with the red car immediately taking its place, or if, after the blue car left, a yellow one took its place that you never saw before the red car arrived. In this metaphor, even if you came back to find a blue car in the slot, you cannot know for certain that it is the same blue car that was there when you left. The value of Equation 13.7 is that these equations account for all of this complexity all on their own.

The Meaning of Time

Two questions that will likely have occurred to you about Equation 13.7 is how we are going to measure time and why in the model that time is confounded with the mutation rate μ (see Figure 13.9). In a few cases, we may have some knowledge of the time at which a pair of species split from fossil data. However, that knowledge is unusual and generally imprecise. Equally importantly, we cannot distinguish with the model of Equation 13.7 between two sequences that have diverged for one million years with a mutation rate μ_1 and two sequences that have diverged for two million years with a mutation rate $\mu_2 = 1/2\,\mu_1$. Hence, for the phylogenetic problems we are about to consider, we will always treat time and the mutation rate as a single parameter of the model. More importantly, rather than needing to know the values of μt from external sources, we will see that it is a parameter that we can estimate as part of the tree inference process.

COMPUTING THE LIKELIHOOD OF A PHYLOGENETIC TREE

Having considered the problem of multiple hits and of the confounding of time and mutation rate, we can now develop an approach that will allow us to use Equation 13.7 to compute the probability or *likelihood* of a given sequence alignment having evolved on a particular phylogenetic tree. I will note that, for practical purposes, the words "likelihood" and "probability" are synonymous here: we use the term *likelihood* simply to note that we need not be concerned if we are off by the same constant factor in our computation of the probabilities of each of the possible trees [31].

Conditional Probability

The first step of our computation is to determine the *conditional probability* of the four DNA bases at the *internal* nodes of the proposed tree. **Figure 13.10** presents

A)

$$P_{X \to X}(\mu t) = \tfrac{1}{4} + \tfrac{3}{4} e^{-\mu t}$$

$$P_{X \to Y}(\mu t) = \tfrac{1}{4} - \tfrac{1}{4} e^{-\mu t}$$

N=	Prob
A	$P^N_A = P_{A \to T}(0.3) \cdot P_{A \to T}(0.2) = 0.003$
C	$P^N_C = P_{C \to T}(0.3) \cdot P_{C \to T}(0.2) = 0.003$
G	$P^N_G = P_{G \to T}(0.3) \cdot P_{G \to T}(0.2) = 0.003$
T	$P^N_T = P_{T \to T}(0.3) \cdot P_{T \to T}(0.2) = 0.70$

Species 1
T
$\mu t_1 = 0.3$
←N
$\mu t_2 = 0.2$
T
Species 2

B)

N=	
A	$P^N_A = P_{A \to G}(0.3) \cdot P_{A \to A}(0.2) = 0.056$
C	$P^N_C = P_{C \to G}(0.3) \cdot P_{C \to A}(0.2) = 0.003$
G	$P^N_G = P_{G \to G}(0.3) \cdot P_{G \to A}(0.2) = 0.037$
T	$P^N_T = P_{T \to G}(0.3) \cdot P_{T \to A}(0.2) = 0.003$

Species 1
G
$\mu t_1 = 0.3$
←N
$\mu t_2 = 0.2$
A
Species 2

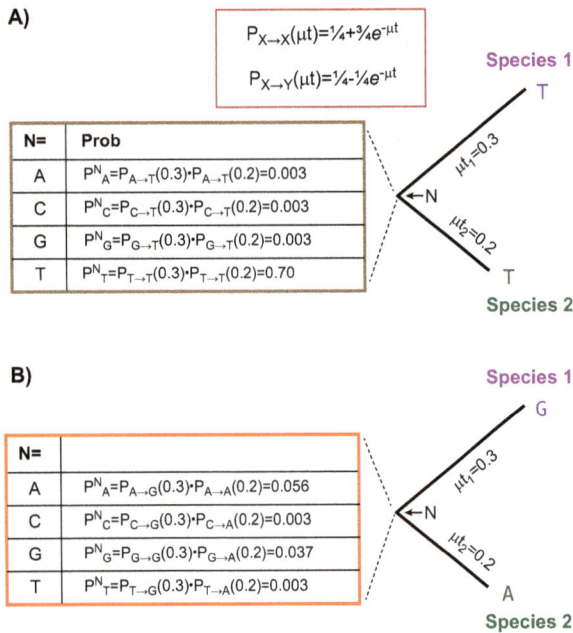

Figure 13.10 Computing *conditional* probabilities at an internal node *N* of a tree. Taking the two bases seen for a pair of "sister" taxa in the tree, we compute the probability of all four bases at the internal node by assuming that internal node has a particular base (e.g., *A*) and then computing the probability of lineage #1 transitioning from that *A* to the observed base in species #1 and similarly for lineage #2. The conditional probability of $A\left(P^N_A\right)$ is then just the product of those two probabilities.

two examples of this problem. We would like to know how probable it is that the internal node was occupied by each of the four bases. To compute these values, we can recall a basic fact from probability theory: the probability of two independent events A and B *both* happening is $P_A \cdot P_B$, namely the product of their individual probabilities. Thus, if we know the DNA bases at the two extant species 1 and 2 are Y and Z, the probability of base X at their shared internal node N is $P_{X \to Y} \cdot P_{X \to Z}$. When we extend this computation to larger trees, we will also have internal nodes that depend on other internal nodes, but this simply involves extending Figure 13.10 to use the conditional probabilities at the child nodes to compute those at the parent.

At this point it is worth comparing maximum likelihood to the parsimony approach from Figure 13.8. Notice that in Figure 13.8 we assumed we knew with certainty what the internal node states were (although in some cases an internal node might have more than one potential state). Essentially, with parsimony, we neglect internal states of lower probability while we retain these probabilities with likelihood-based approaches. Parsimony hence can be expected to perform poorly in cases where neglecting those probabilities is inappropriate.

The Likelihood Computation

With the conditional probability computation of Figure 13.10 in hand, we can now develop an approach for computing the overall likelihood L_i of a single alignment column for a given tree. To do so, we start at an arbitrary pair of tips in the proposed tree and descend down that tree, computing conditional probabilities at all internal nodes. Once we reach a point where we have computed all possible conditional probabilities and have only a single tip left in the tree, we compute the probability of the observed data, given the conditional probabilities at the nearest node and the transition probabilities for the branch leading to that last tip (**Figure 13.11**). Doing so gives us L_i: the likelihood of a single alignment column. The likelihood L of the entire alignment given a particular tree then is just the product of the likelihoods of column $L_1..L_n$:

$$L = \prod_{i}^{n} L_i \qquad\qquad (13.8)$$

A)

N=	Prob
A	$P^N_A=0.003$
C	$P^N_C=0.003$
G	$P^N_G=0.003$
T	$P^N_T=0.70$

Species 1 T

$\mu t_1=0.3$

Species 3 T $\mu t_3=0.18$ ←N

$\mu t_2=0.2$

Species 2 T

N=	Prob
A	$P^3_{\to A}=P_{T\to A}(0.18)\bullet P^N_A=0.0001$
C	$P^3_{\to C}=P_{T\to C}(0.18)\bullet P^N_C=0.0001$
G	$P^3_{\to G}=P_{T\to G}(0.18)\bullet P^N_G=0.0001$
T	$P^3_{\to T}=P_{T\to T}(0.18)\bullet P^N_T=0.610$

$\Sigma_1=0.611$

$$P_{X\to X}(\mu t)=\tfrac{1}{4}+\tfrac{3}{4}e^{-\mu t}$$
$$P_{X\to Y}(\mu t)=\tfrac{1}{4}-\tfrac{1}{4}e^{-\mu t}$$

B)

N=	Prob
A	$P^N_A=0.056$
C	$P^N_C=0.003$
G	$P^N_G=0.037$
T	$P^N_T=0.003$

Species 1 G

$\mu t_1=0.3$

Species 3 G $\mu t_3=0.18$ ←N

$\mu t_2=0.2$

Species 2 A

N=	Prob
A	$P^3_{\to A}=P_{G\to A}(0.18)\bullet P^N_A=0.0023$
C	$P^3_{\to C}=P_{G\to C}(0.18)\bullet P^N_C=0.0001$
G	$P^3_{\to G}=P_{G\to G}(0.18)\bullet P^N_G=0.032$
T	$P^3_{\to T}=P_{G\to T}(0.18)\bullet P^N_T=0.0001$

$\Sigma_2=0.035$

Likelihood $=\Sigma_1\bullet\Sigma_2=0.0214$

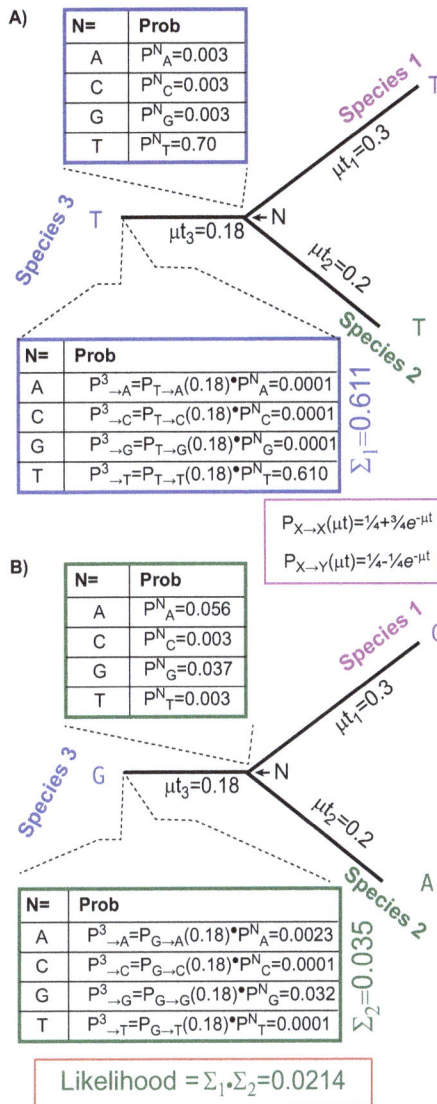

Figure 13.11 Computing the likelihood of two alignment columns (A and B) along a phylogeny of three species using the equations comprising Equation 13.7 and the conditional probabilities we computed in Figure 13.10. Starting from the four conditional probabilities at node *N*, we compute the probability of the transition from the known base for species 3 to those four intermediate bases. When we sum those four probabilities, we get the *likelihood* of that alignment column for that tree. The total likelihood of the tree is the product of the individual column likelihoods.

What Is a Likelihood?

At this point we should consider exactly what this new statistic *L* tells us. To do so, it is probably easier to use a simple problem as an example. In **Figure 13.12** we consider the problem of flipping a coin that we suspect to be unfair a number of times. Supposing we collect data from 1,000 flips of such a coin, yielding 635 heads and 365 tails. What is our best guess of the true probability of flipping a head (P_{head}), given these data?

We can model these flips with the binomial distribution from Chapter 8:

$$\frac{n!}{k!\cdot(n-k)!}\cdot k^{P_{head}}\cdot(n-k)^{(1-P_{head})} \tag{13.9}$$

In this case, we treat P_{head} as an unknown parameter of the model, meaning that in effect we have a whole family of similar models that differ in P_{head}. If we then compute the likelihood of our data for all these possible models, we find that letting $P_{head}=0.635$ gives us the *maximum* likelihood of observing 635 heads (see Figure 13.12C). That value is hence our best guess of the probability of a head from this coin. (We defer the question of whether this value of P_{head} is statistically different from 0.5 for another time.)

We can now abstract this concept to our problem here. Our data, instead of coin flips, are sequence alignments, and our models are the equations comprising

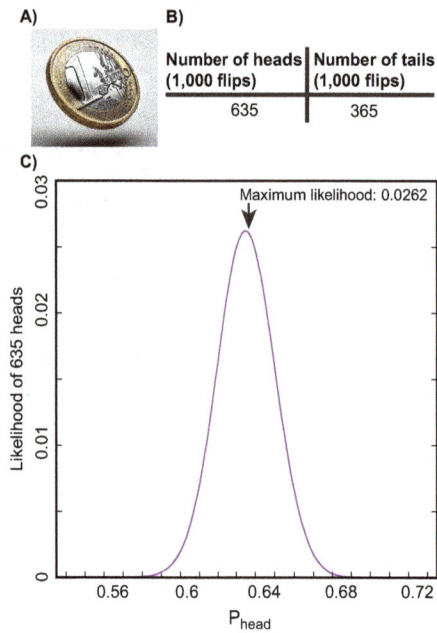

A) B)

	Number of heads (1,000 flips)	Number of tails (1,000 flips)
	635	365

C)

Figure 13.12 Building a likelihood model for the flipping of an unfair coin. This model allows us to compute the likelihood of our coin (A) generating a dataset consisting of 635 heads in 1000 flips (B) under models with different values of the probability of a head (P_{head}, C). If we observe 635 heads over 1000 flips we can use the binomial distribution to compute the probability, or likelihood, of observing those data for a large family of binomial models that differ in the value of P_{head}. When we do so, we find that the value of P_{head} giving the highest likelihood of generating 635 heads is $P_{head} = 0.635$.

Equation 13.7 *plus* our possible phylogenetic tree topologies. Our best estimate of the tree is then the tree that gives the maximum likelihood of observing the alignment, over all possible such trees.

Numerical Problems

It is probably obvious that the procedure just described will result in a likelihood that is very small: looking at Figure 13.11, we can see if we multiplied 0.611 by 0.035 50 times (corresponding to an alignment of 100 bases), the likelihood would be 3.2×10^{-84}. Most computers can, without modification, represent numbers no smaller than about 10^{-300}; once they get smaller, one has to use special techniques that require much longer computing times. For our problem, however, it is sufficient to simply take the natural log of the likelihood of each column. Since multiplication becomes addition in log space, we can then just *add* the log-likelihoods of each column to get the total log-likelihood of the tree. You can see that the natural log of 3.2×10^{-84} is -192.3, which is comfortably within the limits of the computer's numerical representation.

Branch Lengths

Another apparent problem with the computations we have made so far is that I have, without any useful explanation, assumed that I knew the three μt values for the tree in Figures 13.10 and 13.11. In a real analysis, such an assumption is unwarranted. The solution to this problem is conceptually simple but somewhat computationally expensive: we select the combinations of branch lengths for a given tree that maximizes the likelihood of that tree [32]. Though it is beyond the scope of this chapter, it turns out to be possible to compute the derivatives of the likelihood function with respect to the branch lengths [32], which then allows programmers to use approaches like Newton's method [33] to find branch lengths that maximize the likelihood for a given topology. When we search for the maximum likelihood tree, therefore, we are comparing the likelihoods of different topologies, and, for each such topology, we have found the set of branch lengths that gives us the highest likelihood of the data for that topology.

Tree Search

Conceptually, we have now solved the phylogenetics problem we originally posed. For every possible tree, we compute the (log of the) likelihood of the alignment given that tree, using branch lengths that maximize that likelihood. We then retain the tree of highest likelihood. Of course, in practice there are usually too many trees to search all of them, but there are approaches that give generally good performance by searching only part of the tree space [20].

Other Models

Equation 13.7 assumes that the probability of any DNA base changing to any other base is the same. Now, we happen to know that the DNA bases are not all equally chemically similar: C and T are *pyrimidines*, with a single nitrogenous ring in the nucleoside, while A and G are the larger *purines*, with two rings (see **Figure 13.13**). It is reasonable to think that a C changing to a T might disrupt the double helix less than a C changing to an A. Hence, we might prefer a model that allows the pyrimidine-to-pyrimidine changes and the purine-to-purine changes to occur more frequently than changes that involve a purine and a pyrimidine. We can modify Equation 13.7 to allow for such a difference, creating a new model that is known as the "Kimura 2 parameter" (or K2P) model [34]. If we also reflect that the four DNA bases might not be equally common in our data, we could also include the four *base frequencies*, resulting a model called the HKY model, after its three inventors (Hasegawa, Kishino and Yano [35]). The differential equations for the HKY model are:

$$\frac{dP_A}{dt} = -\mu\left(\kappa\pi_G + \pi_C + \pi_T\right)P_A + \mu\pi_C P_C + \mu\kappa\pi_G P_G + \mu\pi_T P_T$$

$$\frac{dP_C}{dt} = \mu\pi_A P_A - \mu\left(\kappa\pi_T + \pi_A + \pi_G\right)P_C + \mu\pi_G P_G + \mu\kappa\pi_T P_T$$

$$\frac{dP_G}{dt} = \mu\kappa\pi_A P_A + \mu\pi_C P_C - \mu\left(\kappa\pi_A + \pi_C + \pi_T\right)P_G + \mu\pi_T P_T$$

$$\frac{dP_T}{dt} = \mu\pi_A P_A + \mu\kappa\pi_C P_C + \mu\pi_G P_G - \mu\left(\kappa\pi_C + \pi_A + \pi_G\right)P_T$$

(13.10)

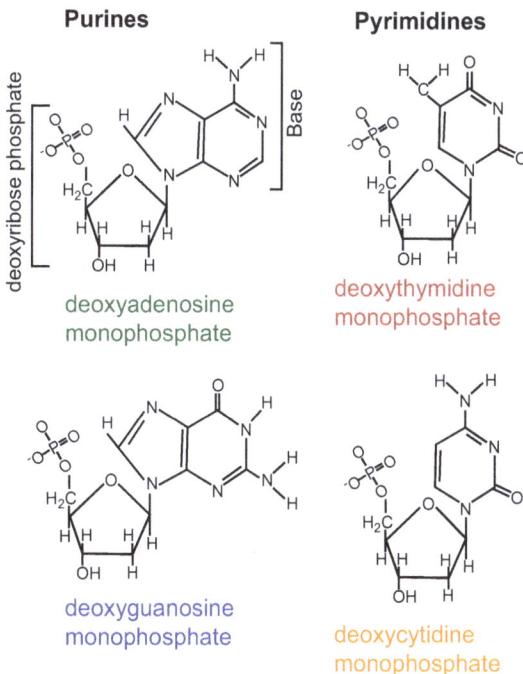

Purines   Pyrimidines

deoxyribose phosphate

deoxyadenosine monophosphate

deoxythymidine monophosphate

deoxyguanosine monophosphate

deoxycytidine monophosphate

Figure 13.13 The chemical structures of the four DNA bases, divided between the larger purines and the smaller pyrimidines.

where π_A, π_C, π_G, and π_T are the base frequencies and κ gives the excess rate of purine-to-purine and pyrimidine-to-pyrimidine changes. We should not be too intimidated by these equations. Instead, we should notice something very powerful about the models built to use with maximum likelihood approaches. If we let $\kappa = 1$ and $\pi_A = \pi_C = \pi_G = \pi_T$, we find that Equation 13.10 collapses back into the equations comprising Equation 13.6. Now, just as with the branch lengths, we can estimate the values of $\pi_A .. \pi_T$ and κ from our sequence alignments. If we do so, we do not need to be too worried that our model has become overcomplicated through the addition of these parameters. The reason is that if the true underlying substitution pattern is actually that of the Juke–Cantor model, we should estimate values of $\pi_A .. \pi_T$ near ¼ and a value of κ near 1. Thus, we can comfortably use the HKY model for our analyses in Exercise 13.4, and potentially even more complex models in other analyses.

> **Exercise 13.2**
>
> Compute the number of possible *unrooted* trees for 10 species.

> **Exercise 13.3**
>
> Confirm that taking the derivatives of the equations in Equation 13.7 yields the original differential equations in Equation 13.6.

Maximum Parsimony versus Maximum Likelihood

We can now return very briefly to the choice we made earlier not to define the maximum parsimony algorithm more carefully. One obvious difference between the parsimony and likelihood approaches is that we have branch lengths in the likelihood model. What these branches are telling us is that there was more time for substitutions to occur along some branches than along others. We can detect this pattern by optimizing the branch lengths via maximum likelihood. What happens if we decide to forgo this temporal information? In other words, what if we assume that each column in the alignment is so different from the other columns that we cannot estimate the branch length for that column using data from the other columns? What we would find is that we have made a new maximum likelihood model where each column has its own set of branch lengths. However, since we do not have enough information in a single alignment column to estimate those branch lengths, we fall back on a model where we treat all the branch lengths as being equal. And when we do that, we create a maximum likelihood model whose optimal tree is always the same as the optimal tree found with maximum parsimony [36]. And *that* means that one way to look at the maximum parsimony approach is as a special kind of maximum likelihood model, justifying our choice to focus instead on the likelihood models.

> **Exercise 13.4**
>
> Retrieve the six aligned DNA sequences for rhodopsin genes posted at http:// qbio.statgen.ncsu.edu/data/Phylogenetics. Use phyml online (http://www. atgc-montpellier.fr/phyml/) or on a local computer with the HYK model to infer a maximum likelihood phylogeny. You can visualize that tree at phylo.io.

Modeling and the Maximum Likelihood Approach

If we step back from the details for the moment, we can also see something else about this maximum likelihood approach. In Chapters 2–6, one of the key difficulties we encountered was how to take a model we had created with some unknown parameters and to match that model to observation by selecting appropriate values for those parameters. In the SIR model, we needed to at least estimate r and a, and yet we saw that slightly changing those estimates might give us very different model predictions. Here, we have created our most complex model so far, with (at least) $2n-1$ branch lengths as parameters. But because we have adopted the maximum likelihood approach, we have a principled way to estimate those parameters: we collect data and select the parameter values that maximize the probability/likelihood that the model would generate the data we collected. As a result, we have unified modeling, data collection, and the end goal of our analysis (a phylogenetic tree topology) into a single computational step. Quite an impressive technique, given where we started in Chapter 3.

PHYLOGENETICS IN PRACTICE

Phylogenetic inference is an enormously important topic and is widely studied, with thousands of scientific papers published in this area every year. There are many different software programs available for inferring phylogenetic trees from sequence data: a few of them are Phylip (the first program for maximum likelihood phylogenetics [24]), PAUP* [26], Phyml [37], and RAxML [38]. In Exercise 13.4 you will infer a phylogenetic tree from the sequences of the rhodopsin gene taken from six vertebrates, asking the question: Are whales and dolphins fish?

Rooting a Phylogenetic Tree

Figure 13.14 presents the tree we infer in Exercise 13.4. As we would expect, it is an unrooted tree. We could, therefore, root that tree at any of its nine branches. As a result, we cannot yet claim that we have evidence that orcas are not fish. The reason for our uncertainty is that if we placed the root along the branch leading to humans, orcas and everything else in the tree would be more closely related to each other than any would be to humans. However, in creating this dataset of six sequences, I have included the *little skate*, a cartilaginous fish related to sharks. Because it lacks a bony skeleton, among other characteristics, we can be fairly certain that skates split from the other five species here before they split from each other. The skate is hence an *outgroup* for these species and serves as an appropriate place to root the tree. As is suggested in Figure 13.14A, we root the tree by conceptually grabbing the unrooted tree by that branch and "pulling" it through our hand to produce the rooted tree in Figure 13.14B. And in that panel we indeed find that humans and orcas are each other's closest living relatives in the tree, indicating that whales and dolphins are not, in fact, fish.

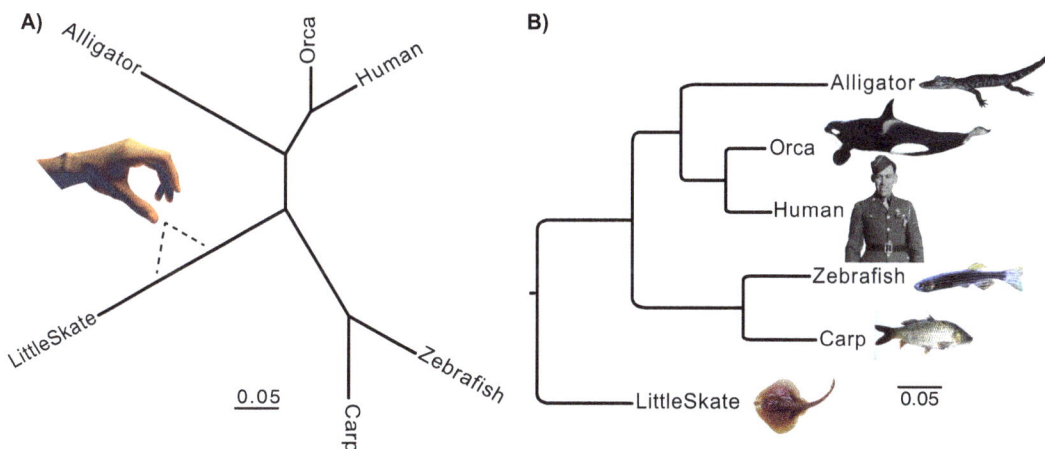

Figure 13.14 *Are whales and dolphins fish?* Shown in **A** is the unrooted maximum likelihood phylogeny inferred in Exercise 13.4, using the aligned sequences of six vertebrate rhodopsin genes. The little skate is a cartilagenous fish known to be evolutionarily distant from the other five species. It can therefore be used as a root for the tree. Conceptually, we grasp the unrooted tree at this branch in our hand. We then "pull" that tree through our hand to produce **B**: the rooted topology that shows that orcas most recently shared a common ancestor with humans and not carp, confirming that orcas are mammals. (Carp image courtesy of George Chernilevsky, on Wikimedia, published under CC BY-SA 3.0 license. Skate image courtesy of MDI Biological Laboratory, Maine, published on Wikimedia and in the public domain.) [49, 50].

REFLECTIONS AND PREVIEWS

In this chapter we first developed an intuition about the phylogenetic trees as theoretical objects we can use to represent the relationships and evolutionary histories of the species that live on this planet. We then considered how hard it might be to find the correct tree among all the possible trees and discussed both the maximum parsimony and maximum likelihood approaches to that tree search problem. Obviously, the problem of phylogenetic inference is a very important one in evolutionary biology. But we have spent a great deal of time on it for another reason as well. Phylogenetic methods unite many of the strands of thought we have been using throughout this book. The model of Equation 13.6 is another system of differential equations like those with which we started the book, except that these equations express probabilities, relating them to the continuous probability density functions we saw in Chapters 8 and 9. But we built the model using the results of our simulations of genetic drift from Chapter 11 and used the sequence alignments of Chapter 12 as the data to be fit to that model. In a way, then, this chapter is an illustration of one of the chief goals of quantitative biology: to use mathematics, computation, and statistics *together* to take biological data and turn those data into inferences, or, more generally, *insight*.

In the next chapter we will take an apparently totally different approach to quantitative biology, exploring *networks*. But I will just point out here that the computational objects we will use there are not as different from evolutionary trees as one might think: in both cases we have nodes; in the next chapter we will call the links between nodes *edges* rather than branches. In both cases, however, we are seeing that computers give us a power to model biology using approaches that go beyond the simple equations with which we started the book.

APPENDIX 13A: SOLVING THE INSTANTANEOUS TRANSITION PROBABILITY EQUATIONS FOR THE JUKES–CANTOR MODEL

The mathematics of solving Equation 13.6 to produce Equation 13.7 are beyond that needed for the rest of the book. However, they are less complex than one might expect. What follows is a sketch of the solution, which has the added benefit of showing a case where a system of differential equations does in fact have an analytic solution, unlike the SIR and Lotka–Volterra models.

We will first rewrite the constants from Equation 13.6 as the matrix A. For simplicity, we have multiplied each equation by 3 to remove the fractions.

$$
\begin{bmatrix} dP_A/dt \\ dP_C/dt \\ dP_G/dt \\ dP_T/dt \end{bmatrix} = \begin{bmatrix} -3\mu & \mu & \mu & \mu \\ \mu & -3\mu & \mu & \mu \\ \mu & \mu & -3\mu & \mu \\ \mu & \mu & \mu & -3\mu \end{bmatrix} \cdot \begin{bmatrix} P_A \\ P_C \\ P_G \\ P_T \end{bmatrix} \tag{13.11}
$$

To make our analysis easier later on, we will next replace the individual P_A, P_C, P_G, and P_T variables with a single *vector* of probabilities $\bar{P}$, which is of length 4, and correspondingly replace its derivative with $\bar{P}'$:

$$
\bar{P}' = \begin{bmatrix} -3\mu & \mu & \mu & \mu \\ \mu & -3\mu & \mu & \mu \\ \mu & \mu & -3\mu & \mu \\ \mu & \mu & \mu & -3\mu \end{bmatrix} \cdot \bar{P} \tag{13.12}
$$

Because μ is a constant relative to time, A in Equation 13.12 can be treated just like the matrices we used in Chapter 6. As we discussed, solutions to Equation 13.12 can then be expected to be some form of exponential function, since the coefficients of the derivatives of $\bar{P}$ are constants with respect to time. Let us propose, for the moment, that one solution to Equation 13.12 might be a function like [39]:

$$\bar{P} = \xi \cdot e^{rt} \tag{13.13}$$

Here r is an unknown real number and ξ is a vector of unknown constants of length 4. In other words, the P_A, P_C, P_G, and P_T probability functions have the same exponential form, but each has a different constant out in front (the ξ values). We could question this type of solution, but let us first see where the idea leads. Taking the derivative of Equation 13.13 gives us:

$$\bar{P'} = r \cdot \xi \cdot e^{rt} \tag{13.14}$$

If we now substitute Equations 13.14 and 13.13 back into the left and right sides of Equation 13.12, respectively, we have:

$$r \cdot \xi \cdot e^{rt} = \begin{bmatrix} -3\mu & \mu & \mu & \mu \\ \mu & -3\mu & \mu & \mu \\ \mu & \mu & -3\mu & \mu \\ \mu & \mu & \mu & -3\mu \end{bmatrix} \cdot \xi \cdot e^{rt} \tag{13.15}$$

When we cancel the e^{rt} term from both sides, we have:

$$r \cdot \xi = \begin{bmatrix} -3\mu & \mu & \mu & \mu \\ \mu & -3\mu & \mu & \mu \\ \mu & \mu & -3\mu & \mu \\ \mu & \mu & \mu & -3\mu \end{bmatrix} \cdot \xi \tag{13.16}$$

As an aside, I will mention that ξ is what is called an *eigenvector* of the matrix A [40]. An eigenvector of a matrix is a vector that, when multiplied by that matrix, gives a resulting vector that is just the original eigenvector scaled by a constant (here r). Eigenvectors are extremely useful in linear algebra: here they will assist us in solving the differential equations comprising Equation 13.6.

We would like some way of finding the values of the vector ξ that satisfy Equation 13.16. To do so, we need first to convert the constant r into a matrix that can be combined with the matrix on the right side. We do this by multiplying r by what is called the *identity* matrix I, which has 1s on the diagonal and 0s everywhere else:

$$\begin{bmatrix} r & 0 & 0 & 0 \\ 0 & r & 0 & 0 \\ 0 & 0 & r & 0 \\ 0 & 0 & 0 & r \end{bmatrix} \cdot \xi = \begin{bmatrix} -3\mu & \mu & \mu & \mu \\ \mu & -3\mu & \mu & \mu \\ \mu & \mu & -3\mu & \mu \\ \mu & \mu & \mu & -3\mu \end{bmatrix} \cdot \xi \tag{13.17}$$

We can next move this new rI matrix to the other side by subtracting each of its elements from the corresponding element of A. The result is to produce a new matrix $A - rI$, that, when multiplied by ξ, gives:

$$\begin{bmatrix} -3\mu-r & \mu & \mu & \mu \\ \mu & -3\mu-r & \mu & \mu \\ \mu & \mu & -3\mu-r & \mu \\ \mu & \mu & \mu & -3\mu-r \end{bmatrix} \cdot \xi = 0 \tag{13.18}$$

The equations comprising Equation 13.18 have a solution if and only if the *determinant* of $A - rI$ is 0 [39]. To compute values of r that allow the determinant to be 0, we will first let $x = -3\mu - r$ for convenience. The determinant of $A - rI$ is then given by the following recursive formula:

$$\det\begin{bmatrix} x & \mu & \mu & \mu \\ \mu & x & \mu & \mu \\ \mu & \mu & x & \mu \\ \mu & \mu & \mu & x \end{bmatrix} = 0 = x \cdot \det\begin{bmatrix} x & \mu & \mu \\ \mu & x & \mu \\ \mu & \mu & x \end{bmatrix} - \mu \cdot \det\begin{bmatrix} \mu & \mu & \mu \\ \mu & x & \mu \\ \mu & \mu & x \end{bmatrix}$$

$$+ \mu \cdot \det\begin{bmatrix} \mu & x & \mu \\ \mu & \mu & \mu \\ \mu & \mu & x \end{bmatrix} - \mu \cdot \det\begin{bmatrix} \mu & x & \mu \\ \mu & \mu & x \\ \mu & \mu & \mu \end{bmatrix} \quad (13.19)$$

In other words, the determinant of a 4×4 matrix is just the determinant of the four 3×3 matrices formed by omitting the first row and each successive column and multiplying those four determinants by the value of the first row at that column. The determinant of a 3×3 matrix is computed similarly:

$$x \cdot \det\begin{bmatrix} x & \mu & \mu \\ \mu & x & \mu \\ \mu & \mu & x \end{bmatrix} = x \cdot \left(x \cdot \det\begin{bmatrix} x & \mu \\ \mu & x \end{bmatrix} - \mu \cdot \det\begin{bmatrix} \mu & \mu \\ \mu & x \end{bmatrix} + \mu \cdot \det\begin{bmatrix} \mu & x \\ \mu & \mu \end{bmatrix} \right) \quad (13.20)$$

and the same for the other three terms of Equation 13.19. The determinant of a 2×2 matrix is just the product of the two diagonal elements minus the product of the two off-diagonal elements, meaning that we can rewrite Equation 13.20 as:

$$x \cdot \left(x \cdot \left(x^2 - \mu^2 \right) - \mu \cdot \left(\mu x - \mu^2 \right) + \mu \cdot \left(\mu^2 - \mu x \right) \right) \quad (13.21)$$

which simplifies to:

$$x^2 \cdot \left(x^2 - \mu^2 \right) - 2\mu^2 x \cdot \left(x - \mu \right) \quad (13.22)$$

The other three terms of Equation 13.19 simplify to:

$$-\mu^2 \cdot \left(x^2 - \mu^2 \right) + 2\mu^3 \cdot \left(x - \mu \right)$$
$$\mu^3 \cdot \left(x - \mu \right) - \mu^2 x \cdot \left(x - \mu \right) \quad (13.23)$$
$$\mu^3 \cdot \left(x - \mu \right) - \mu^2 x \cdot \left(x - \mu \right)$$

respectively. Combining Equations 13.20 and 13.23 gives:

$$\left(x^2 - \mu^2 \right)^2 - 4\mu^2 \cdot \left(x - \mu \right)^2 = 0 \quad (13.24)$$

We can rewrite the term $\left(x^2 + \mu^2 \right)^2$ as $\left(x - \mu \right)^2 \left(x + \mu \right)^2$, giving:

$$\left(x - \mu \right)^2 \cdot \left[\left(x + \mu \right)^2 - 4\mu^2 \right] = 0 \quad (13.25)$$

If we expand and substitute $x = -3\mu - r$, we get:

$$\left(-4\mu - r \right)^2 \cdot \left[\left(-2\mu - r \right)^2 - 4\mu^2 \right] = 0 \quad (13.26)$$

or:

$$\left(-4\mu - r \right)^2 \cdot \left[4\mu^2 + 4\mu r + r^2 - 4\mu^2 \right] = 0 \quad (13.27)$$

which simplifies to:

$$r\left(-4\mu-r\right)^3 = 0 \tag{13.28}$$

If we solve for r, we find either $r = 0$ or $r = -4\mu$. Looking back at Equation 13.13 tells us that solutions of the form:

$$\bar{P} = \xi \cdot e^{-4\mu t} \tag{13.29}$$

should satisfy Equation 13.12 once we solve for ξ in a moment. We will return to the $r = 0$ solution shortly.

For the general case, finding the values of ξ would require some tools of linear algebra beyond the scope of this book. However, if we plug $r = -4\mu$ back into Equation 13.18, we find:

$$\begin{bmatrix} u & \mu & \mu & \mu \\ \mu & u & \mu & \mu \\ \mu & \mu & u & \mu \\ \mu & \mu & \mu & u \end{bmatrix} \cdot \xi = 0 \tag{13.30}$$

Reminding ourselves that ξ is just a vector of four elements,

$$\begin{bmatrix} u & \mu & \mu & \mu \\ \mu & u & \mu & \mu \\ \mu & \mu & u & \mu \\ \mu & \mu & \mu & u \end{bmatrix} \cdot \begin{bmatrix} \xi_1 \\ \xi_2 \\ \xi_3 \\ \xi_4 \end{bmatrix} = 0 \tag{13.31}$$

we can see that the equations are satisfied when all four elements of ξ are equal in absolute value and two are positive and two are negative. There are hence a whole range of values for ξ. Examples include:

$$\begin{bmatrix} 1 \\ -1 \\ 1 \\ -1 \end{bmatrix} or \begin{bmatrix} 1 \\ 1 \\ -1 \\ -1 \end{bmatrix} or \begin{bmatrix} 1 \\ -1 \\ -1 \\ 1 \end{bmatrix} \tag{13.32}$$

What about $r = 0$? That solution returns our original A matrix, meaning we want a ξ that satisfies:

$$\begin{bmatrix} -3\mu & \mu & \mu & \mu \\ \mu & -3\mu & \mu & \mu \\ \mu & \mu & -3\mu & \mu \\ \mu & \mu & \mu & -3\mu \end{bmatrix} \cdot \xi = 0 \tag{13.33}$$

But evidently, any ξ where all the elements are equal does this, so we can just use a vector of all 1s for simplicity.

INITIAL CONDITIONS

We now have a set of solutions to our original differential equations from Equation 13.12. However, to apply them to the phylogenetics problem, we need to ensure that our solutions follow the rules of probability, namely that:

$$\sum_{Y=A,C,G,T} P_{X \to Y}(t) = 1.0 \tag{13.34}$$

for any value of t.

The general approach to this problem is somewhat complex, but we can see that we should expect two kinds of solutions, namely $P_{X \to X}(t)$ (the base is the same at the start and the end of the branch) and $P_{X \to Y}(t)$ (the base changes: $X \neq Y$). Because we have several possible ξ vectors for the $r = 4\mu$ case, our solutions can combine different constants in front of the $e^{-4\mu t}$ term. We also know that both solutions must share the same constant solution from the $r = 0$ case above, and we will call this constant c_1. Let's claim for the moment:

$$P_{X \to X}(t) = c_1 + c_2 \cdot e^{-4\mu t}$$
$$P_{X \to Y}(t) = c_1 + c_3 \cdot e^{-4\mu t}$$

$$(13.35)$$

We can now use two initial conditions to try to find values for $c_1..c_3$. We know that $P_{X \to X}(0) = 1$ because there has been no time for the base to undergo a substitution away from X. We also know that $P_{X \to Y}(0) = 0$ for the same reason. Since $e^{m0} = 1$ for any m, we can rewrite Equation 13.35 as:

$$1 = c_1 + c_2$$
$$0 = c_1 + c_3$$

$$(13.36)$$

We can now replace c_1 with $1 - c_2$ and c_3 with $-c_1 = c_2 - 1$, since $c_1 = 1c_2$ and $c_3 = c_2 - 1$. To find c_2, we return to Equation 13.34, which implies that $P_{X \to X}(t) + 3P_{X \to Y}(t) = 1$. Using Equation 13.35 and substituting c_1 and c_3, we have:

$$1 - c_2 + c_2 \cdot e^{-4\mu t} + 3\left((1 - c_2) + (c_2 - 1) \cdot e^{-4\mu t}\right) = 1.0$$

$$(13.37)$$

We can simplify this to:

$$4(1 - c_2) + c_2 \cdot e^{-4\mu t} + (3c_2 - 3) \cdot e^{-4\mu t} = 1.0$$

$$(13.38)$$

or:

$$4(1 - c_2) + 4c_2 \cdot e^{-4\mu t} - 3e^{-4\mu t} = 1.0$$

$$(13.39)$$

or:

$$4c_2 \cdot e^{-4\mu t} - 4c_2 = 3e^{-4\mu t} - 3$$

$$(13.40)$$

Factoring gives:

$$4c_2\left(e^{-4\mu t} - 1\right) = 3\left(e^{-4\mu t} - 1\right)$$

$$(13.41)$$

Meaning that $c_2 = \frac{3}{4}$. We then compute $c_1 = 1 - \frac{3}{4} = \frac{1}{4}$ and $c_3 = -\frac{1}{4}$. Our new solutions are therefore:

$$P_{X \to X}(t) = \frac{1}{4} + \frac{3}{4} \cdot e^{-4\mu t}$$
$$P_{X \to Y}(t) = \frac{1}{4} - \frac{1}{4} \cdot e^{-4\mu t}$$

$$(13.42)$$

Which are just the solutions we found in Equation 13.7 save for a factor of 4 in the exponent that comes from how we structured Equation 13.6 and the multiplication by 3 with which we started this section. Since we estimate the confounded μt from our data by maximum likelihood, this factor of 4 will wash out of the computation.

REFERENCES

1. Darwin C (1859) *On the origin of species by means of natural selection* (John Murry, London).
2. Gould SJ (1980) The Panda's Thumb. *The Panda's Thumb* (W. W. Norton, New York), pp. 19–26.
3. The Chimpanzee Sequencing and Analysis Consortium (2005) Initial sequence of the chimpanzee genome and comparison with the human genome. *Nature* 437(7055):69–87.
4. Lodish H, Baltimore D, Berk A, Zipursky SL, & Matsudaira P (1995) *Molecular cell biology*. 3rd Edition (Scientifc American Books, New York), p. 1344.
5. Santos MA, Moura G, Massey SE, & Tuite MF (2004) Driving change: the evolution of alternative genetic codes. *TRENDS in Genetics* 20(2):95–102.
6. Haig D & Hurst LD (1991) A quantitative measure of error minimization in the genetic code. *Journal of Molecular Evolution* 33:412–417.
7. Wong JF (1980) Role of minimization of chemical distances between amino acids in the evolution of the genetic code. *Proceedings of the National Academy of Sciences of the USA* 77(2):1083–1086.
8. Błażej P, Wnętrzak M, Mackiewicz D, Gagat P, & Mackiewicz P (2019) Many alternative and theoretical genetic codes are more robust to amino acid replacements than the standard genetic code. *Journal of Theoretical Biology* 464:21–32.
9. Bouquet M (1996) Family trees and their affinities: the visual imperative of the genealogical diagram. *Journal of the Royal Anthropological Institute*:43–66.
10. Futuyma DJ (1998) *Evolutionary biology*. 3rd Edition (Sinauer Associates, Inc, Sunderland, MA).
11. Larsen BB, Miller EC, Rhodes MK, & Wiens JJ (2017) Inordinate fondness multiplied and redistributed: the number of species on earth and the new pie of life. *The Quarterly Review of Biology* 92(3):229–265.
12. Mora C, Tittensor DP, Adl S, Simpson AG, & Worm B (2011) How many species are there on Earth and in the ocean? *PLoS Biology* 9(8):e1001127.
13. Curtis TP, Sloan WT, & Scannell JW (2002) Estimating prokaryotic diversity and its limits. *Proceedings of the National Academy of Sciences* 99(16):10494–10499.
14. Baum DA, Smith SD, & Donovan SS (2005) The tree-thinking challenge. *Science* 310(5750):979–980.
15. Gould SJ (1983) *The Stinkstones of Oeningen. Hen's teeth and horse's toes: Further reflections in natural history*, (WW Norton & Company, New York), pp. 94–106.
16. Edman P, Högfeldt E, Sillén LG, & Kinell P-O (1950) Method for determination of the amino acid sequence in peptides. *Acta Chemica Scandinavica* 4(7):283–293.
17. Sanger F, Air GM, Barrell BG, Brown NL, Coulson AR, Fiddes CA, Hutchison CA, Slocombe PM, & Smith M (1977) Nucleotide sequence of bacteriophage phi X174 DNA. *Nature* 265(5596):687–695.
18. Rogers RL & Slatkin M (2017) Excess of genomic defects in a woolly mammoth on Wrangel island. *PLoS Genetics* 13(3):e1006601.
19. Camin JH & Sokal RR (1965) A method for deducing branching sequences in phylogeny. *Evolution*:311–326.
20. Hillis DM, Moritz C, & Mable BK (1996) *Molecular systematics*. 2nd Edition (Sinauer Associates, Sunderland, MA).
21. Felsenstein J (1978) Cases in which parsimony or compatibility methods will be positively misleading. *Systematic Zoology* 27:401–410.
22. Conant GC & Lewis PO (2001) Effects of nucleotide composition bias on the success of the parsimony criterion in phylogenetic inference. *Molecular Biology and Evolution* 18(6):1024–1033.
23. Felsenstein J (1978) The number of evolutionary trees. *Systematic Zoology* 27:27–33.
24. Felsenstein J (1993) *PHYLIP (phylogeny inference package) (distributed by the author)*, The University of Washington, Seattle, WA), 3.5.
25. Lewis PO (1998) A genetic algorithm for maximum-likelihood phylogeny inference using nucleotide sequence data. *Molecular Biology and Evolution* 15(3):277–283.
26. Swofford DL (2002) *PAUP** (Sinauer, Sunderland, MA), 4.0b10.
27. Stamatakis A, Ludwig T, & Meier H (2005) RAxML-III: a fast program for maximum likelihood-based inference of large phylogenetic trees. *Bioinformatics* 21(4):456–463.
28. Kumar S, Filipski A, Swarna V, Walker A, & Hedges SB (2005) Placing confidence limits on the molecular age of the human–chimpanzee divergence. *Proceedings of the National Academy of Sciences* 102(52):18842–18847.
29. Jukes TH & Cantor CR (1969) Evolution of protein molecules. *Mammalian protein metabolism*, ed Munro HN (Academic Press, New York), pp. 21–132.
30. Lewis PO (1998) Maximum likelihood as an alternative to parsimony for inferring phylogeny using nucleotide sequence data. *Molecular systematics of plants II: DNA sequencing* (Springer), pp. 132–163.
31. Lewis PO (2001) Phylogenetic systematics turns over a new leaf. *Trends in Ecology and Evolution* 16(1):30–37.
32. Felsenstein J (1981) Evolutionary trees from DNA sequences: A maximum likelihood approach. *Journal of Molecular Evolution* 17:368–376.
33. Larson RE & Hostetler RP (1986) *Calculus with analytic geometry*, 3rd Edition (D. C. Heath and Company, Lexington, MA), p. 1013.
34. Kimura M (1980) A simple method for estimating evolutionary rates of base substitutions through comparative studies of nucleotide sequences. *Journal of Molecular Evolution* 16:111–120.
35. Hasegawa M, Kishino H, & Yano T (1985) Dating of the human-ape splitting by a molecular clock of mitochondrial DNA. *Journal of Molecular Evolution* 22:160–174.
36. Tuffley C & Steel M (1997) Links between maximum likelihood and maximum parsimony under a simple model of site substitution. *Bulletin of Mathematical Biology* 59(3):581–607.
37. Guindon S, Dufayard J-F, Lefort V, Anisimova M, Hordijk W, & Gascuel O (2010) New algorithms and methods to estimate maximum-likelihood phylogenies: assessing the performance of PhyML 3.0. *Systematic Biology* 59(3):307–321.
38. Stamatakis A (2014) RAxML version 8: a tool for phylogenetic analysis and post-analysis of large phylogenies. *Bioinformatics* 30(9):1312–1313.
39. Boyce WE & DiPrima RC (1992) *Elementary differential equations and boundary value problems* (John Wiley and Sons, New York), p. 680.
40. Lay DC (1994) *Linear Algebra and its applications* (Addison-Wesley, Reading, MA), p. 445.
41. Eldredge N (2005) *Darwin's Other Books:"Red" and "Transmutation" Notebooks,"Sketch,""Essay," and Natural Selection*. Public Library of Science, San Francisco, USA.
42. Pennisi E (2003) *Modernizing the tree of life*. American Association for the Advancement of Science.
43. NIAID Visual and Medical Arts (9/5/2024) Mountain Lion Silhouette. in *NIAID BIOART Source*. https://bioart.niaid.nih.gov/bioart/368
44. NIAID Visual and Medical Arts (9/10/2024) Raptor Silhouette. in *NIAID BIOART Source*. https://bioart.niaid.nih.gov/bioart/428

45. NIAID Visual and Medical Arts (9/12/2024) Wood Mouse Silhouette. in *NIAID BIOART Source*. https://bioart.niaid.nih.gov/bioart/20

46. NIAID Visual and Medical Arts (6/27/2024) Chimp Outline. in *NIAID BIOART Source*. https://bioart.niaid.nih.gov/bioart/76

47. NIAID Visual and Medical Arts (8/28/2024) Boy Silhouette. in *NIAID BIOART Source*. https://bioart.niaid.nih.gov/bioart/59

48. NIAID Visual and Medical Arts (9/16/2024) Hare Silhouette. in *NIAID BIOART Source*. https://bioart.niaid.nih.gov/bioart/195

49. Chernilevsky G (2008) Common carp or European carp (Cyprinus carpio). https://commons.wikimedia.org/wiki/File:Cyprinus_carpio_2008_G1_(cropped).jpg

50. MDI Biological Laboratory M (2009) Skate (Raja erinacea). https://en.wikipedia.org/wiki/File:Skate91-300.jpg

Life in a Net

Network Tools for Modeling Complex Systems from a Cell to an Ecosystem

"Does aught befall you? It is good... All that befalls you is part of the great web."
—Marcus Aurelius, Meditations

INTERACTIONS GIVE STRUCTURE TO LIVING PROCESSES ACROSS BIOLOGICAL SCALES

Figure 14.1 show three diagrams depicting interactions between biological entities, one at the scale of an ecosystem, one at that of an organism, and one at that of a cell. If we think back to Chapter 3, we find that the predator–prey model we built there would represent a single line in Figure 14.1A, where fish species #21 preys on species #20. Likewise, in Chapter 7, when we discussed the central dogma of molecular biology, one detail we omitted was the *transcription factors*. Transcription factors are proteins that can specifically bind to an upstream DNA region of a gene. That sequence is commonly called a *promoter*. By binding to the promoter, the transcription factor either prevents or enables binding of the RNA polymerase to that gene. Hence, the binding of transcription factors serves to control whether a particular gene is turned on or off. Knowing these properties, we could expand the kinetic model of the central dogma we built in Chapter 7 to include these transcription factors. If we had done so, this would describe a single pair of connected dots in Figure 14.1C, which correspond to a transcription factor (red) and one of the genes it regulates.

We should probably draw two rough inferences from Figure 14.1, then. First, biological systems often draw their functions from the *interactions* between entities: pairs of species in an ecosystem, pairs of neurons in a brain, pairs of proteins or reactions in the cell, and so forth. Second, there are often an enormous number of these interactions. As we saw in Chapter 6, developing models of systems with very large numbers of distinct parts is often difficult, not only for computational reasons but also because we need to know the various kinetic parameters of those models. These parameters could correspond to, for instance, the predation rates in the predator–prey model, $V_{\max}$ and K_m for every possible metabolic reaction, or the binding affinity of every transcription factor to every gene.

The conclusion we drew in Chapter 6, then, applies potentially with even more force in these other systems. The mathematical models we used in the first half of this book are not ideal for full-scale biological systems, because we lack both the computing power to model—say, the molecular dynamics of all the proteins in the cell—and the scientific resources to measure all the necessary kinetic parameters for such a model.

The *network* approaches or models illustrated in Figure 14.1 provide us with a set of tools for studying these systems of many parts that do not require such detailed information. We do need to remember that, as we saw with the flux–balance models in Chapter 6, more abstract models come with the potential cost of yielding predictions that are more qualitative in nature. With that caveat in mind, and before we delve any deeper into this topic, we will explore these network objects in a slightly more familiar context.

DOI: 10.1201/9781003687504-14

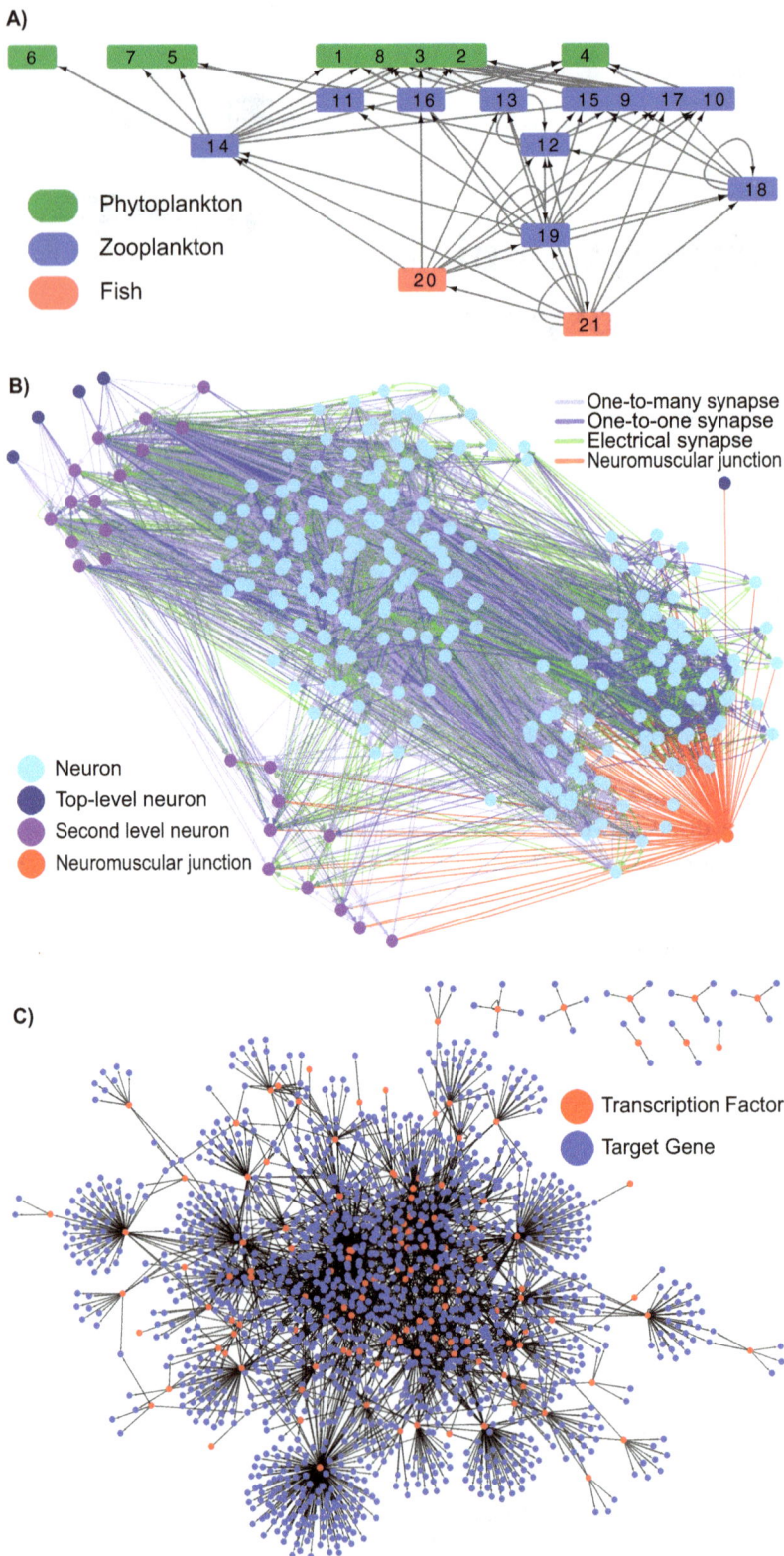

Figure 14.1 Three biological *networks* drawn from decreasing scales of biological organization. (**A**) A food web from an aquatic system, where the nodes (colored boxes) are "metaspecies" occupying a similar ecological niche [4]. Metaspecies that prey on other metaspecies are indicated with arrows. (**B**) The neuronal network of a nematode, showing the connections in between different neurons and how those neurons enervate the animal's muscles [9]. (**C**) The transcriptional regulatory network of baker's yeast, showing its transcription factors (*red*) and the genes they regulate [10]. All networks were visualized with Cytoscape [7].

Six Degrees of Kevin Bacon

Figure 14.2 gives a diagram of a game that enjoyed a certain vogue when I was in college in the mid-1990s: the Kevin Bacon game. The goal of the game was to connect a named actor or actress with Kevin Bacon through as few co-stars as possible [11].

The specifics of this game are perhaps of less interest today: our concern is the representation of the relationships involved, as shown in Figure 14.2. The first

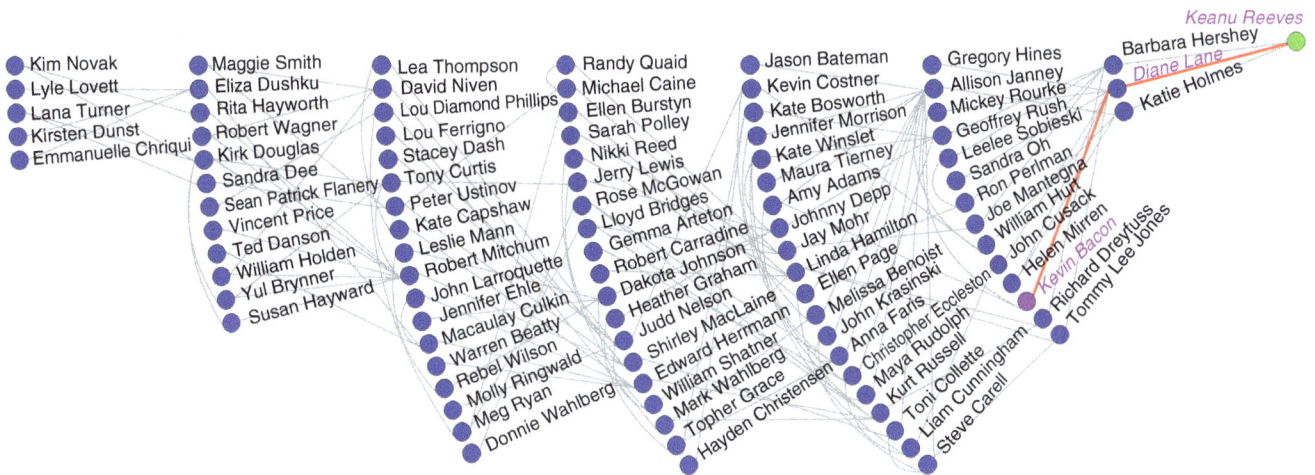

Figure 14.2 The Kevin Bacon game, for a small subset of actors and actresses. I have selected, more or less at random, 100 actors and actresses with which to play the game. I used Keanu Reeves as the start of this game: he co-starred with Diane Lane in *My Dog Skip* and then Diane Lane co-starred with Kevin Bacon in *Hard Ball*. The diagram, however, also illustrates other co-starring relationships among this group, with distance from Keanu Reeves shown as the concentric partial rings in the diagram. You can regenerate this diagram at http://qbio.statgen.ncsu.edu/PlayBacon/ (Exercise 14.1). Note that the data shown are taken from 2016; newer films do not appear.

point to notice is that we can represent data of very different kinds with this network layout, a fact that is obvious from comparing Figures 14.1 and 14.2. Another thing to notice is that even though networks are very different from other mathematical approaches we have used, there are still some common properties. One useful one is that of *distance*: in Figure 14.2, we notice that Katie Holmes and Keanu Reeves were co-stars and are therefore close in the network, while, at least when we consider only this set of film stars, Kirsten Dunst is very distant from Mr. Reeves.

A surprising thing about the movie star network, at least when we consider all films, is that the average distances are actually very small. If we take a sample of about 200,000 actors and actresses from 2016, connected with about 425,000 co-staring relationships, we find that the average distance from any of those actors to Kevin Bacon is 5.3 connections. This result makes us think at least a bit, as the set of people in the comparison comprises film stars going all the way back to the silent film era in the 1920s, and yet all these people connect to each other rather easily.

We will call these distances *path lengths*. It is actually quite easy to build a network with short path lengths: simply connect every node to every other one, so that the mean (and only) path length is 1. However, such a system has a cost: many, many connections or *links*. In fact, the number of links L required to connect every pair of nodes in a group of N such nodes is:

$$L = \frac{N(N-1)}{2} \tag{14.1}$$

We could express this more succinctly as $L \approx N^2 / 2$.

Yet, as just mentioned, the movie star network has nothing like the twenty billion or so links needed to be completely connected: the number of links in this case is only about twice the number of nodes. We will now explore some ways to achieve short path lengths in much less connected networks.

NETWORKS AND GRAPHS: NEW APPLICATIONS OF OLD IDEAS

While they may appear unfamiliar at first, networks are an old idea in mathematics, going back to Euler in the 18th century [12]. In the mid-20th century, Paul Erdős and Alfréd Rényi made key discoveries regarding how the *distribution* of the

A)

Lattice

B)

Small-
world

C)

Uniform
random

Mean path length	2.33	1.92	1.81
Clustering coefficient	0.60	0.41	0.30

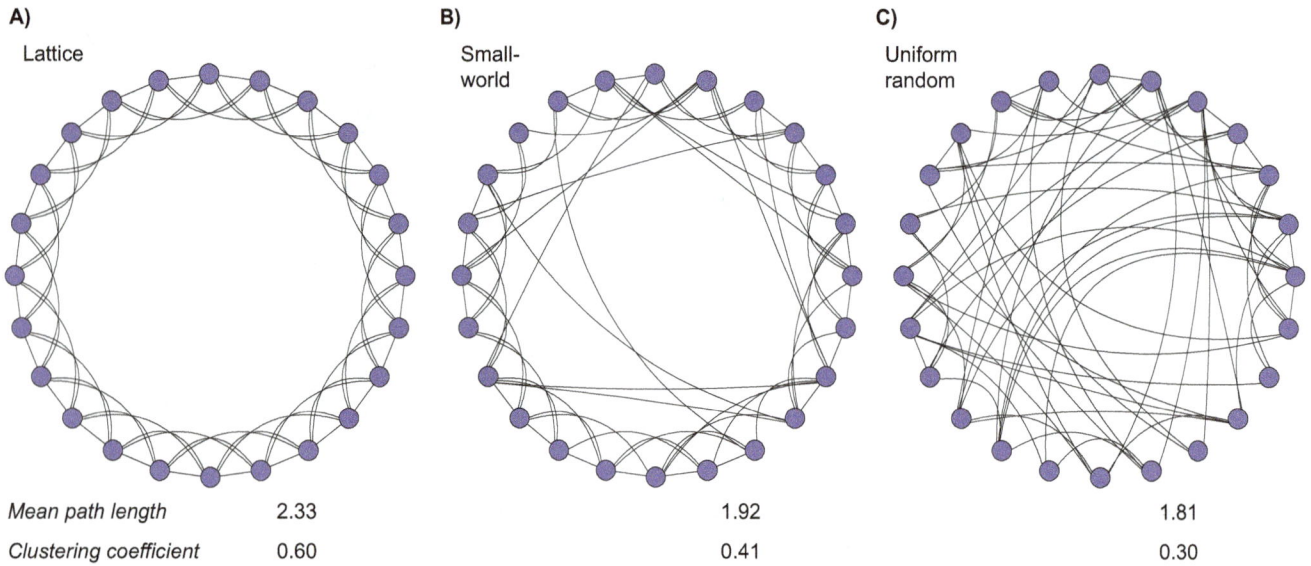

Figure 14.3 Comparisons of a regular lattice (A) where each node is connected to its six nearest neighbors to a nearly completely random network (C) and to a "small-world" network (B). Each network has the same 24 nodes and 72 links. To produce **B** and **C**, we randomly selected a link and connected one side of it to another random node. In **B** we conducted this randomization for 20% to the total edges, while for **C** we performed it as many times as there were edges (although we did not prevent a previously randomized link from be re-randomized). Clustering coefficients and path lengths for each are given. (Adapted from Watts DJ & Strogatz SH (1998) *Nature* 393:440–442. With permission from Springer Nature.)

links (which we will call *edges* henceforth, to match the terminology in the literature) had a critical role in the global properties of the network [12]. One of their discoveries involved graphs in which each node was connected to just a few other nodes completely at random. They compared such random graphs to *lattices*: graphs in which every node is uniformly connected to a fixed number of neighbors (**Figure 14.3**). (The completely connected graph we just mentioned is the limiting case of such a lattice.) What Erdős and Rényi found was quite striking: random graphs had much shorter *average* path lengths than did lattices (cf. Figure 14.3C with Figure 14.3A).

Networks and the Coming of Universal Computing

In keeping with a pattern we have seen in this book, what brought networks to a wider audience of scientists was the appearance of new, large scientific datasets suitable for analyses with them. In the late 1990s, a series of papers began to take information about a range of real systems and describe the structure of the networks they defined. In a pioneering paper, Duncan Watts and Steven Strogatz described *small world networks* [6]. Such networks have local structure, meaning that if node A interacts with node B and with node C, it is relatively likely that B and C will also interact with each other. These two authors named this propensity the *clustering coefficient*: Figure 14.3 shows how the clustering coefficients and the path lengths differ between a lattice (see Figure 14.3A) and a random graph similar to those considered by Erdős and Rényi (see Figure 14.3C).

Computing the Clustering Coefficient

In **Figure 14.4** I give a diagram outlining Watts and Stogatz's clustering coefficient for a focal node [6]. Formally, we can compute the clustering coefficient C for a node F as follows. Let $N_0..N_i$ be the i nodes that are neighbors of F (in other words, F has i edges) and let E be the set of all edges in the network. Then we have:

$$C = \frac{\sum_{j,k \leq i} \begin{vmatrix} 1 & \in E_{N_j,N_k} \\ 0 & Otherwise \end{vmatrix}}{i \cdot (i-1)} \qquad (14.2)$$

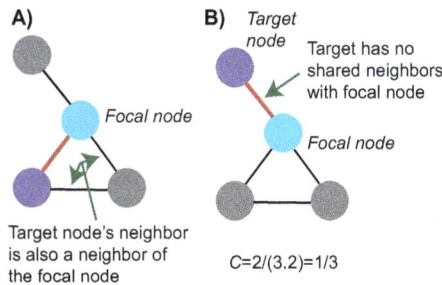

Figure 14.4 Principles of finding the clustering coefficient *C* for a focal node (*light blue*). (**A**) When we consider a target node (*dark blue*) whose neighbors are also neighbors of the target, this adds to *C* (Equation 14.2). (**B**). When the target node has no neighbors that are also neighbors of the focal node, that target node does not contribute to the numerator of *C* (Equation 14.2). In Exercise 14.2, you can compute the average clustering coefficient of this small network.

In other words, we count all the cases where an edge exists between a pair of nodes N_j and N_k that are both neighbors of F and then divide by the number of neighbor pairs possible for F (i.e., $i(i-1)$). In Figure 14.3, the reported clustering coefficients are just the average of C across all nodes, where nodes with a single edge have $C = 1$.

Small-World Networks: Clustered and Connected

The key insight that Watts and Stogatz gave us is that networks with relatively small perturbations from the regular lattice can have the short path lengths characteristic of Erdős and Rényi's random networks while at the same time maintaining some of the local nature of the regular lattice. In Figure 14.3B I give an example of such a perturbed network, in which I randomly selected and reconnected 20% of the edges in a lattice. This network is intermediate between Figure 14.3A and Figure 14.3C in both clustering coefficient and mean path length.

Watts and Stogatz demonstrated that networks as diverse as the film star network (see Figure 14.2), the neural network of *C. elegans* (see Figure 14.1B) [9], and the electrical grid of the western United States showed this type of intermediate structure. Thus, these networks have path lengths not too dissimilar to random networks yet have clustering coefficients that might approach those of lattices [6]. Watts and Stogatz referred to such networks as *small-world networks*, in homage to the Disney song.

Node Degree Distributions: Hubs, Attacks, and Long Tails

The next piece of the network puzzle was presented by Albert, Jeong, and Barabási [13]. Implicit in Figure 14.3 is the assumption that every node has approximately the same number of edges. Yet when these authors studied actual networks, the distribution of the number of edges per node was found to be strikingly uneven. Figure 14.4 shows how a network where most nodes have roughly the same number of edges (panel A) differs from one where a few nodes have a very larger number of edges (panel B). This second type of network has often been referred to as a *scale-free* network [13]. This name derives from the claim that the degree distribution follows a *power-law degree probability distribution*. Such power-law distributions are scale free because they retain their shape when their measurement scale is changed, which is not the case for the probability distributions we considered in Chapters 8 and 9 [14]. However, I will use the term *long-tailed* network here in preference to scale free because it is not the case that all biological networks with long-tailed degree distributions are truly scale free [15].

Long-Tailed Networks, Robustness, and Emergence

Albert, Jeong, and Barabási [13] considered the network structure of both the computers wired together to form the Internet and the pages comprising the World Wide Web. They pointed out that because both networks have this long-tailed distribution of edges, the removal of a random node from either network will most often have minimal effect on statistics such as the path length. The reason for this

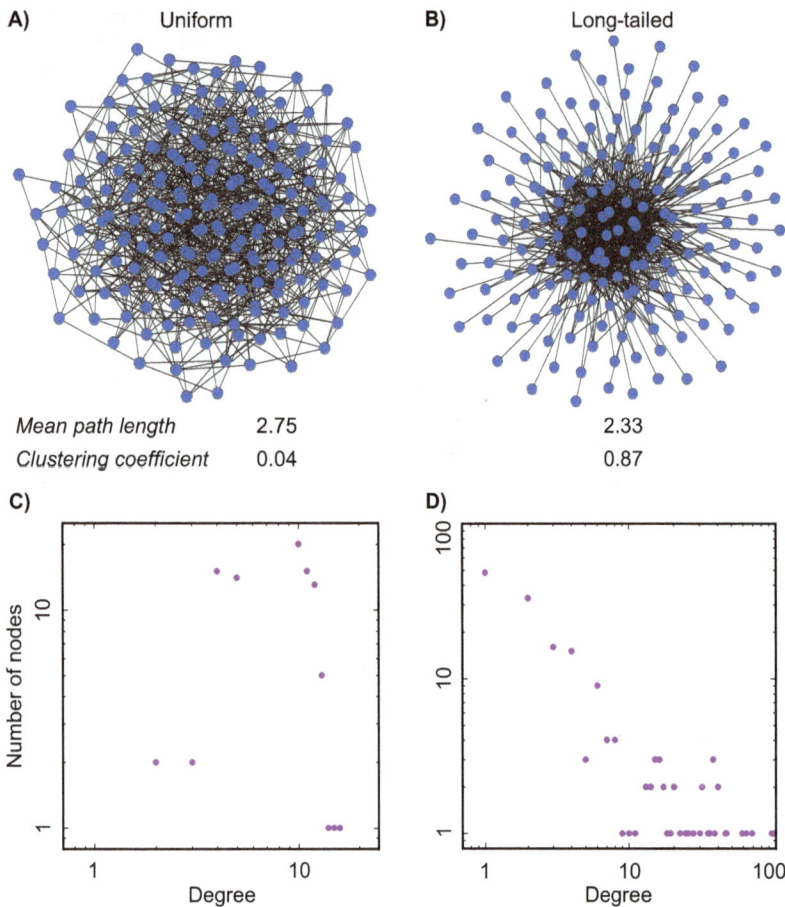

A) Uniform

Mean path length 2.75
Clustering coefficient 0.04

B) Long-tailed

2.33
0.87

C)

D)

Figure 14.5 Comparisons of networks where the number of links per node is distributed approximately normally (A) and one where the number of links per node approximately follows a power-law distribution (B). In both cases, the networks have 200 nodes and 800 links, but the *long-tailed* network in **B** has a higher clustering coefficient and a lower mean path length than does the *uniform* network of **A**. Visualized with Cytoscape [7]. (**C**) Distribution of number of edges per node, plotted on a log-log scale for the network in **A**. [**D**] As for **C** but for the network in **B**.

insensitivity, or *robustness*, is that most nodes in such a long-tailed network have a low degree: in **Figure 14.5B**, the modal edge count is actually one. On the other hand, such networks are very sensitive to the removal of those nodes with a high degree: one can see this implicitly by imagining the removal of www.google.com from the World Wide Web.

Barabási and Albert also considered how such long-tailed networks might emerge. One mechanism that they proposed was *growth by preferential attachment*. In this mechanism, a small ancestral network might add new nodes by connecting them to the existing nodes. Such a process bears a strong resemblance to those seen in evolution, whereby new genes and proteins, created by processes such as gene duplication, connect into existing nodes. This type of growth process can generate degree distributions like those in **Figure 14.5**, with the implication that older nodes in the network are more highly connected [16].

BIOLOGICAL NETWORKS: SCALE FREE AND SMALL WORLD

While these advances were being made in the computational analyses of networks, biology was entering what has been called the genomic era. The first bacterial genome was sequenced and published in 1995 [17], the first eukaryote in 1996 [18], and the human genome was released in 2001 [19, 20]. At first, these genomes could mostly give researchers an idea of the "parts list" of an organism: all its genes and the proteins those genes coded for. However, often little was known about what most of those genes did. Hence, networks were a very useful tool to ask questions about the global structure of an organism's genetic network(s) when the pieces were known but where the information about those pieces was relatively limited.

We have already encountered one such network approach in the book: the *metabolic network* that was shown in Figure 6.1. In that figure the nodes are metabolites, and pairs of metabolites are connected if they co-occur in at least one reaction. While that depiction served the goal of understanding the matter flow in the cell, a more common depiction of a metabolic network would be to make the nodes represent biochemical reactions and have edges connect reactions that have at least one metabolite in common with each other [21].

Figure 14.6 presents this alternative view of the *Synechocystis* sp. PCC 6803 metabolic network. The edge distribution (see Figure 14.6B) is clearly long-tailed, with a few highly connected reactions. To reinforce the small-world, long-tailed nature of this network, Figure 14.6C shows the distribution of path lengths and clustering coefficients for a sample of uniformly connected random networks (see Figure 14.5A) and for an ordered lattice (see Figure 14.3A), both with the same number of nodes and edges as Figure 14.6A. The true metabolic network is much more clustered than the uniform network yet has path lengths that are much closer to it than to a lattice. We can therefore think of the real network as having something akin to the best of both worlds: the short path lengths of a random network with the clustering of a lattice. In other words, it is like the small-world networks of Watts and Stogatz.

We can also wonder whether the structure we see in Figure 14.6A gives us another view of the robustness we discussed in Chapter 6. In particular, the highly connected nature of the network might be another way to think about the fact that the network showed robustness to the loss of many reactions in the same way that the World Wide Web does not show strong disconnection when random web pages are removed [13]. Indeed, it has been shown for a large number of metabolic networks that their structure shows the kind of robustness to random node loss that is seen in other types of network [22]. Perhaps even more interestingly, if the preferential attachment hypothesis is correct, those highly connected reactions in the metabolic network could be some of the earliest biochemical reactions. Under that hypothesis, they gained their current highly connected status through a long evolutionary history of newer reactions connecting to them.

Another interesting fact about metabolic networks, however, is that they are not unique among biological networks. Several other types of biological network have surprisingly similar degree distributions, clustering, and tolerance to node removal. We will consider some of these networks next.

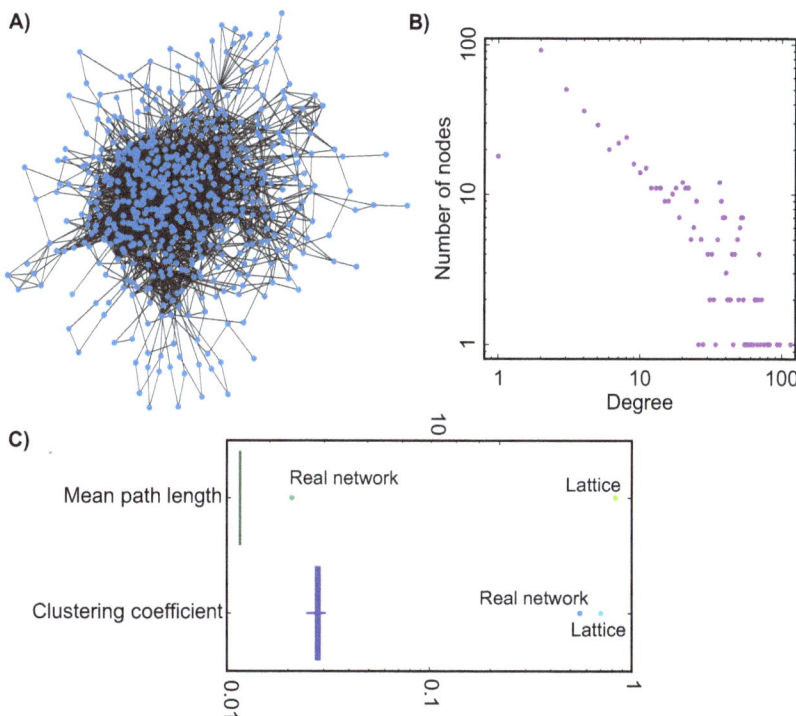

Figure 14.6 Representing an organism's metabolic map as a network. (**A**) The metabolic network of *Synechocystis* sp. PCC 6803 [1]. Nodes are metabolic reactions and edges join pairs of reactions sharing a metabolite. Metabolites occurring in more than 50 reactions are omitted from the edge list. Visualized with Cytoscape [7]. (**B**) Distribution of the number of edges per node from **A**, plotted on a log-log scale. (**C**) The mean network path length and clustering coefficient for the actual network in **A** are both lower than a lattice network with the same number of nodes and edges ("Lattice"). However, when we compare these values to the distribution seen for 100 simulated uniform random networks, we find that the real network has higher clustering and longer path lengths.

Protein Interaction Networks

One of the first kinds of biological networks that were studied extensively were *protein interaction networks*. These networks were inferred using the results of two kinds of large-scale experiments invented in the early 2000s. In the first, a technique called a yeast two-hybrid experiment [23, 24] can rapidly assess whether or not two proteins are able to bind to each other. In the second, techniques involving mass spectrometry and affinity purification allow research to extract full protein complexes and identify their constituent proteins [25, 26]. In this context, all we mean by "complex" is a set of two or more proteins that are all bound together at some point into a single entity.

It is worth reflecting back to Chapter 5 at this point because there we considered the parameter K_m for an enzymatic reaction. This parameter, in part at least, describes the relative affinity of a protein enzyme for a metabolite. We can compute similar affinities for pairs of proteins (think of the actin and myosin in your muscles, for example). But just as we discussed in Chapter 6, experimentally determining such values for every possible pair of proteins would be intractable. Hence, a protein interaction network represents a sort of intermediate level of knowledge, where we know that a degree of binding exists between a pair of proteins, but we cannot quantify it to the accuracy of the models in Chapters 5–7.

Figure 14.7 illustrates a large set of protein interactions from the bakers' yeast, *Saccharomyces cerevisiae*, taken from the BIOGRID database of such interactions [2].

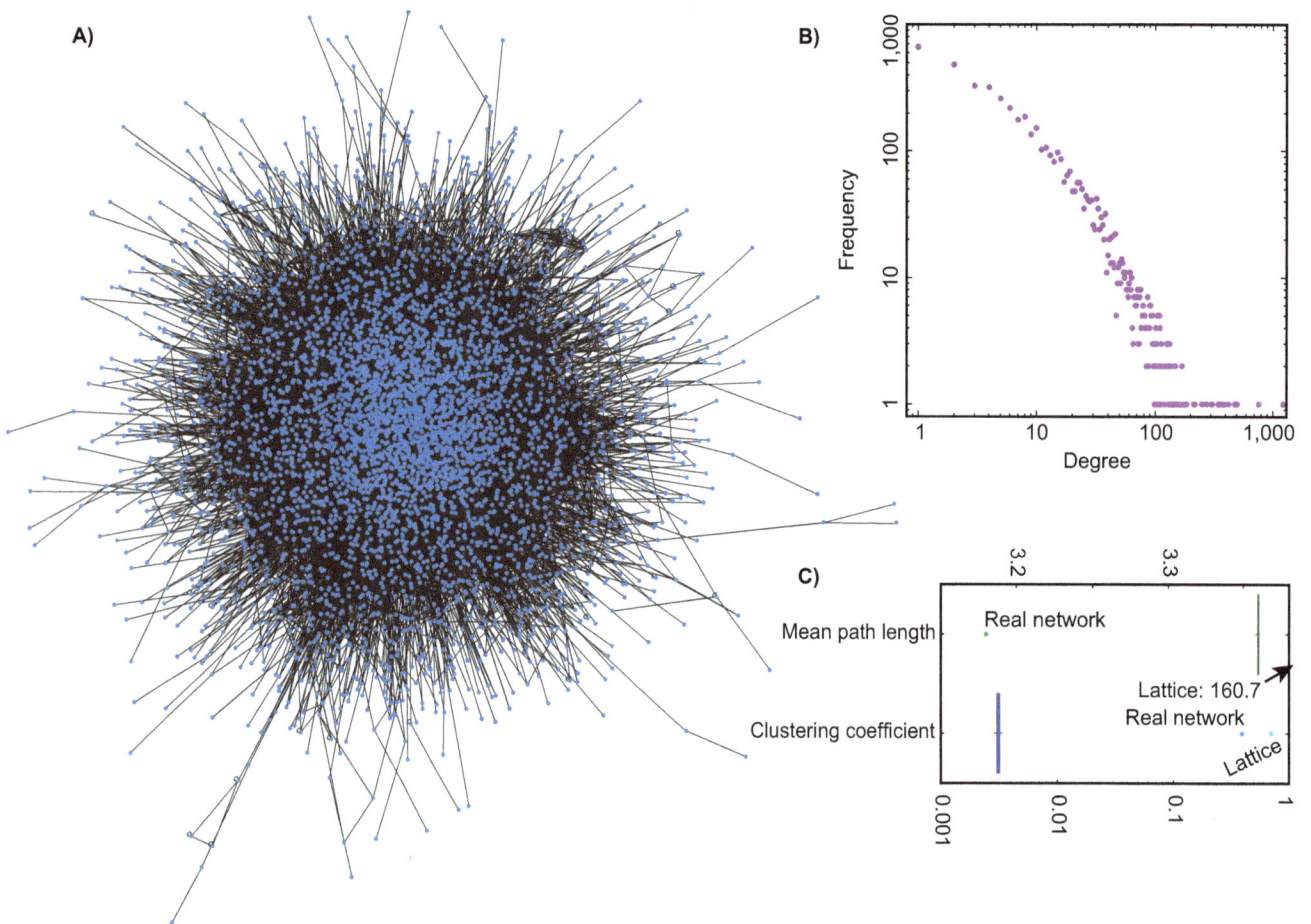

Figure 14.7 The network view of protein interactions. (A) The protein interaction network of bakers' yeast (*Saccharomyces cerevisiae*) [1, 2]. Nodes are proteins, and two nodes are connected if they share a protein interaction. Those interactions were identified either through a yeast two-hybrid experiment or an affinity-capture experiment that employed either a Western blot or mass-spectrometry for identification (see text). The network has 5,325 nodes and 44,322 interactions. It was visualized with Cytoscape [7]. **(B)** Distribution of the number of edges per node from **A**, plotted on a log-log scale. **(C)** The network is **A** has lower clustering and (vastly) shorter path lengths than does a lattice with the same number of nodes and edges. The network in **A** also has shorter path lengths than are seen across 100 random networks with uniform connectivity. The clustering coefficient of **A** is also much larger than for random networks and approaches the level seen in the lattice (note the log-scale for the clustering coefficient measurements).

The network in Figure 14.7 does not show a straight-line relationship between degree-rank and degree on a log-log plot, arguing that the degree distribution does not follow a power-law [27]. However, we should not read too much into this fact: the data in Figure 14.7 incorporate different kinds of experimental data, and such pooled data can alter the network structure [28]. It is, in any case, clearly long-tailed.

One question we could ask is whether the patterns of robustness and sensitivity for the loss of nodes of low and high degree, respectively, seen in the networks studied by Albert, Jeong, and Barabási [13], also apply to protein interaction networks. And in fact, some of these same authors conducted just such an analysis [29]. To do so, they used some very valuable experimental data on the effects of genetically deleting every gene in bakers' yeast [30]. This large series of experiments showed that the deletion from the genome of approximately 19% of the genes caused the carrying yeast cells to die; these genes are termed *essential*. Notably, the loss of the remaining 81% of the genes was tolerated in the conditions tested (think back to the knockout analyses we did in Figure 6.9). Jeong and coauthors determined the location of the essential genes' protein products in the protein interaction network. They found, perhaps not surprisingly, that essential proteins tend to have more interactions and lie nearer to the center of the network than do other proteins [29]. In other words, the protein interaction network is sensitive to the loss of highly connected nodes, just as the World Wide Web is.

Insights into Cell Structure from Combining Networks

If you are wondering how these network concepts enlighten us about other aspects of biology, an example might be helpful. In this case, we can uncover interesting behaviors by *merging* the two types of biological networks we have just discussed, namely metabolic and protein interaction networks.

Metabolic Channeling

In Chapters 6 and 7, our biochemical models implicitly assumed that all of the cell's metabolites and enzymes were dissolved and freely diffusing. That assumption greatly simplifies the mathematics of the models but does not exactly represent the situation in real cells. Instead, cells often use *metabolic channeling*. In such channeling, the cell uses several mechanisms to prevent intermediate metabolites from defusing freely around the cell. The most straightforward of these mechanisms is simply keeping the sequential enzymes of the pathway physically close to each other, such that intermediate metabolites encounter the next enzyme in the pathway soon after they are produced [31, 32]. In more complex cases, a pair of sequential enzymes can even form a physical tunnel between each other to prevent the intermediate from escaping before it can be delivered to the second enzyme [33, 34].

Cells use metabolic channeling for at least two reasons. First, some metabolic intermediates are toxic, so the cell does not want them circulating [35]. Second, even for nontoxic intermediates, the cell would prefer not to be constrained by having to wait until each intermediate diffuses to the correct enzyme and can be metabolized [31]. Instead, to return to the factory analogy in Chapter 6, it would be more efficient for cells to keep enzymes for the same pathway close to each other in the cell, reducing the time needed for the metabolites to move from one enzyme to the next.

This question of the spatial organization of the cell is obviously a complex one [36], and we will limit ourselves to just one example involving protein–protein interactions.

Protein Interactions Can Facilitate Metabolic Channeling

One effect of a protein–protein interaction is so obvious that it is sometimes overlooked. Such an interaction means that, at least on occasion, the two proteins involved are in the same place at the same time. As a result, one way evolution may

A)

B)

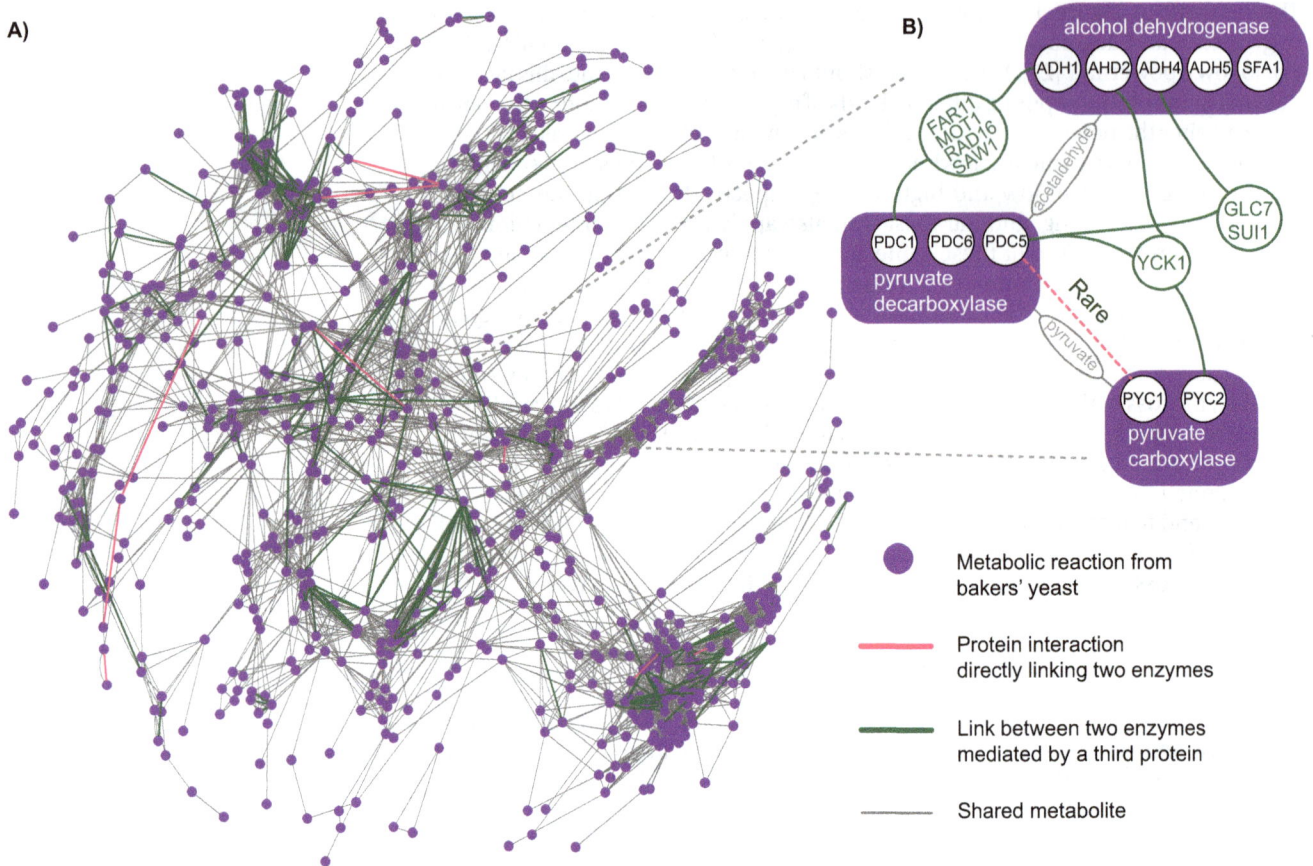

Figure 14.8 Metabolic channeling through shared protein–protein interactions. (**A**) The metabolic network of bakers' yeast [3], with enzymes (nodes) with shared metabolites connected by gray edges. (Rare) cases where such connected enzymes are also connected by a protein interaction are shown in pink. Cases where a pair of connected enzymes mutually interact with a third, nonenzymatic protein are shown as green links. These green lengths are statistically overrepresented in this network relative to random ones [8]. (**B**) A small region of **A**, illustrating the concept of a mediator protein (like YCK1) that can interact with a number of enzymes that are part of a single pathway, potentially bringing those enzymes into physical proximity in the cell. (Adapted from Pérez-Bercoff Å, McLysaght A & Conant GC (2011) *Molecular BioSystems* 7:3056–3064.)

have chosen to co-localize two enzymes in a pathway is by using protein interactions [37]. Some colleagues of mine and I analyzed the metabolic and protein interaction networks of bakers' yeast. We found that cases where the sequential enzymes of a pathway directly shared a protein interaction with each other were relatively rare [8]. Instead, we found it to be more common for both enzymes to each interact with a third *mediator* protein. That protein was not itself an enzyme; instead, it helped organize those enzymes in space.

In **Figure 14.8** I illustrate the metabolic network of yeast, with shared protein interactions between linked enzymes layered on top. In the inset panel I show several examples of these mediator proteins that bring sequential enzymes into physical proximity, potentially allowing them to efficiently channel metabolites. Our network analyses showed that these enzyme–mediator interactions were more common than could be explained by chance not only in yeast but also in *E. coli* and humans [8]. In yeast and *E. coli*, reactions involving a mediator protein also carried higher metabolic flux (see Chapter 6) than did other reactions (we did not conduct the equivalent test for humans).

While there are other approaches to studying metabolic channeling, the simple use of networks arguably uncovers this key organizing principle of cellular organization in an almost trivial way, without the need to add the complex mathematics of channeling to models like those we used in Chapter 5.

Networks at Higher Levels of Biological Organization

In Figure 14.1 we saw two networks from larger-scale biological systems, namely the nervous system and an ecosystem. In the case of the brain of the nematode worm *Caenorhabditis elegans* [9], we are fortunate that development in these animals is highly stereotyped across individuals, meaning that the fate of every cell can be tracked [38]. As a result, researchers could draw the network of Figure 14.1B, showing which neurons connect to which others for any *C. elegans* individual. As an aside, such a diagram would be nearly impossible to draw for humans, as we have vastly more neurons and the connections between them are different in each of us.

One of the slightly surprising features of this network is that we can begin to see outlines of function in the diagram: we have "top-level" neurons that are presumably receiving information from the outside world, information that is processed through the intermediate levels of the network until a decision is made. That decision is then implemented by some action by the animal, initiated by nerve impulses to the neuromuscular junction (red in the figure).

Neural Networks and Computation

It is worth pausing for a moment, because we have found something rather surprising: if we add some mathematics to the edges of this network, we have actually created a particular type of computer—a *neural network* [39]. If we imagine, for instance, that we have some difficult computational task such as examining all the pixels in a black-and-white photograph and deciding if they represent a tree, we could represent each input pixel as an input node (**Figure 14.9**). The intermediate neurons are called the *hidden* layer or layers of the network. The output node might then output a "0" if the picture is not a tree and a "1" if it is (see Figure 14.9). Of course, just drawing such a network will not produce a classifier of any value; in general, we will need training data: cases where the inputs and the desired output are known. There are then computational approaches for adjusting the mathematics of the hidden layers to give accurate classifications for the training data, and hopefully for new input data as well [39]. The new artificial intelligence approaches that we are beginning to see in many places are most often built off extensions of this idea of a neural network [5].

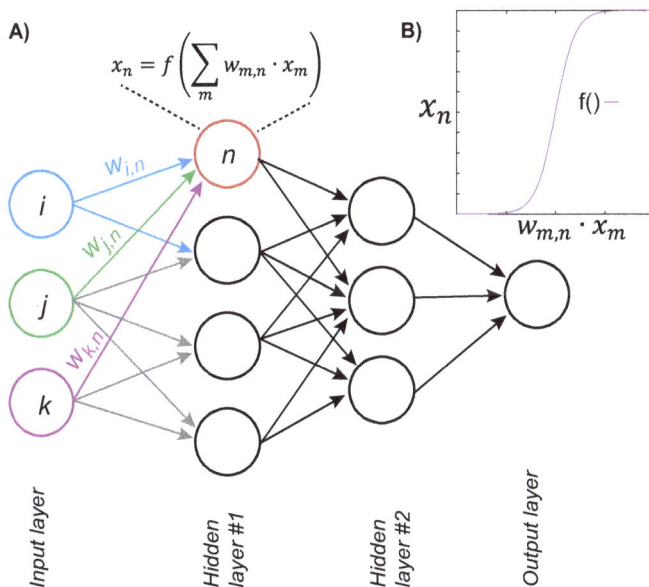

Figure 14.9 Artificial neural networks for computation. (A) The network is divided into layers, including an input layer that receives data, several hidden layers, and an output layer containing the desired result of the computation. Here we show that three input nodes i, j and k "innervate" node n: the signal from each is multiplied by a weight $w_{i,n}$ indicating its influence on n. The sum of the weighted inputs is then generally transformed by some function f, often a sigmoidal one (**B**) before determining the output of n. By controlling the connections and weights, arbitrary data classifiers can be built. (Adapted from LeCun Y, Bengio T & Hinton G (2015) *Nature* 521:436–444. With permission from Springer Nature.)

Directed Networks

Looking closely at Figure 14.1, we can see another feature of these networks that differs from Figure 14.2: the edges have arrows. What do they mean? They imply directionality to the relationships. This concept is most obvious for Figure 14.1A: saying that a lion preys on a hyena is quite different from saying that a hyena preys on a lion. For the neural network, we biologically imply that one neuron has an axon that *innervates* the dendrite of the second neuron [40]: in the computational framework we just mentioned, it might be clearer to simply say that information is flowing from the first neuron to the second one.

In fact, this notion of information flow is quite powerful because it can be applied beyond neural networks. If we look at Figure 14.1C, we can also imagine that information is flowing between transcription factors and their target genes. (Remember that transcription factor proteins bind to the promoters of those targets and activate or inhibit the production of mRNA from those genes.) In this situation we can posit that some external stimulus in the cell activates a particular transcription factor that in turn activates a group of genes to respond to that stimulus [41]. A transcription factor tied to cell temperature thus might turn genes that the cell uses to respond to excess heat (or *heat shock*). Hence, we can consider a transcriptional regulatory network as another type of computational engine that life uses to respond to its environment. We might almost call it the brain of a single-celled organism.

We will return to this idea of computation, information, and life in the last chapter. For the moment, we are going to consider some other features of transcriptional regulatory networks. In doing so, we will uncover a surprising link back to phenomena we saw in Chapter 7.

NETWORK MOTIFS AND THE TRANSCRIPTIONAL REGULATORY NETWORK

The ideas we will consider are built on network *motifs*. The idea of a network motif simply asks us to consider the different patterns of connection one can see in small sets of three or four nodes. In an undirected network like the protein interaction network, these patterns are not terribly interesting, as there are only two possible patterns where there exists a path between three nodes (**Figure 14.10A**). However,

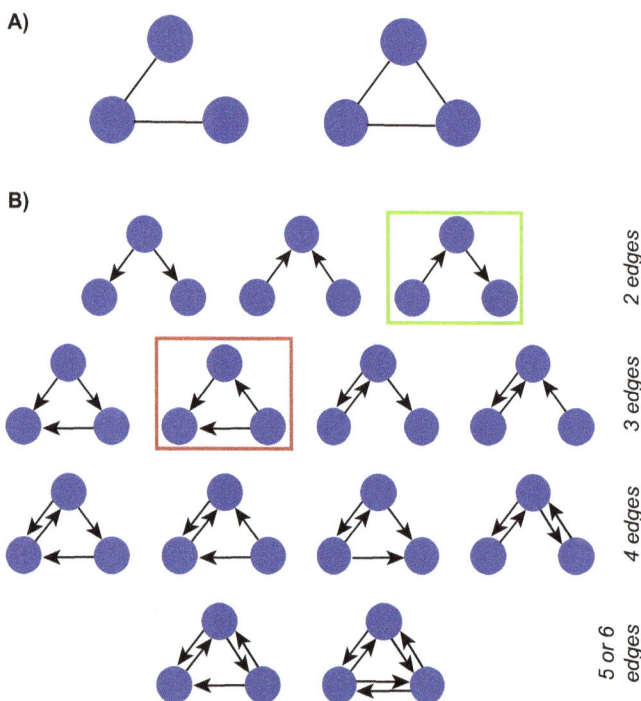

Figure 14.10 **Motifs in undirected and directed networks.** (**A**) In an undirected network there are only two "motifs" of three nodes where all three nodes are connected. (**B**) In a directed graph, there are 13 ways to connect a set of three nodes such that each node is reachable from the other two. Of these 13 motifs, three have two edges, four have three edges, four have four edges, one has five edges, and one has six edges. The "feedforward loop" motif is boxed in red and the "three-chain" motif in green.

in the case of a directed network, there are 13 possible patterns that can completely connect three nodes (**Figure 14.10B**). A clever analysis by Milo and colleagues [42] examined how common these 13 motifs were across a wide range of different directed networks, including food webs, regulatory networks, and neural networks. These authors then compared those frequencies to the motif frequencies seen in random networks with the same connection patterns. Now, this idea of "random" networks is not as obvious as it may seem. Comparisons to random networks of differing edge distributions would give different motif frequencies. So, in essence, what these authors did is to break all the links in the networks. They then reconnected those broken edges at random. By doing so they produced random networks in which each node retained the same number of links but with different connections.

In food webs, they found that a motif they referred to as a "three chain" (see the green box in Figure 14.10B) was more common than expected. That motif could be most easily described as "A eats B that eats C," and it is rather obvious why chains of this kind might be found in such networks.

In both the gene regulatory network and the neural networks, a different motif was found to be overly common: the *feedforward loop* (see the red box in Figure 14.10B). Strikingly, this same motif was found in a human-designed information-processing device: electronic circuits.

Regulatory Behavior of the Feedforward Loop

Figure 14.12 illustrates one potential pattern by which a feedforward loop might regulate a target gene. In contrast to the simple regulation of such a gene by a single transcription factor in **Figure 14.11**, with a feedforward loop, the target gene requires the simultaneous presence of a pair of transcription factors (TF1 and TF2) at its promoter for that target gene to be transcribed [43]. Because it is a feedforward loop, transcription of the second transcription factor is also activated by the binding of the first transcription factor to the promoter of the second transcription factor.

Figure 14.12 looks quite unnecessarily complicated relative to Figure 14.11. Why might evolution have chosen to regulate a gene in this fashion? One answer is to repress the gene expression noise we saw in Chapter 7. In Exercise 14.3 you can model the two regulatory circuits shown in Figures 14.11 and 14.12. When we compare the noise level for the target gene's mRNA between the simple regulation of Figure 14.11 and the feedforward loop, we find that that noise is reduced with

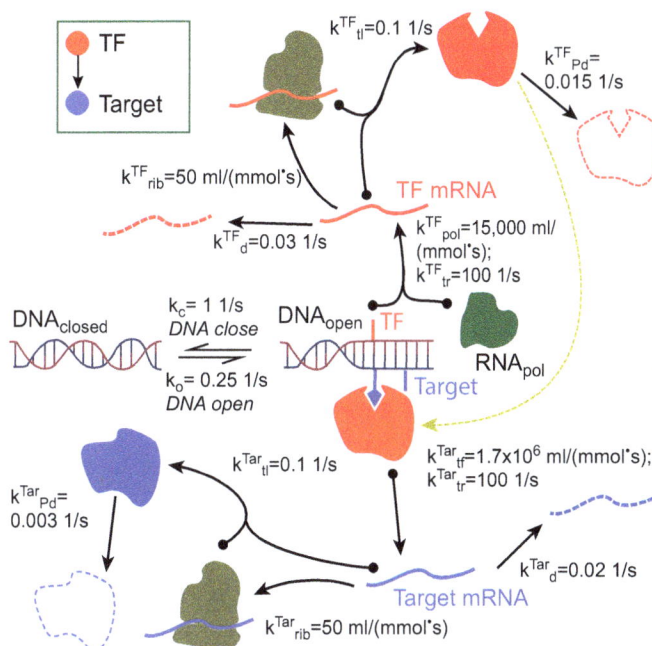

Figure 14.11 Regulation of a target gene by a single transcription factor, where the protein product of the transcription factor gene causes the expression of the target gene by binding to its DNA (*yellow arrow*). The kinetic parameters given can be used to model this system in Exercise 14.3. (For diagrammatic clarity, the transcription factor and target genes are shown on the same DNA segment, but the model does not assume they are close to each other in the genome).

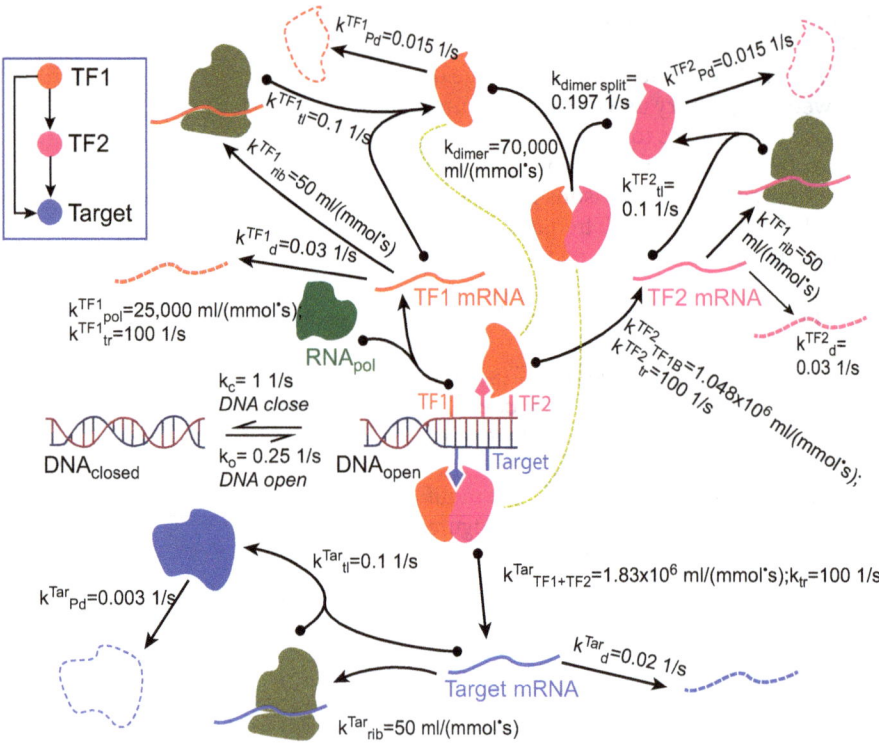

Figure 14.12 Regulation of a target gene by a pair of transcription factors TF1 and TF2. Both TF1 and TF2 must bind to each other (becoming a *dimer*) and then bind to the regulatory site of the target gene to allow transcription of that target. Meanwhile, TF1 is also required for the transcription of TF2's gene (*second yellow arrow*). Kinetic parameters are again for the models in Exercise 14.3. The behavior of this system is otherwise similar to Figure 14.11.

the feedforward loop (see Figure 14.12). In fact, several authors have discussed this idea in more detail [44–46].

What is the source of this noise reduction? In essence, we assume that the cell requires the presence of both TF1 and TF2 to transcribe the target gene. The levels of both of these proteins are subject to noise. However, if we assume that the noise in TF2 is independent of that in TF1, then, at any given instant, it is unlikely that both TF1 and TF2 will be unusually high or low. As a result, the level of the TF1 + TF2 unit seen by the target gene is less variable than the level of either protein on its own. We are in fact seeing the same behavior we considered in our discussion of the central limit theorem in Chapter 9. The cell is using the increased "sample size" of having two transcription factors to reduce the variability in the level of their sum to better control noise in its target genes. This example illustrates how quantitative biology informs on itself: because we had considered both network approaches and the kinetic models of gene expression in Chapter 7, we were able to discover something about the cell that would not have been obvious from either approach alone.

REFLECTIONS AND PREVIEWS

In this chapter we saw the power of an apparently abstract approach to quantitative biology: representing biology systems as networks of entities (nodes) and their connections. We described how these methods are especially useful in cases where there are many entities but our information about those entities is incomplete. Biological networks tend to show local structure in their connections, as shown by high clustering coefficients. This clustering likely represents biological entities working together for a common purpose, such as enzymes in a pathway. We also saw that these networks had short path lengths, implying that they are indeed *systems* and that changes in one part of the network will have effects all over the network.

Nearly every topic we have touched on in this book can be considered from a network perspective, from the biochemical networks of Chapters 5–7 to the food webs (see Figure 14.1A) that are implied by extending the Lotka–Volterra models of Chapter 3. Networks extend even to the infectious disease models of Chapter 2, where using information about individuals' social network can improve over the random contact model we employed [47]. Likewise, in phylogenetics, phylogenetic *networks* help address the problem of species not being completely genetically isolated from each other [48]. And networks, or *graphs*, as they are often referred to, come up again and again as basic computational tools in bioinformatics, where they are used to represent genome assembly problems [49], accelerate genetic analyses [50], and allow us to find optimal solutions to alignment problems [51]. As we become better at quantitative biology, these linkages across fields of biology will give us new understanding of the kinds of complex biological systems that surround us.

We also saw how networks extended across the *spatial* scale of biological questions, being applicable from systems within a single cell (metabolic, protein interaction, and regulatory networks) all the way up to the ecosystem scale (food webs). In the next chapter we will take up this question of scale in earnest, asking how processes like metabolism change as organisms get smaller and larger.

Exercise 14.1

Use the website at http://qbio.statgen.ncsu.edu/PlayBacon/ to play the Kevin Bacon for few different actors and actresses.

Exercise 14.2

Compute the *average* clustering coefficient $\bar{C}$ for the network in Figure 14.4, recalling that for nodes with a single edge $C = 1$.

Exercise 14.3

Using the two Copasi files provided (http://qbio.statgen.ncsu.edu/data/FeedForwardLoop/), compute time courses similar to Figure 14.13.

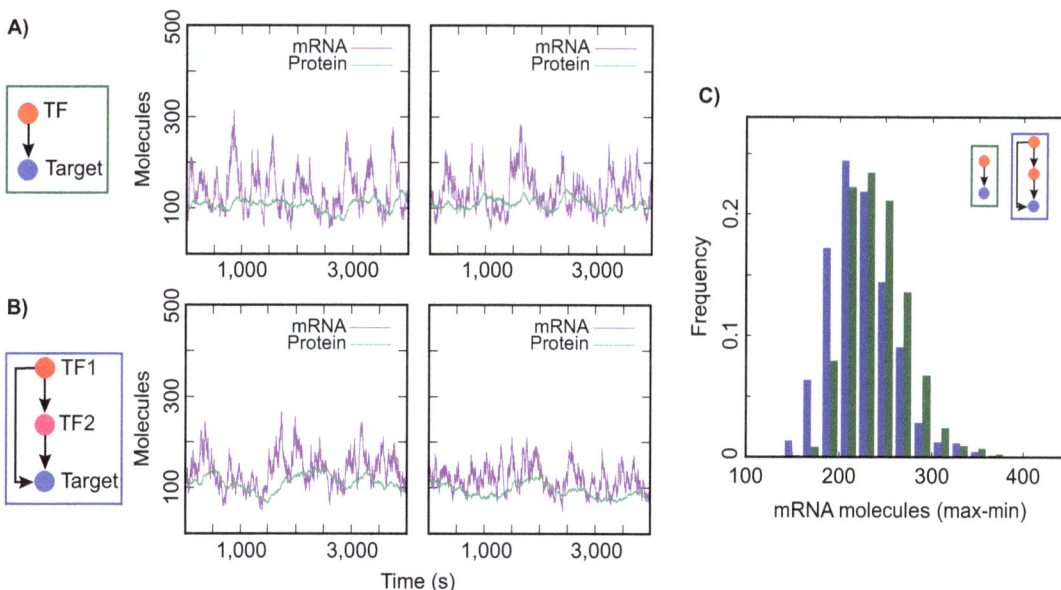

Figure 14.13 The feedforward loop and noise in gene expression. (A) Two simulated time courses for the regulation of a target gene by a single transcription factor (Figure 14.11 and Exercise 14.3). **(B)** Two time course simulations for a target gene regulated by a feedforward loop (Figure 14.12 and Exercise 14.3). The noise in the mRNA levels of the target gene appears reduced relative to **A**. **(C)** Shown are the difference between the maximum and minimum mRNA level seen across 1000 simulated time courses from **A** (green) and **B** (blue), showing that, indeed, the noise in mRNA levels is reduced by regulating a gene with a feedforward loop.

REFERENCES

1. Knoop H, Gründel M, Zilliges Y, Lehmann R, Hoffmann S, Lockau W, & Steuer R (2013) Flux balance analysis of cyanobacterial metabolism: the metabolic network of Synechocystis sp. PCC 6803. *PLoS Computational Biology* 9(6):e1003081.

2. Stark C, Breitkreutz BJ, Chatr-Aryamontri A, Boucher L, Oughtred R, Livstone MS, Nixon J, Van Auken K, Wang X, Shi X, Reguly T, Rust JM, Winter A, Dolinski K, & Tyers M (2011) The BioGRID interaction database: 2011 update. *Nucleic Acids Research* 39:D698–D704.

3. Duarte NC, Herrgård MJ, & Palsson BØ (2004) Reconstruction and validation of *Saccharomyces cerevisiae* iND750, a fully compartmentalized genome-scale metabolic model. *Genome Research* 14:1298–1309.

4. Jonsson T, Cohen JE, & Carpenter SR (2005) Food webs, body size, and species abundance in ecological community description. *Advances in Ecological Research* 36:1–84.

5. LeCun Y, Bengio Y, & Hinton G (2015) Deep learning. *Nature* 521(7553):436–444.

6. Watts DJ & Strogatz SH (1998) Collective dynamics of 'small-world' networks. *Nature* 393:440–442.

7. Shannon P, Markiel A, Ozier O, Baliga NS, Wang JT, Ramage D, Amin N, Schwikowski B, & Ideker T (2003) Cytoscape: a software environment for integrated models of biomolecular interaction networks. *Genome Research* 13(11):2498–2504.

8. Pérez-Bercoff Å, McLysaght A, & Conant GC (2011) Patterns of indirect protein interactions suggest a spatial organization to metabolism. *Molecular BioSystems* 7:3056–3064.

9. Varshney LR, Chen BL, Paniagua E, Hall DH, & Chklovskii DB (2011) Structural properties of the Caenorhabditis elegans neuronal network. *PLoS Computational Biology* 7(2):e1001066.

10. Harbison CT, Gordon DB, Lee TI, Rinaldi NJ, Macisaac KD et al., (2004) Transcriptional regulatory code of a eukaryotic genome. *Nature* 431(7004):99–104.

11. Collins JJ & Chow CC (1998) It's a small world. *Nature* 393(6684):409–410.

12. Barabási A-L (2002) *Linked: how everything is connected to everything else and what it means for business, science and everyday life* (Basic Books, New York).

13. Albert R, Jeong H, & Barabasi AL (2000) Error and attack tolerance of complex networks. *Nature* 406(6794):378–382.

14. Newman ME (2005) Power laws, Pareto distributions and Zipf's law. *Contemporary Physics* 46(5):323–351.

15. Broido AD & Clauset A (2019) Scale-free networks are rare. *Nature Communications* 10(1):1017.

16. Barabási A-L & Albert R (1999) Emergence of scaling in random networks. *Science* 286:509–512.

17. Fleischmann RD, Adams MD, White O, Clayton RA, Kirkness EF et al., (1995) Whole-genome random sequencing and assembly of *Haemophilus influenzae Rd*. *Science* 269(5223):496–512.

18. Goffeau A, Barrell BG, Bussey H, Davis RW, Dujon B, Feldmann H, Galibert F, Hoheisel JD, Jacq C, Johnston M, Louis EJ, Mewes HW, Murakami Y, Philippsen P, Tettelin H, & Oliver SG (1996) Life with 6000 genes. *Science* 274(5287):546,563–567.

19. Lander ES, Linton LM, Birren B, Nusbaum C, Zody MC et al., (2001) Initial sequencing and analysis of the human genome. *Nature* 409(6822):860–921.

20. Venter JC, Adams MD, Myers EW, Li PW, Mural RJ et al., (2001) The sequence of the human genome. *Science* 291(5507):1304–1351.

21. Wagner A & Fell DA (2001) The small world inside large metabolic networks. *Proceedings of the Royal Society of London, Series B* 268(1478):1803–1810.

22. Jeong H, Tombor B, Albert R, Oltvai ZN, & Barabási A-L (2000) The large-scale organization of metabolic networks. *Nature* 407:651–654.

23. Ito T, Chiba T, Ozawa R, Yoshida M, Hattori M, & Sakaki Y (2001) A comprehensive two-hybrid analysis to explore the yeast protein interactome. *Proceedings of the National Academy of Sciences, USA* 98(8):4569–4574.

24. Uetz P, Giot L, Cagney G, Mansfield TA, Judson RS, Knight JR, Lockshon D, Narayan V, Srinivasan M, Pochart P, Qureshi-Emili A, Li Y, Godwin B, Conovert D, Kalbfeisch T, Vijayadamodar G, Yang M, Johnston M, Fields S, & Rothbery JM (2000) A comprehensive analysis of protein-protein interactions in *Saccharomyces cerevisiae*. *Nature* 403:623–627.

25. Ho Y, Gruhler A, Heilbut A, Bader GD, Moore L et al., (2002) Systematic identification of protein complexes in Saccharomyces cerevisiae by mass spectrometry. *Nature* 415(6868):180–183.

26. Gavin AC, Bosche M, Krause R, Grandi P, Marzioch M et al., (2002) Functional organization of the yeast proteome by systematic analysis of protein complexes. *Nature* 415(6868):141–147.

27. Newman ME (2006) Modularity and community structure in networks. *Proceedings of the National Academy of Sciences USA* 103(23):8577–8582.

28. Hahn MW, Conant GC, & Wagner A (2004) Molecular evolution in large genetic networks: Connectivity does not equal constraint. *Journal of Molecular Evolution* 58(2):203–211.

29. Jeong H, Mason SP, Barabási A-L, & Oltvai ZN (2001) Lethality and centrality in protein networks. *Nature* 411:41–42.

30. Winzeler EA, Shoemaker DD, Astromoff A, Liang H, Anderson K et al., (1999) Functional characterization of the *S. cerevisiae* genome by gene deletion and parallel analysis. *Science* 285:901–906.

31. Srere PA (2000) Macromolecular interactions: tracing the roots. *Trends Biochem Sci* 25(3):150–153.

32. Gaertner FH (1978) Unique catalytic properties of enzyme clusters. *Trends in Biochemical Sciences* 3(1):63–65.

33. Hyde CC, Ahmed SA, Padlan EA, Miles EW, & Davies DR (1988) Three-dimensional structure of the tryptophan synthase alpha 2 beta 2 multienzyme complex from Salmonella typhimurium. *Journal of Biological Chemistry* 263(33):17857–17871.

34. Yanofsky C & Rachmeler M (1958) The exclusion of free indole as an intermediate in the biosynthesis of tryptophan in Neurospora crassa. *Biochim Biophys Acta* 28(3):640–641.

35. Winkel BS (2004) Metabolic channeling in plants. *Annual Review of Plant Biology* 55:85–107.

36. Ellis RJ (2001) Macromolecular crowding: obvious but underappreciated. *Trends in Biochemical Sciences* 26(10):597–604.

37. Huthmacher C, Gille C, & Holzhutter HG (2008) A computational analysis of protein interactions in metabolic networks reveals novel enzyme pairs potentially involved in metabolic channeling. *Journal of Theoretical Biology* 252(3):456–464.

38. Sulston JE & Horvitz HR (1977) Post-embryonic cell lineages of the nematode, Caenorhabditis elegans. *Developmental Biology* 56(1):110–156.

39. Flake GW (1998) *The computational beauty of nature: Computer explorations of fractals, chaos, complex systems, and adaptation* (The MIT Press, Cambridge, MA).

40. Marieb EN (1991) *Human anatomy and physiology* (The Benjamin/Cummings Publishing Company, Redwood City, CA), p. 1040.

41. Luscombe NM, Babu MM, Yu H, Snyder M, Teichmann SA, & Gerstein M (2004) Genomic analysis of regulatory network dynamics reveals large topological changes. *Nature* 431 (7006):308–312.

42. Milo R, Shen-Orr S, Itzkovitz S, Kashtan N, Chklovskii. D, & Alon U (2002) Network motifs: simple building blocks of complex networks. *Science* 298:824–827.

43. Pires JC & Conant GC (2016) Robust yet fragile: expression noise, protein misfolding and gene dosage in the evolution of genomes. *Annual Review of Genetics* 50(1):113–131.

44. Prill RJ, Iglesias PA, & Levchenko A (2005) Dynamic properties of network motifs contribute to biological network organization. *PLoS biology* 3(11):e343.

45. Klemm K & Bornholdt S (2005) Topology of biological networks and reliability of information processing. *Proceedings of the National Academy of Sciences, USA* 102(51):18414–18419.

46. Alon U (2007) Network motifs: theory and experimental approaches. *Nature Reviews Genetics* 8(6):450–461.

47. Keeling MJ (1999) The effects of local spatial structure on epidemiological invasions. *Proceedings of the Royal Society of London. Series B: Biological Sciences* 266(1421):859–867.

48. Huson DH & Bryant D (2006) Application of phylogenetic networks in evolutionary studies. *Molecular Biology and Evolution* 23(2):254–267.

49. Pop M (2009) Genome assembly reborn: recent computational challenges. *Brief Bioinform* 10(4):354–366.

50. Conant GC, Plimpton SJ, Old W, Wagner A, Fain PR, Pacheco TR, & Heffelfinger G (2003) Parallel Genehunter: Implementation of a linkage analysis package for distributed-memory architectures. *Journal of Parallel and Distributed Computing* 63:674–682.

51. Conant GC & Wagner A (2004) A fast algorithm for determining the longest combination of local alignments to a query sequence. *BMC Bioinformatics* 5:62.

Scale
Metabolic Rate, Body Size, and Fractal Geometry

"To see a World in a Grain of Sand And a Heaven in a Wild Flower
Hold Infinity in the palm of your hand And Eternity in an hour"
—William Blake, *"Auguries of Innocence"*

Figure 15.1 depicts a fact you might not have considered: how much difference there is in the sheer size of the various living things. An *E. coli* cell is something like *one hundred million* times smaller in its length than the largest single organism we know, a giant sequoia. Scientists speak of these powers of 10 shown in Figure 15.1 as *orders of magnitude*.

Moreover, biological processes occur at even smaller and larger scales. **Figure 15.2** illustrates the components of cells, including proteins and DNA molecules: these occupy sizes intermediate between that of cells and of the atomic scale. Of course, all of the processes in Figures 15.1 and 15.2 are occurring in ecosystems that exist on the scale of tens to hundreds of kilometers. Moreover, as we discussed in Chapter 5, all these processes reshape the environment on a global scale. One such global change was the introduction of high levels of oxygen into

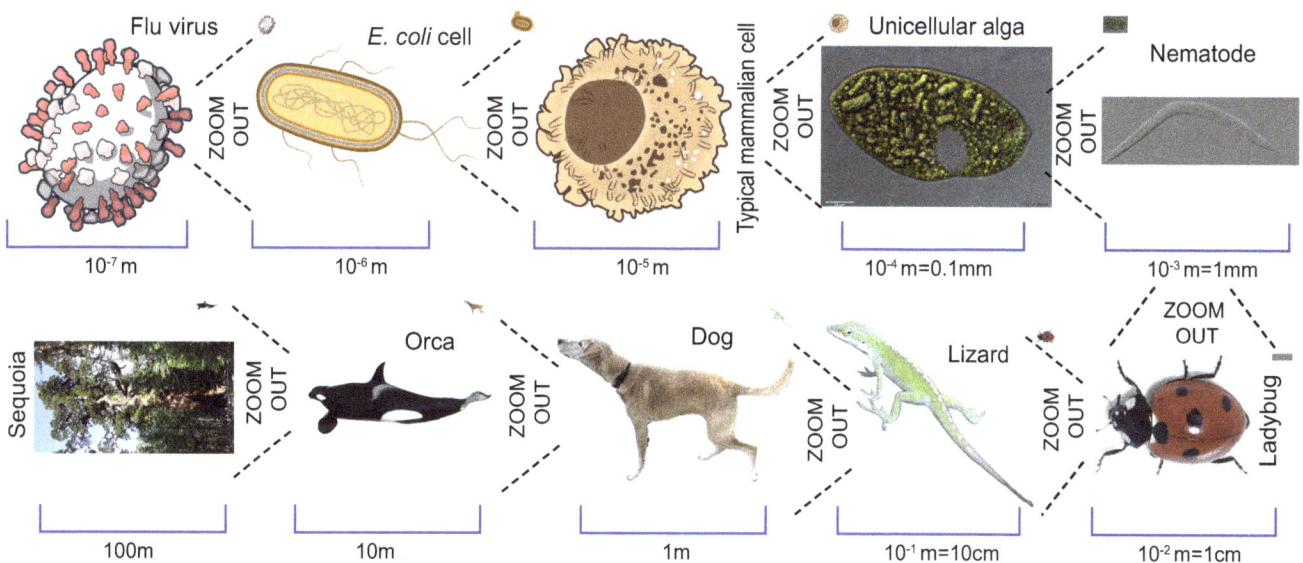

Figure 15.1. Organisms range in their length across approximately 10 orders of magnitude. Starting from arguably the smallest complete biological entity, a virus, we can find organisms (or parts of an organism, in the case of a mammalian cell) that span sizes from the micrometer (*E. coli*) up to a tenth of a kilometer in the case of a giant sequoia. (Virus, bacteria, and mast cell art all courtesy of NIAID. Pellicle of Euglena image courtesy of David Shykind, on Wikimedia, released into the public domain. *C. elegans* image courtesy of Kbradnam, on Wikimedia, published under CC BY-SA 2.5 license. Giant sequoia image courtesy of Mike Murphy, on Wikimedia, published under CC BY-SA 2.0 license. Image references: virus [40], E. coli (edited [41]), mammalian cell [42], alga [43], nematode [44], sequoia [45].)

DOI: 10.1201/9781003687504-15

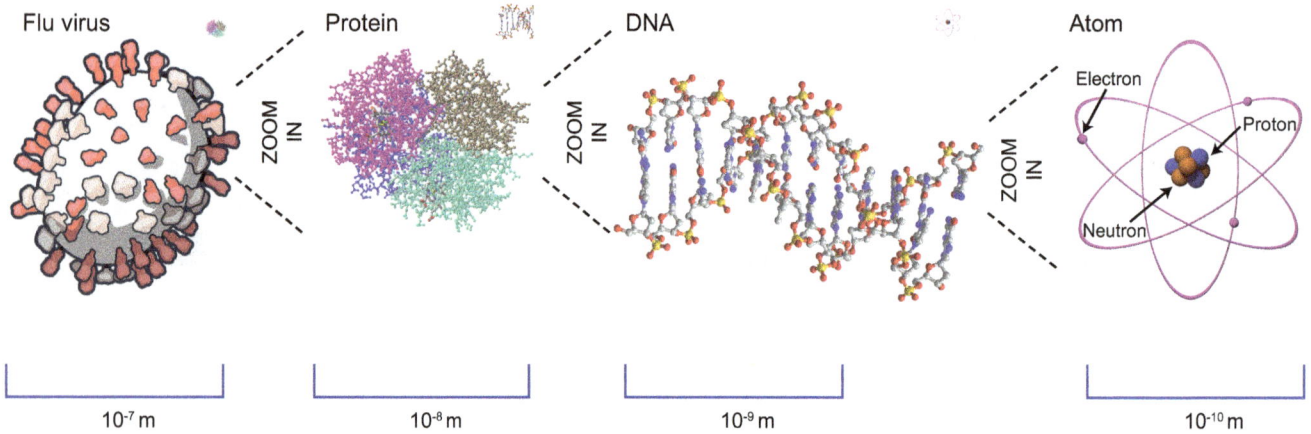

Figure 15.2 **Biological processes extend** *down* **in scale several more orders of magnitude from a virus [40], running from protein molecules (shown is fetal hemoglobin, PDB ID 4MQJ) to bases of DNA (PDB ID: 1BNA) to atoms.** (Virus art courtesy of NIAID.)

the atmosphere. That new abundance of molecular oxygen allowed the lifeforms that use large quantities of it to engage in new biochemical activities and lifestyles that were not possible prior to its widespread availability [1].

BASAL METABOLIC RATE IN DIFFERENT ORGANISMS

We have returned to this point about oxygen in the atmosphere because it invites us to consider more carefully the energy needs of animals and plants of differing sizes. For animals such as ourselves, oxygen forms the final electron acceptor in our use of various organic (i.e., carbon-rich) molecules as biochemical fuel [2]. If we place a human being in a room with a comfortable temperature (perhaps 25°C) and ask her or him to simply relax, how much energy does it take for that person to just "do nothing"? This rate of energy consumption is called the *basal metabolic rate*, and for human beings, it is, very approximately, 85W [3]. Just for perspective, this is the power required to continuously run about 1½ old-style incandescent lightbulbs.

A very natural question that arises is how this power level differs between animals of different sizes. That question has interested biologists for nearly two centuries now [4, 5]. Before we explore those researchers' discoveries, however, we will take a brief detour to see what progress we can make on this problem on our own.

We will start with the Etruscan shrew (*Suncus etruscus*), the mammal with the smallest known mass: about 2 grams [6]. Although it is intuitive to consider metabolic rates quantified in watts, researchers more often measure the oxygen uptake rate [7], and so we will use these values for comparing species. Doing so will also avoid the assumptions necessary in converting such values to power units. In these terms, the Etruscan shrew consumes 14.4 ml of O_2 per hour [7].

A natural assumption would be that the metabolic rate increases in direct proportion to the volume of the animal in question. We would make such a prediction on the presumption that metabolic processes are distributed relatively evenly through that body. Fortunately, animals are typically rather similar in their body composition, meaning that, very roughly, we can use body *mass* as a reasonable proxy for volume [8], under the equality we saw in Chapter 7 of 1 ml of water having a mass of 1 gram. You can explore the relationship of mass to metabolic rate in Exercise 15.1.

Exercise 15.1

Using the metabolic rate of the Etruscan shrew (14.4 ml O_2/hour) and its mass (2.4 g), use Excel to predict the metabolic rate of a meerkat (mass of 850g) and an elk (mass of 325 kg). Notice that the shrew has a metabolic rate to body mass ratio of 14.4/2.4 = 6. Compare your predictions to the observed values for the two animals (meerkat: 310 ml O_2/h; elk: 51,419 ml O_2/h). See **Figure 15.3**.

A)

B)

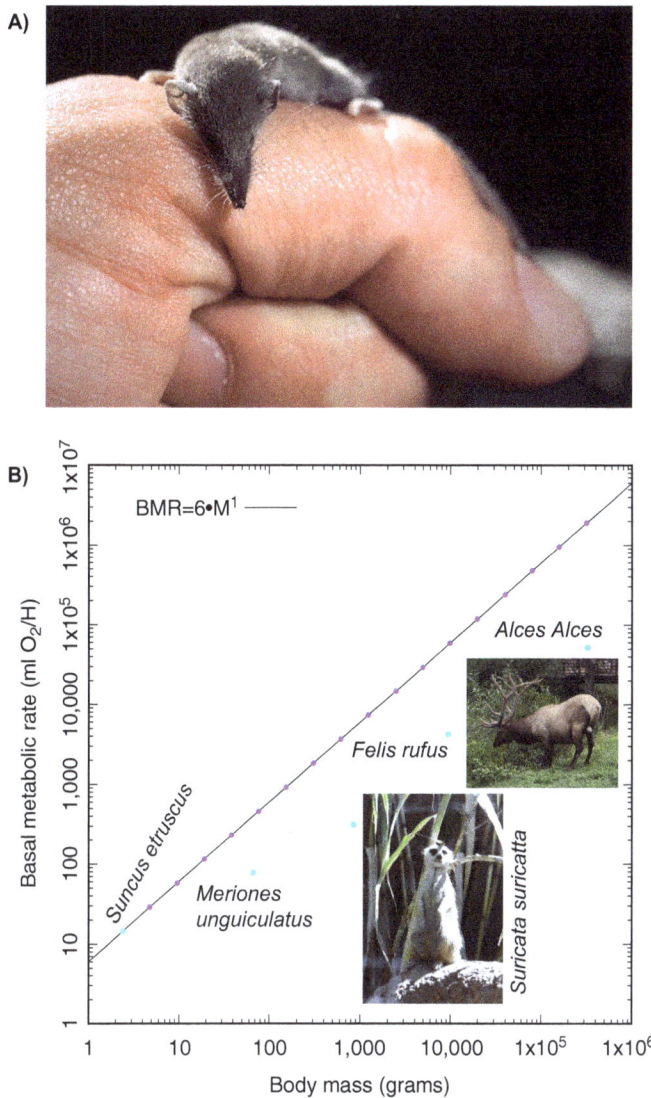

Figure 15.3 *Scaling* **of the metabolic rate with body size in mammals.** (**A**) The smallest known mammal, the Etruscan shrew, *Suncus etruscus*, [46], has a mass of about 2.5 grams. It consumes about 14 ml of O_2 per hour in a resting state [6, 7]. (**B**) If we directly scale this ratio of 6 ml O_2/hour per gram of body mass, our predictions of the metabolic rates of several other mammals (gerbils, meerkats, bobcats, and elk) are quite poor. However, the metabolic rates of these species do appear to follow a rough linear trend on the log-log axes of **B**. (Data from White CR & Seymour RS (2003) *PNAS* 100:4046–4049. A, courtesy of Trebol-a, on Wikimedia, published under CC BY-SA 3.0 license.)

Figure 15.3 shows the predicted metabolic rates for a few mammals under the assumption that that rate scales *directly* with body mass: in other words, that a doubling of body mass produces a doubling of metabolic rate. Quite obviously, our hypothesized direct relationship does not explain observed metabolic rates very well.

Why might our prediction be so far off? A primary reason is that our argument about the distribution of metabolic activity in animals' bodies is incomplete. While it makes sense to imagine metabolic activity distributed across the animal's mass, not all the metabolic activities of the animal can be thought of in this way. Consider the problems of absorbing nutrients and disposing of waste to the environment. These processes involve moving matter into or out of the organism. For a three-dimensional animal, that movement occurs across a two-dimensional surface, such as the lining of the intestines, the lungs, or the tubules of the kidneys [3].

As we remind ourselves with **Figure 15.4**, the relationship between the surface area and the volume of a simple geometrical object like a sphere is nonlinear. The surface area of a sphere of radius r is given by:

$$SA = 4\pi r^2 \tag{15.1}$$

while its volume is given by:

$$V = \frac{4}{3}\pi r^3 \tag{15.2}$$

A)

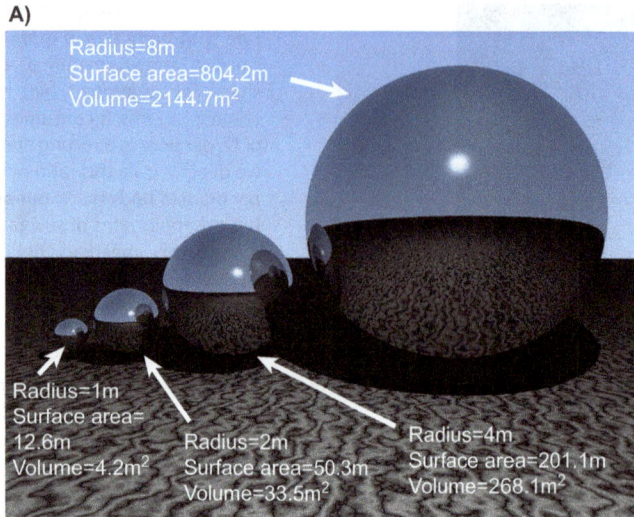

B)

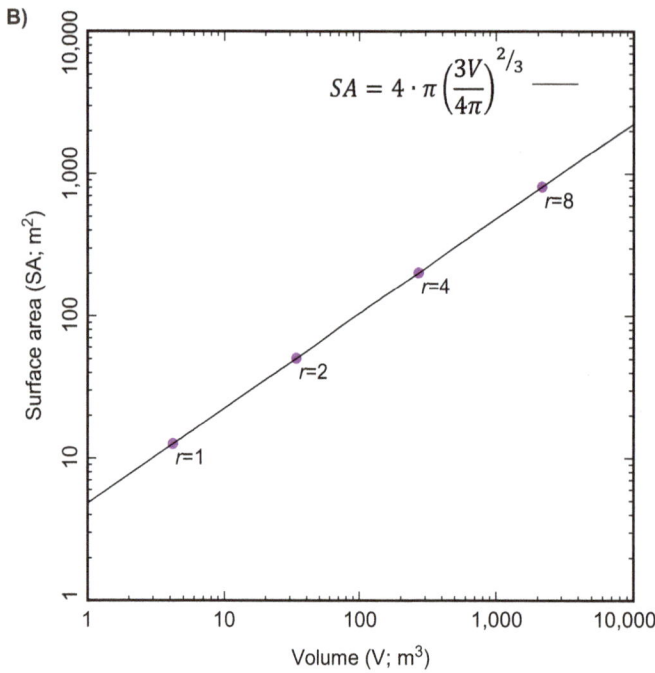

Figure 15.4 **The relationship of surface area to volume in objects of differing size.** (**A**) The relationship of surface area to volume for a series of spheres of increasing volume. (**B**) The ratio of the volume of a sphere to its surface area, plotted on a log-log scale. In Exercise 15.2, you can compute the surface area and volume of these four spheres and plot these values on a log-log scale to verify that they match 15.4B.

Hence, as we change the unit dimension r, the volume increases more quickly than does the surface area.

We can express the surface area of a sphere as a function of its volume by first solving for r in Equation 15.2 to yield:

$$r = \left(\frac{3V}{4\pi}\right)^{1/3} \tag{15.3}$$

reminding ourselves that raising a number to the power of 1/3 is equivalent to taking the cubic root. If we now substitute the value of r from Equation 15.3 into Equation 15.1, we get:

$$SA = 4\pi\left(\left(\frac{3V}{4\pi}\right)^{1/3}\right)^2 \tag{15.4}$$

or:

$$SA = 4\pi \left(\frac{3V}{4\pi} \right)^{2/3} \tag{15.5}$$

In Figure 15.4B, we have plotted Equation 15.5 on a log-log scale, meaning that both the x and y axes are logarithmic. We see that in this log-log framework, the relationship of volume to surface area follows a straight line. By taking the log of both sides of Equation 15.5, we see both why this is the case and what the slope of that line is:

$$\log_{10}(SA) = \log_{10}\left(4\pi \left(\frac{3V}{4\pi} \right)^{2/3} \right) \tag{15.6}$$

By the rules of logarithms, $\log(x \cdot y) = \log(x) + \log(y)$, allowing us to write:

$$\log_{10}(SA) = \log_{10}(4\pi) + \log_{10}\left(\left(\frac{3V}{4\pi} \right)^{2/3} \right) \tag{15.7}$$

Because it is also the case that $\log(x^y) = y\log(x)$, we can further simplify this expression to:

$$\log_{10}(SA) = \log_{10}(4\pi) + \frac{2}{3}\log_{10}\left(\frac{3V}{4\pi} \right) \tag{15.8}$$

In other words, the function described by Equation 15.5 is indeed a line in log-log space and, more interestingly, that line has a slope of 2/3. Looking back at Figure 15.3B, we see that the metabolic rate does appear to follow a rough line on this log-log plot and that the slope of that line is indeed less than 1.0. So perhaps the necessity of feeding a three-dimensional body over two-dimensional surfaces is what dictates the slope of the mass to metabolic rate relationship.

Two-Thirds versus Three-Fourths: What Is the Metabolic Rate to Body Size Slope?

Of course, we are hardly the first to consider the question of the scaling of metabolic rate with body size. As illustrated in **Figure 15.5**, researchers have considered this problem for more than a century [5], using plots similar to that of Figure 15.3. These researchers were keenly aware of the potential importance of the surface area to volume problem of Figure 15.4. They were, therefore, in some sense expecting a 2/3 slope in their plots [5, 9]. However, as suggested by Figure 15.5, it was common to observe a higher slope, often close to ¾ [10]. Two natural questions thus arise. First, using the data in these figures, how do we place the line and compute its slope? This problem looks especially challenging in cases like Figure 15.5C, D, where there are many points that do not fall naturally onto a single line. Second, is there some biological rule that could explain this arguably larger than expected scaling exponent?

DIGRESSION 1: LEAST-SQUARES REGRESSION

We will tackle the first problem first. How do we fit a line through clouds of points such as those in Figure 15.5C, D? To answer this question will require us to take a brief digression back into the matrix mathematics we first explored in Chapter 6.

A)

B)

C)

D)

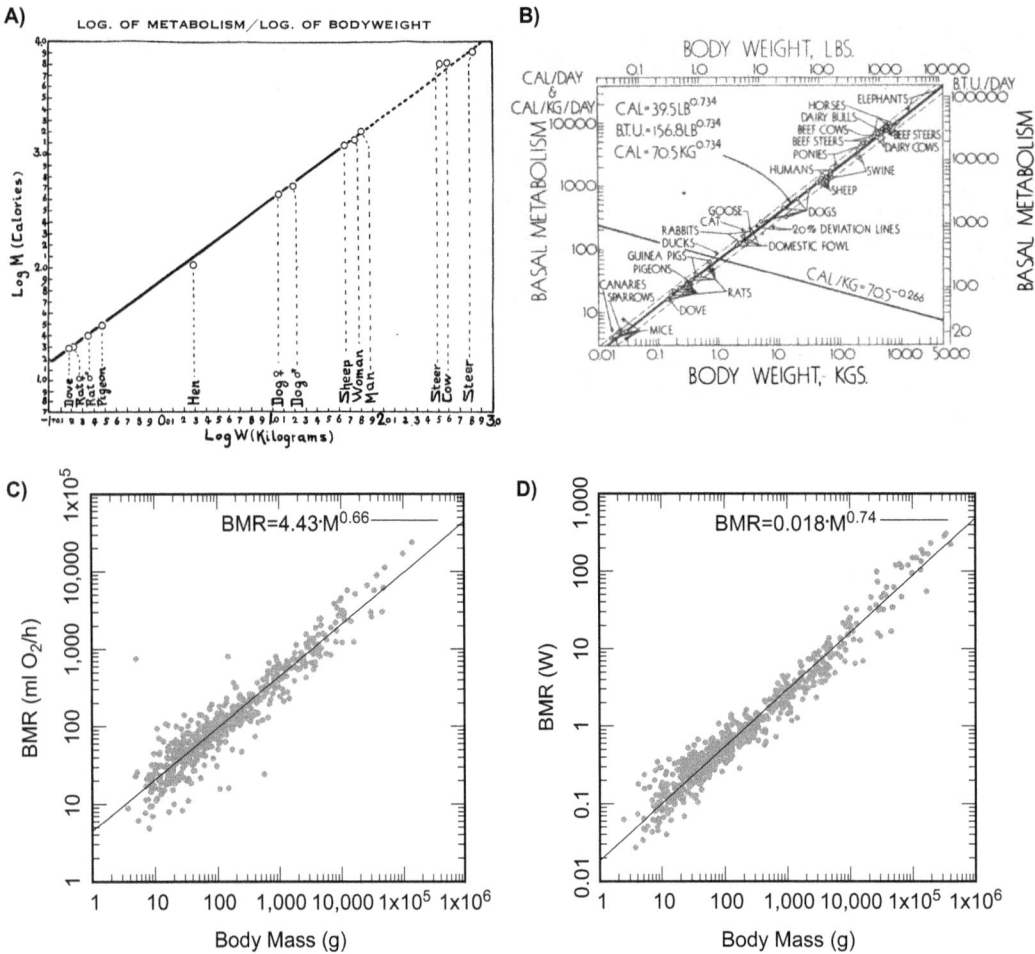

Figure 15.5 **More than 70 years of comparisons of body mass versus basal metabolic rate.** (**A**) Data from Kleiber in 1932: the slope of the line on the log-log plot is approximately 0.74 [9, 14]. (**B**) From Brody in 1945: note the 0.73 exponent/slope [5]. (**C**) Data from White and Seymour in 2005 [12]. (**D**) From Savage et al., showing a slope/exponent of 0.737 [10].

Matrices and Vectors: An Alternative Representation of Lines and Planes

The first concept we need for our problem of fitting a line through points is that of a *vector* in some N-dimensional space. In **Figure 15.6** I illustrate such vectors in three dimensions: in other words, a line connecting the origin at 0,0,0 to another point in $x - y - z$ space. If we accept this concept, we can see that two vectors that do not fall on the same line are sufficient to delimit a plane that passes through 0,0,0 (Figure 15.6). If this rule appears unclear to you, imagine asking if you could lay a piece of cardboard on the vectors and have it balance. A single vector would allow the cardboard to rotate, but two pieces would hold it stably. Since we can write down a vector in the notation of Chapter 6 as:

$$\bar{v} = \begin{vmatrix} x \\ y \\ z \end{vmatrix} \tag{15.9}$$

we can see that every point in $x - y - z$ space can be thought of as representing a vector connecting that point to the origin. What does it mean for two vectors $\bar{v}_1$ and $\bar{v}_2$ to fall on the same line in this framework? Simply that $\bar{v}_1 = c\bar{v}_2$: the two vectors differ by multiplication by a constant.

In the section that follows I will confine my examples to two and three dimensions for clarity. However, all the ideas I will describe extend naturally to more

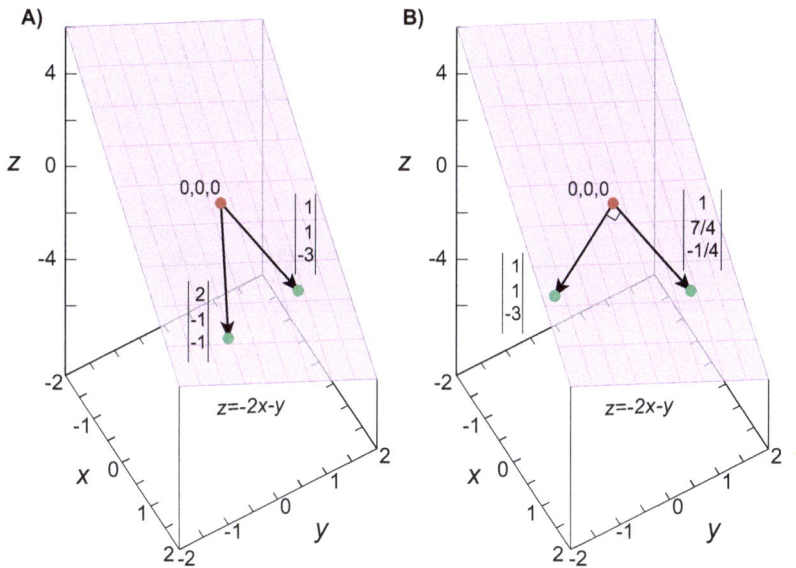

Figure 15.6 Points in space can be thought of as representing vectors connecting a point to the origin (red dot at $x = 0$, $y = 0$, and $z = 0$). (A) Two *independent* vectors (i.e., vectors that do not lie along the same line) are sufficient to define a plane [11, 14], in this case $z = -2x + y$. **(B)** If the two vectors chosen are perpendicular to each other, we call them *orthogonal*. In Exercise 15.3, you can verify that the dot product of the two vectors in 15.6B is 0.

dimensions. In that case, we would speak of a *hyperplane* instead of a plane. However, I think the value of using such general language is probably less than the distraction it introduces, so I will avoid it.

Representing Planes as Combinations of Vectors

Two vectors $\bar{u}$ and $\bar{v}$ that *do not* fall on the same line are said to be independent. Because they define a plane, we can represent any other point in the plane that they define as a combination of those two vectors, each multipled by a different constant. We can represent this process by listing the two vectors as columns of a matrix, which we multiply by an unknown, two-element vector $\bar{c}$. In $\bar{c}$, the two unknown constants c_1 and c_2 can take on values $-\infty < c_1, c_2 < \infty$. For the vectors in Figure 15.6A, this computation would have the form:

$$\begin{vmatrix} 2 & 1 \\ -1 & 1 \\ -1 & -3 \end{vmatrix} \cdot \begin{vmatrix} c_1 \\ c_2 \end{vmatrix} = \begin{vmatrix} x \\ y \\ z \end{vmatrix} \tag{15.10}$$

which we evaluate by the rules of matrix-vector multiplication we discussed in Chapter 6. Any point on the plane $z = -2x - y$ can be represented by picking appropriate values of c_1 and c_2.

Dot Products and Vector Projection

In addition to multipling vectors by matrices or constants, we can also multiply vectors by each other. The most common form of this multiplication is called a *dot product* and is illustrated in **Figure 15.7**. To form the dot product of two vectors $\bar{u}$ and $\bar{v}$ we take what is called the *transpose* of $\bar{u}$. The transpose of a matrix or vector is simply the result of turning all of the rows of the vector (or matrix) into columns. Alternatively, we can think of rotating that vector or matrix 90° counterclockwise. For example, if $\bar{u}$ is given by:

$$\bar{u} = \begin{vmatrix} u_1 \\ u_2 \\ u_3 \end{vmatrix} \tag{15.11}$$

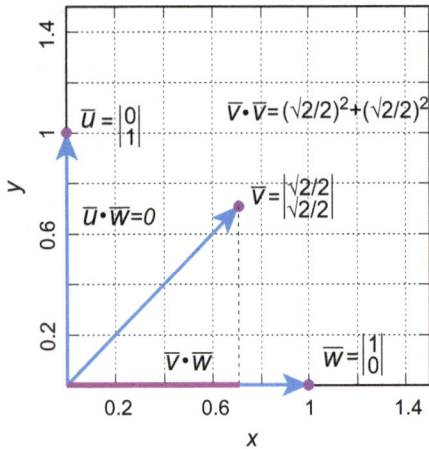

Figure 15.7 Vectors, vector dot products, and orthogonality. The three vectors $\bar{u}, \bar{v}$, and $\bar{w}$ are all of unit length, meaning that the sum of the squares of their x and y coordinates is 1. Vector $\bar{v}$ lies on the line $y = x$: when we take the dot product of it with $\bar{w}$ (or $\bar{u}$), we get a *scalar* value corresponding to the part of $\bar{v}$ that falls on the x-axis (or the y-axis for $\bar{u}$). The length of that projection is $\sqrt{2} / 2$. Vectors $\bar{u}$ and $\bar{w}$ are perpendicular, or *orthogonal*, to each other; their dot product is therefore 0.

then the transpose is named $\bar{u}^T$ and is written:

$$\bar{u}^T = \begin{vmatrix} u_1 & u_2 & u_3 \end{vmatrix} \tag{15.12}$$

It is first instructive to consider the dot product of $\bar{u}$ with itself:

$$\bar{u} \cdot \bar{u} = \begin{vmatrix} u_1 & u_2 & u_3 \end{vmatrix} \begin{vmatrix} u_1 \\ u_2 \\ u_3 \end{vmatrix} \tag{15.13}$$

By the rules of matrix-vector multiplication, this product works out to:

$$\bar{u} \cdot \bar{u} = u_1^2 + u_2^2 + u_3^2 \tag{15.14}$$

Looking at the right-hand side of Equation 15.14, we can see that it is the three-dimensional version of the Pythagorean theorem [11], meaning that the dot product of a vector with itself gives the square of the length of that vector. Notice that in Figure 15.7, all three vectors produce a dot product of 1.0 when multiped by themselves: such *unit-length* vectors are useful when we wish to compare the direction of vectors without worrying about their length.

When we take the dot product of two nonidentical vectors, we find something interesting: the dot product *projects* the part of the total length of first vector that is in the direction as the second. We can make this result particularly useful if we assume that the second vector, $\bar{v}$, is of unit length. In that case, the scalar value produced by computing $\bar{u} \cdot \bar{v}$ gives the part of $\bar{u}$'s length that falls in the direction of $\bar{v}$. We can then write the *projection* of $\bar{u}$ in the direction of $\bar{v}$ as:

$$(\bar{u} \cdot \bar{v}) \bar{v} \tag{15.15}$$

Since the $\bar{u} \cdot \bar{v}$ dot product is a scalar, this equation simply multiples the vector $\bar{v}$ by this scalar, giving the part of the vector $\bar{u}$ that falls in the direction of $\bar{v}$.

Importantly, this concept can be extended to the case of multiple vectors. If we assume we have vectors $\bar{v}$ and $\bar{w}$ that are orthogonal to each other and are each of unit length, then it is clear that between them they define a plane. We can then project another vector $\bar{u}$ onto that plane to produce the new vector u_p. This computation would have the form:

$$\bar{u_p} = (\bar{u} \cdot \bar{v}) \bar{v} + (\bar{u} \cdot \bar{w}) \bar{w} \tag{15.16}$$

Figure 15.8 gives a geometrical illustration of this principle. What we have done is in fact more interesting that just projecting $\bar{u}$ into the $\bar{v} \times \bar{w}$ plane: we have also, in making that projection, found the closest point to $\bar{u}$ in that plane. Knowing this,

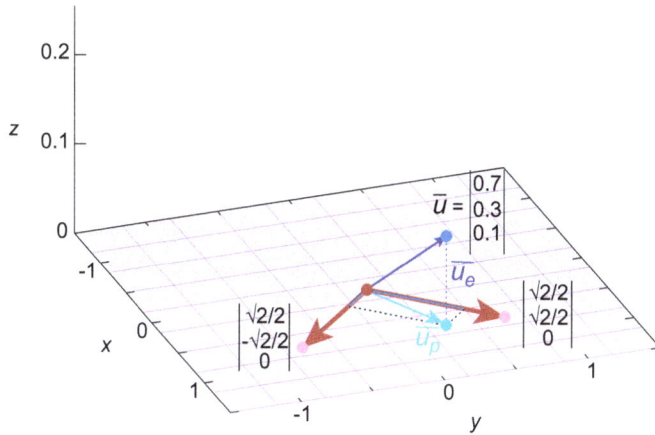

Figure 15.8 Using dot products to *project* a vector onto a plane. The vector $\bar{u}$ rises out of the $x - y$ plane into the z dimension. To find the closest point to that vector in the $x - y$ plane, we first create a pair of unit-length, orthogonal vectors in the $x - y$ plane (the red lines following $y = x$ and $y = -x$). We take the dot product of $\bar{u}$ with each of these vectors to give its projection along that vector (blue lines on top of the unit vectors). The sum of those two projections is $\bar{u}_p$—the point in the $x - y$ plane closest to $\bar{u}$, namely $(0.7, 0.3, 0)$. The "remainder" of the vector $\bar{u}$ projects perpendicular to the plane and can be thought of as the "error" in the projection, $\bar{u}_e$.

we also have the ability to compute the *distance* between $\bar{u}$ and the $\bar{v} \times \bar{w}$ plane. To do that, we subtract $\bar{u}_p$ from $\bar{u}$ to form the *error* vector $\bar{u}_e$. Notice that $\bar{u}_e$ is orthogonal to $\bar{v}$ and $\bar{w}$, which makes sense because any part of $\bar{u}$ that is *not* orthogonal to the $\bar{v}$ x $\bar{w}$ plane is part of $\bar{u}_p$ rather than $\bar{u}_e$. Moreover, if we transform $\bar{u}_e$ to unit length, then the dot product of $\bar{u}$ with $\bar{u}_e$ gives the distance from $\bar{u}$ to the $\bar{v} \times \bar{w}$ plane.

Orthogonality and Dot Products

There is one last property of dot products that is implicit in this projection idea but that we should make explicit. The dot product of two orthogonal (i.e., perpendicular) vectors is 0. We can therefore see that $\bar{v} \cdot \bar{w}$ in Figure 15.8 is 0, which is the test of their orthogonality. Similarly, since both $\bar{v}$ and $\bar{w}$ have 0s for their z component, they are both orthogonal to the z vector $(0,0,1)$. We will use both this property of the dot product and its ability to project between vectors in our technique to fit lines to a set of points.

The Least Squares Approach and the Best-Fit Line

Having arranged these mechanics of vectors and matrices, we can return to the problem at hand. **Figure 15.9** depicts some x, y data for which the y values appear to depend, at least in part, on the value of x. Circling back to our metabolic rate versus body size questions, we would like to fit the best possible line through this cloud of points. We will define *best possible* to mean that the sum of the squared distances between the data points and this line should be as small as possible. Figure 15.9 depicts the calculation of two such squared distances between points and the best-fit line.

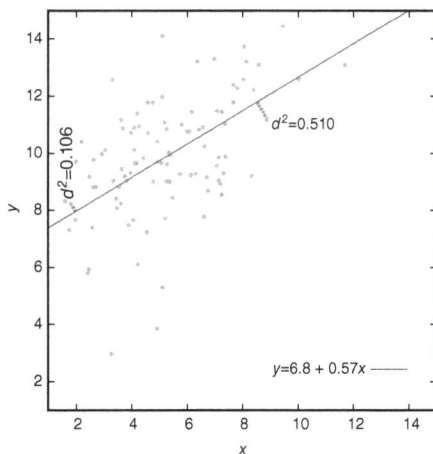

Figure 15.9 The line of least-squared distance to a set of *x*, *y* points. The points were generated from a bivariate normal distribution as described in Exercise 15.4. The equation of the line of best fit is given, and for two of the points, the squared Euclidian distance (d^2) between that point and the best-fit line is shown with dashed lines.

If necessary, we could use a programming approach such as the ones in Chapter 10 to fit this line. With that approach we could try many different lines and find the one with the minimal squared distance. However, the vector tools we have just discussed allow a much neater solution.

First we need to frame our problem as one of matrix-vector multiplication, similar to what we did in Chapter 6. We will begin by storing our x, y pairs (i.e., points) as two vectors of length n:

$$\bar{x} = \begin{vmatrix} x_1 \\ x_2 \\ \vdots \\ x_n \end{vmatrix}, \bar{y} = \begin{vmatrix} y_1 \\ y_2 \\ \vdots \\ y_n \end{vmatrix} \tag{15.17}$$

Were it the case that the x, y pairs fell precisely on a line, we could write $y = mx + b$ for all pairs. Representing this idea in matrix-vector form could be written as:

$$\begin{vmatrix} x_1 & 1 \\ x_2 & 1 \\ \vdots & \vdots \\ x_n & 1 \end{vmatrix} \begin{vmatrix} m \\ b \end{vmatrix} = \begin{vmatrix} y_1 \\ y_2 \\ \vdots \\ y_n \end{vmatrix} \tag{15.18}$$

Here we have added a column of 1s to the vector of $x's$ to form a new matrix A. We have also added a vector whose two components are the unknown slope m and the unknown intercept b. The new 1s in A represent the as-of-yet undetermined y-intercept of our equation (b in the vector of length 2). We can now rewrite Equation 15.18 as:

$$A \begin{vmatrix} m \\ b \end{vmatrix} = \bar{y} \tag{15.19}$$

Hence, in our ideal world where every point falls on the same line, our problem is simply to use the tools of linear algebra to find values of m and b.

Were we to have only two x, y pairs, they would necessarily fall on the same line, and we could solve Equation 15.18 exactly. However, with more pairs, the system of equations comprising Equation 15.18 will in general not be solvable because there are no values of m and b that perfectly map every x value onto its corresponding y value.

However, we can now flip our perspective to find an approach. The scheme of Figure 15.8 applies not only to three dimensions but to as many dimensions as we wish (even if we cannot visualize them). The two columns of the matrix A comprise two nonidentical vectors in such an n-dimensional space. We can, therefore, assume that there exist two orthogonal unit vectors $\bar{v}$ and $\bar{w}$ that describe the same *hyper-plane* as does the matrix A. Now, the vector $\bar{y}$ *does not* fall into the hyperplane of A, again because the y values of our data do not all exactly fall onto a line. However, if we were to take the dot product of $\bar{y}$ with $\bar{v}$ and $\bar{w}$, we would be projecting $\bar{y}$ onto the plane of A in exactly the approach of Figure 15.8. We will refer to this projection as $\bar{y_p}$. Even though there is no solution to Equation 15.18 for the actual vector $\bar{y}$, because $\bar{y_p}$ is a projection of $\bar{y}$ into the hyper-plane of A, we can completely accurately write:

$$A \begin{vmatrix} m \\ b \end{vmatrix} = \bar{y_p} \tag{15.20}$$

We can now express an error term of "what is left of" $\bar{y}$ besides $\bar{y_p}$ with:

$$\bar{y_e} = \bar{y} - \bar{y_p} \tag{15.21}$$

We can rewrite this expression for $\overline{y}_e$ by using Equation 15.20 to substitute for $\overline{y}_p$:

$$\overline{y}_e = \overline{y} - A\begin{vmatrix} m \\ b \end{vmatrix} \tag{15.22}$$

But there is something important about $\overline{y}_e$. As we saw in Figure 15.8, it is necessarily orthogonal to the hyper-plane of A because it contains only those elements of $\overline{y}$ that could not be projected into that hyper-plane. Because $\overline{y}_e$ is an n-dimensional column vector, it follows that we can compute the dot project of $\overline{y}_e$ with both columns of A. Recall that to do so, we would need to take the transpose of A, which we can write as A^T. If we multiply both sides of Equation 15.22 by A^T, we get:

$$A^T \overline{y}_e = A^T\left(\overline{y} - A\begin{vmatrix} m \\ b \end{vmatrix}\right) \tag{15.23}$$

But, by the rules of dot products, taking the dot product of two orthogonal vectors gives 0, and we know that $\overline{y}_e$ is orthogonal to any combination of vectors in the hyper-plane of A. Then, since $A^T \overline{y}_e = 0$, it must also be true that:

$$0 = A^T\left(\overline{y} - A\begin{vmatrix} m \\ b \end{vmatrix}\right) \tag{15.24}$$

by the definition of $\overline{y}_e$. In other words, by multiplying $\overline{y}$ by A^T we essentially remove the parts of $\overline{y}$ that are not in our projection. If we rearrange Equation 15.24, we have:

$$A^T A\begin{vmatrix} m \\ b \end{vmatrix} = A^T \overline{y} \tag{15.25}$$

What Equation 15.25 tells us is that we can fit the best line through a set of points by first forming the matrix A, then multiplying both A itself and $\overline{y}$ by the transpose of A (i.e., A^T), and finally solving for m and b in the resulting system of equations [11]. Notice that by the rules of matrix-matrix multiplication, the matrix $A^T A$ will have size 2, as will $A^T \overline{y}$. In Exercise 15.4, you can use R to perform exactly these matrix calculations for data similar to that in Figure 15.9. The result will be the best-fit line to your data. We can also give a name to this fitting procedure: *least-squares regression*.

Exercise 15.2

Compute the surface area to volume ratio for the four spheres in Figure 15.4A to verify that they fall on the line in Figure 15.4B.

Exercise 15.3

Compute the dot product of the two vectors in Figure 15.6B and verify that it is 0.

Exercise 15.4

Using R, fit the least-squares line to data drawn from a bivariate normal distribution with means $\mu_1 = 5$, $\mu_2 = 10$, variances $\sigma_1^2 = 5$, $\sigma_2^2 = 6$, and a covariance coefficient of $\rho = 0.548$. Useful R commands:

- `library(MASS)`
- `bvnData <- mvrnorm(n=100, mu=c(5, 10),`
  `Sigma=matrix(c(3, 5),ncol=2))`

```
• xvals<-bvnData[,1]
• yvals<-bvnData[,2]
• inter <- rep(1,100)
• A<-matrix(c(inter, xvals),ncol=2)
• At<-t(A)
• AmA=At%*%A
• Amb <- At%*%yvals
• solve(AmA,Amb, fractions=TRUE)
```

SLOPE OF THE METABOLIC RATE TO BODY MASS CURVE: COMPARING RESULTS

We can now return to the problem of determining the slope of the metabolic rate to body mass curve. However, we do need to make a few quick alterations to do this. First, in Figure 15.5, it is log(Metabolic rate) that is being compared to log(Mass). From the perspective of the regression approach we just developed, this transformation might seem to be a problem, but in fact it is not. We simply apply the computation of Equation 15.25 to the logarithmic transformations of our proposed x and y values (i.e., mass and metabolic rate).

A more serious question is that of our confidence in the estimated slope. If we estimate $m = 0.69$, how confident are we that that value is statistically different from 2/3 or ¾? Given how many data points we have, how much error can we expect in our estimates of m?

We will skip the second question here, except to note that very often we can assume that our x and y values follow what is known as a joint normal distribution, allowing you to compute confidence intervals or P values regarding m and b using standard tools. The reason we are avoiding this question of the statistical properties of m here is that the data shown in Figure 15.5 have spawned considerable scientific controversy, with some researchers strongly declaring that a value of $m = 2/3$ best describe these data and others equally strongly declaring that $m = 3/4$ [7, 10, 12]. Both groups can point to regression analyses supporting their position, and so, in this particular case, the statistical tests of the value of m are not sufficient, at least in my view, to resolve the question. What then can we do?

Is There a Model for $m = 3/4$?

One of the key themes of this book is that models express the sum of our scientific understanding of a particular question. So rather than expect that regression alone can resolve this question, it is worth asking whether we can incorporate other knowledge to build our expectation for m.

Now, we actually have a rather strong theoretical model as to why metabolism should scale at 2/3 body mass: because metabolism is seeking to fuel a three-dimensional body over two-dimensional surfaces. So, at the moment, $m = 2/3$ better expresses our understanding of the biology than does $m = 3/4$. But is that the end of the story?

DIGRESSION 2: FRACTAL GEOMETRY

Figure 15.10 shows a pattern called the "Koch snowflake" [13, 14]. It is formed by recurrently applying the rule: "Break each straight line segment into three parts. Then replace the center segment with 2 new segments of equal length, rotated by 120° and 60° with respect to the original segment and connected at the top." Each time we produce a new snowflake, we apply the same rule to all the lines within it, theoretically forever. Moreover, if we zoom into just 1/3 of the snowflake, we will see that it is a perfect copy of the entire curve. We can refer to this property as "self-similarity": images that have this property are called *fractals* [14]. Figure 15.11 gives two more examples of such curves. The Peano curve is particularly interesting because it is *space-filling*: when fully iterated, it is a line that fills a section of a plane [14].

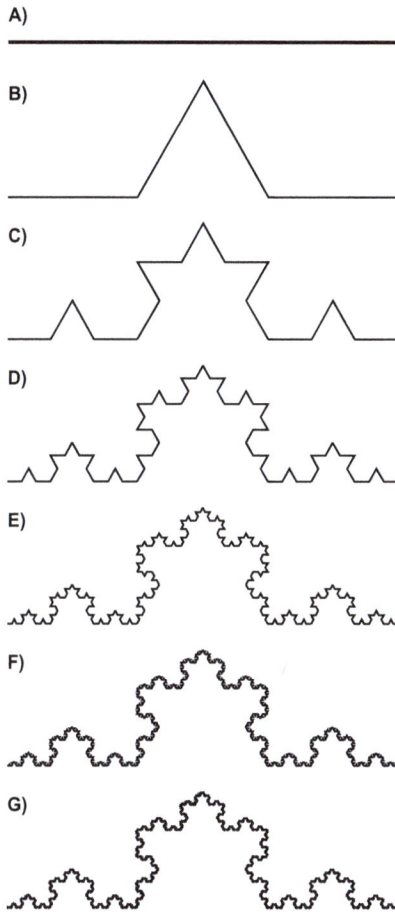

Figure 15.10 The Koch "snowflake" [13]. Starting with a straight line of unit length (**A**), we successively break each line segment into thirds and create a "peak" in the middle third at an angle of 120° to the original line. We then apply this same rule to the resulting curve for as long as we wish to (**B–G** for n = 1 to 6 steps). After six applications of the rule, the resulting figure has 4,096 segments. You can produce animated versions of this snowflake at: http://qbio.statgen.ncsu.edu/Koch_curve/

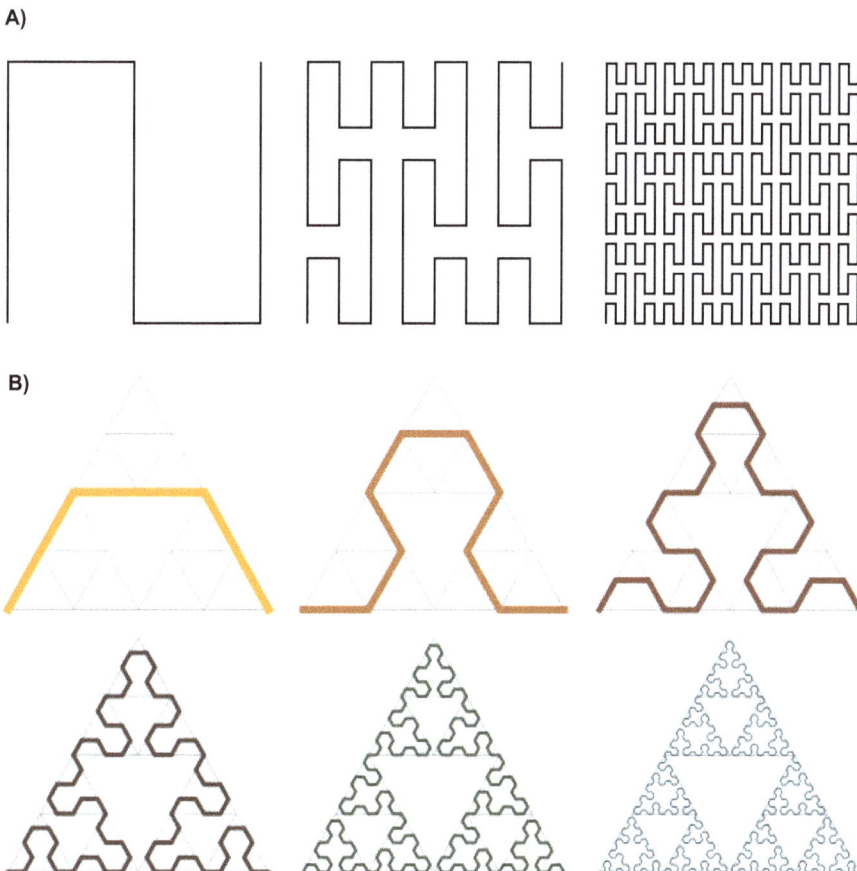

Figure 15.11 Two more fractal curves: the Peano curve (A) and the Sierpinski arrowhead (B). The Peano curve is *space-filling*, meaning that it is a line that completely covers its containing area when fully iterated [14]. (A, Courtesy of Antonio Miguel de Campos, on Wikimedia, published under CC BY-SA 3.0 license. B, Courtesy of Robert Dickau, on Wikimedia, published under CC BY-SA 3.0 license.)

Curve Length and Fractal Dimension

One question we could ask about the curves in Figure 15.10 is: How are long are they? In particular, if we let the number of steps n increase, we see that at every step we add four new segments and scale all segments down by a factor of 3. Hence at step n, there are 4^n pieces each of a length of $(1/3)^n$, making the length

$$\lim_{n \to \infty} \left(\frac{4}{3} \right)^n = \infty \tag{15.26}$$

This result is intriguing in the sense that we have an object that is bounded and yet very long. Obviously, the way the Koch snowflake packs this length into the unit line between 0 and 1 is to extend *up*: in other words, to move out of a unidimensional space into a two-dimensional one. Looking back at Figure 15.11A, we see this concept taken to its logical extreme: a line that seems to take up the same space as a plane. Could we formalize the idea that there might be figures whose number of dimensions does not fall neatly into the one, two, and three dimensions we are comfortable with?

Figure 15.12 gives an approach that applies to systems composed of self-similar objects, such as our fractals in Figures 15.10 and 15.11. In the scheme of Figure 15.12, we consider how what we "see" for an object changes as our resolution increases. We will define s as our *scale*: the size of the smallest self-similar object we can detect. We define the *dimensionality D* as the factor by which n, the number of self-similar objects we see, changes as we change s [15]:

$$n = \frac{1}{s^D} \tag{15.27}$$

To solve for D, first replace $1/s^D$ with $\left(s^D \right)^{-1}$. Taking the log of both sides yields:

$$\ln(n) = \ln\left(\left(s^D \right)^{-1} \right) \tag{15.28}$$

When we move the $-D$ term out of the logarithm we get:

$$\ln(n) = -D \cdot \ln(s) \tag{15.29}$$

or:

$$D = -\frac{\ln(n)}{\ln(s)} \tag{15.30}$$

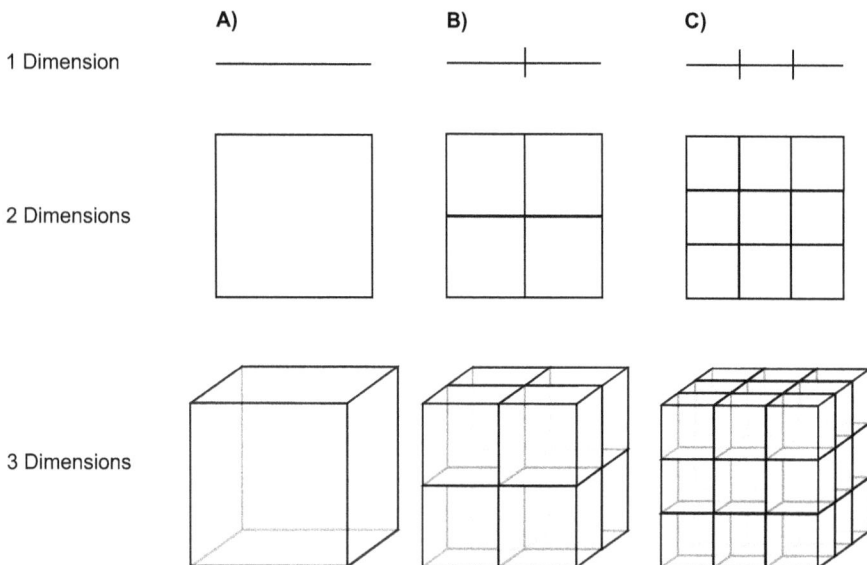

Figure 15.12 Dimensionality and self-similarity. Consider the effect of an increase in resolving power s on our ability to perceive self-similar objects. For one-dimensional objects, the number of self-similar objects (namely lines) increases directly with s: notice how s changes from 1 to 3 in **A–C**. In two dimensions the number of self-similar squares increases as s^2, while in three dimensions the number of cubes increases as s^3.

Equation 15.30 gives us the relationship between the change in scale and the change in the number of objects we see. When we use $s = 1/2$ for Figure 15.12, we find that $n = 4$ for a two-dimensional object like a plane. Similarly, $n = 8$ for a three-dimensional cube. Plugging into Equation 15.30, we get the expected $D = 2$ and $D = 3$ for these cases; we get the same result for $s = 1/3$, where $n = 9$ for a plane and 27 for a cube.

Returning to the Koch snowflake, we see that, unlike simple geometric objects, we need to take some care with our selection of s. In particular, $s = 1/3$ is a natural choice, since each third of the figure reproduces the whole. Plugging in $s = 1/3$, we get $D = \ln(4)/\ln(3) = 1.261$. This result is perhaps a curious one: we have found an object whose *dimensionality* is somewhere between a line and a plane.

In fact, were we to compute D for the Peano curve in Figure 15.11A, we would find $D = 2$, meaning that if that curve is allowed to grow indefinitely, it will in a sense become as large as a plane, even though it consists of segments of one-dimensional lines [14]. Hence, we can learn two things. First, there are self-similar objects with non-integer dimensions. Second, some of these self-similar objects can "reach up" a dimension, building a plane from line segments, or perhaps a sphere from circles.

Fractals in Nature

A curious feature of many fractal curves is that they seem to have a more natural appearance than do the pure lines and curves from conventional mathematics or computation [14]. There are a few ways we could pursue this familiarity; for the moment we will do so by returning to the question of biological surfaces. In **Figure 15.13** I illustrate five biological surfaces at which key metabolic events occur. For instance, in **Figure 15.13A** we see the blood vessels bringing nutrients to the brain of a rat [16] while Figure **15.13B** shows the alveoli in the lungs where

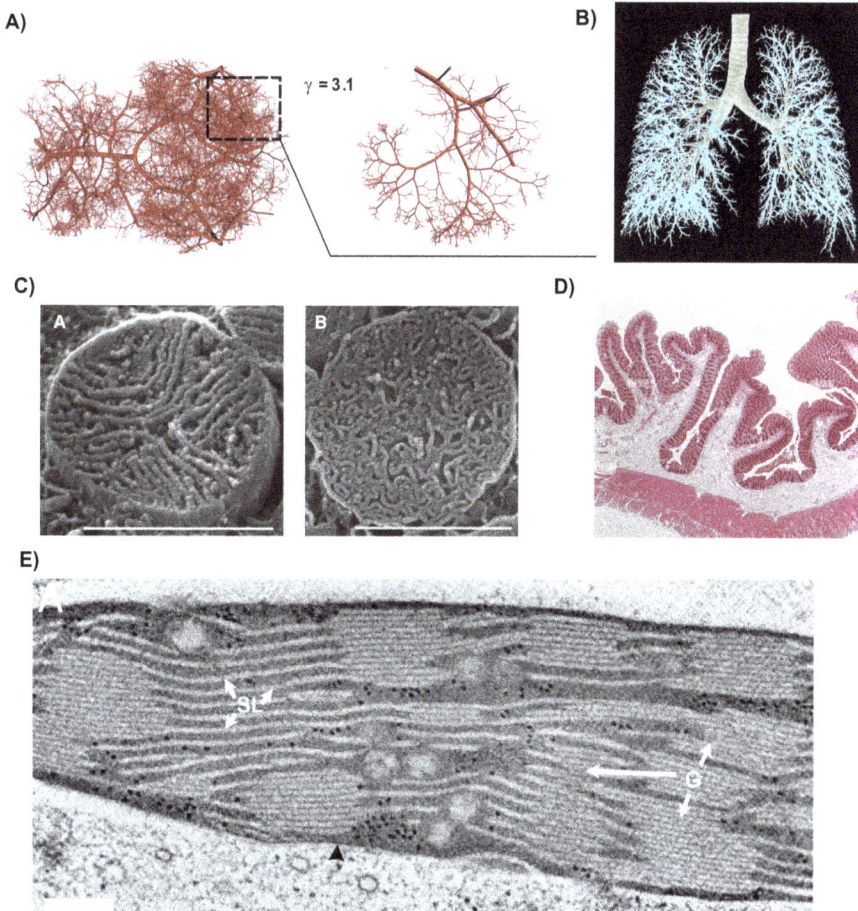

Figure 15.13 Fractal-like surfaces in biological systems. (**A**) A computational model of the branching of the blood vessels in the brain of a rat [16]. (**B**) Branching patterns in the lungs: this image shows a computational model of the deposit of inhaled particles in the mammalian lung. Image courtesy of Pacific Northwest National Laboratory. (**C**) Mitochondrial cristae from cardiac myocytes (muscle cells) of a rat [49]. (**D**) Villi lining the intestine [3]. (**E**) Thylakoid membranes in the chloroplast of lettuce. (A, From Staehelin LA & Paolillo DJ (2020) *Photosynthesis Research* 145:237–258. With permission from Springer Nature. B, Kopylova V et al. (2018) *Journal of Physics: Conference Series*, IOP Publishing. With permission. C, Mannella CA et al. (2013) *Journal of Molecular and Cellular Cardiology* 62:51–57. With permission from Elsevier.)

oxygen absorption occurs [3]. **Figure 15.13C** illustrates the cristae that result from the dense folding of the inner membrane of the mitochondrion: into that membrane are embedded the components of the electron transport chain that catalyze the final steps of respiration in our cells [17]. The rough surface of the lining of the intestine where nutrients are absorbed is shown in **Figure 15.13D**, while the thylakoids of the inner membrane of the chloroplasts of green plants is shown in **Figure 15.13E**. The chlorophyll antennae that absorb the light for photosynthesis are embedded in these membranes [17].

Perhaps the clearest thing we could say about the various biological surfaces in Figure 15.13 is that they are not smooth. Upon reflection, this fact should not surprise us. If organisms are truly faced with the problem of fueling three-dimensional bodies over two-dimensional surfaces, it would make sense that those surfaces should be as large as possible. For the transport networks of Figure 15.13A, B in particular, we might also say that there appear to be fractal-like properties: branching patterns that seem to repeat themselves as we zoom in in space.

With Figure 15.11A as an example, we could conjecture that evolution has sought to escape surface area limitations by employing fractal surfaces whose fractal dimension approaches three. These surfaces would arguably grow just as rapidly as their object's volume. Such surfaces ought, theoretically, to give a 1:1 scaling of metabolic rate to body mass. The flaw in this argument, however, is that fractal surfaces can be truly space-filling only if the fractal recursion is infinite. In real systems, we are absolutely limited by the dimensions of the atoms involved, and probably limited by biological structures at an even larger scale. If we abuse terminology for a moment, what kind of scaling can we achieve with "finite" fractals?

DERIVING THE ¾ SCALING EXPONENT

We are now ready to return to the question posed above. Can we find a physical or biological theory that would predict a ¾ scaling exponent between body mass and metabolic rate? In an elegant series of papers, Geoffrey West, James Brown, and Brian Enquist did exactly this [18]. They first proposed that most biological systems could only be scaled down as far as a particular *invariant*, which would be some basic feature of life whose size could not be changed. Some example invariants are the size of the Rubisco protein in plants [17] or the diameter of red blood cells (and hence capillaries) in mammals [3, 18].

The first step of their argument is to note that the surface area a and volume v of the organism both represent some complex function of the lengths $l_{0..i}$ inside, say, the transport network of the organism. For convenience, they defined the invariant we just mentioned (e.g., capillary diameter) to have length l_0. Rather than dealing with all the other lengths independently, we can take one, l_1, as being representative and describe all the others in terms of their ratios to it:

$$a(l_0, l_1, l_2, \cdots) = l_1^2 \cdot \phi\left(\frac{l_0}{l_1}, \frac{l_2}{l_1}, \cdots\right)$$

$$v(l_0, l_1, l_2, \cdots) = l_1^3 \cdot \psi\left(\frac{l_0}{l_1}, \frac{l_2}{l_1}, \cdots\right)$$

$$(15.31)$$

If we were to perfectly "resize" the organism in l_1 by a constant factor λ, we could represent that change by a single constant raised to the appropriate power in both cases:

$$a(l_0, l_1, l_2, \cdots) = \lambda^2 \cdot l_1^2 \cdot \phi\left(\frac{l_0}{l_1}, \frac{l_2}{l_1}, \cdots\right)$$

$$v(l_0, l_1, l_2, \cdots) = \lambda^3 \cdot l_1^3 \cdot \psi\left(\frac{l_0}{l_1}, \frac{l_2}{l_1}, \cdots\right)$$

$$(15.32)$$

In such a case, surface area would indeed scale as a 2/3 power of volume. However, by our assumption, l_0 is invariant to scaling, meaning that we need to correct for this fact:

$$a'\left(l_0,l_1,l_2,\cdots\right) = \lambda^2 \cdot l_1^2 \cdot \phi\left(\frac{l_0}{\lambda \cdot l_1},\frac{l_2}{l_1},\cdots\right)$$
$$v'\left(l_0,l_1,l_2,\cdots\right) = \lambda^3 \cdot l_1^3 \cdot \psi\left(\frac{l_0}{\lambda \cdot l_1},\frac{l_2}{l_1},\cdots\right)$$

(15.33)

In other words, the objects of length l_0 did not change in size with the rest of the organism. If this is the case, we can no longer assume that a scales as 2/3 v as it does in Equation 15.32.

Instead, we are going to look back at Figure 15.10. So far, we have imagined our fractals as growing ever more complicated within a fixed length or area, such that the inner dimensions increase without limit but are bounded by fixed outer dimensions. What if, instead, we imagined that Figure 15.10B was a building block of a larger system? In that view, the base curve of the Koch snowflake becomes the object measured by l_0. To make a bigger organism, we imagine each of the successive levels of the snowflake as having been built by adding on more component copies of Figure 15.10B, arranged in the fractal pattern.

If organisms were built in this manner, so long as l_0 was much smaller than a and v, we would expect that, no matter their size, we would see fractal-like patterns in their distribution networks similar to what we see in Figure 15.13, because as the organism grows, it simply expands its network with more copies of the fractal unit. Mathematically, we could then claim that the scaling function ϕ has a fractal dimension, meaning that there is some unknown exponent ϵ_a that rescales the rest of the lengths relative to l_0. We can then we rewrite our ratio function to remove the complication of the l_0 invariant:

$$\phi\left(\frac{l_0}{\lambda \cdot l_1},\frac{l_2}{l_1},\cdots\right) = \lambda^{\epsilon_a} \cdot \phi\left(\frac{l_0}{l_1},\frac{l_2}{l_1},\cdots\right)$$

(15.34)

So that now when we transform the area by λ, we account for this invariant, giving:

$$a\left(l_0,l_1,l_2,\cdots\right) = \lambda^{2+\epsilon_a} \cdot l_1^2 \cdot \phi\left(\frac{l_0}{\lambda \cdot l_1},\frac{l_2}{l_1},\cdots\right)$$

(15.35)

We could assume similar behavior with the volume, and its scaling coefficient ϵ_v. However, we can also propose that length has a similar coefficient ϵ_l, allowing us to rewrite ϵ_v as $\epsilon_v = \epsilon_a + \epsilon_l$. Our rescaled volume then becomes:

$$v\left(l_0,l_1,l_2,\cdots\right) = \lambda^{3+\epsilon_a+\epsilon_l} \cdot l_1^3 \cdot \phi\left(\frac{l_0}{\lambda \cdot l_1},\frac{l_2}{l_1},\cdots\right)$$

(15.36)

Omitting the common l_1 factor, our surface area–to–volume ratio from Equations 15.35 and 15.36 is then:

$$\frac{a}{v} \propto \frac{\lambda^{2+\epsilon_a}}{\lambda^{3+\epsilon_a+\epsilon_l}}$$

(15.37)

Under the assumption that organisms might wish to optimize their surface areas to fuel their three-dimensional volumes, we should seek to maximize Equation 15.37, for $0 \leq \epsilon_a, \epsilon_l \leq 1$. These bounds on ϵ_a and ϵ_l argue that fractal-like behavior can increase an object's dimension by at most 1, corresponding to the space-filling fractal of Figure 15.11A or its two-dimensional equivilent. The maximum for Equation 15.37 clearly occurs when $\epsilon_l = 0$ and $\epsilon_a = 1$. Plugging these optimal values into Equation 15.37 yields:

$$\frac{a}{v} \propto \frac{\lambda^3}{\lambda^4} \propto \lambda^{3/4} \tag{15.38}$$

In other words, the assumption that organisms seek to maximize their input surface areas relative to their volumes, subject to the constraint of some internal invariant unit, predicts the ¾ scaling rule that emerges from Figure 15.5. Hence, while the discussion of the fit of these figures and the estimated slope of the regression line might seem difficult to resolve, it is no longer the case that there is no theoretical expectation for a ¾ exponent. Coupled to the values in Figure 15.5 and the apparent complex structures of Figure 15.13, the value of ¾ no longer seems unexpected. And there is even evidence that this scaling might act below the level of single cells in the very structure of the cell itself [19].

USES OF SCALING RULES: EXAMPLE FROM YEAST EVOLUTION

I am going to close this chapter with the slight indulgence of discussing some of my own research. I am doing this to illustrate the reach of the ideas above into domains you might not expect.

The bakers' and brewers' yeast *Saccharomyces cerevisiae* is probably the most important domesticated microorganism, a domestication that is particularly interesting because the ancient humans that performed it had no idea they were doing so, given that yeast cells are invisible to the naked eye [20]. The genome of *S. cerevisiae* was the first eukaryote to have its genome sequenced in 1996 [21]. Soon after this sequence was released, Ken Wolfe and Dennis Shields discovered the remnants of an ancient *genome duplication* in that sequence [22].

Genome duplications, also known as *polyploidies*, most commonly represent the outcome of the hybridization of two closely related species. When this happens, each of these two species brings two complete copies of its genome, effectively forming a new species with four total copies of the genome in its nucleus [23]. These events have been quite important over evolutionary time, including for the evolution of vertebrates [23, 24].

Even though we refer to these events as genome duplications, many or most of duplicated gene copies created by these events are quickly lost [25]. We have already discovered the reason for this rapid loss. The genetic drift we modeled in Chapter 11 also applies to duplicated genes, because as long as one functional copy of the gene survives, any mutations in the other will be selectively neutral [26]. As a result, mutations that delete that gene (or its function) will behave according to the simulation rules we created in Chapter 11. Because of this prevalence of drift, when we look across many different polyploidies, we see the rapid loss of at least 50% of duplicated genes created by those polyploidy events [27]. In some sense, then, we become interested in genes that *did not* lose their duplicate copies. One of the main reasons that a duplicate gene copy might be kept over evolutionary time is that having two copies of the gene allows the cell to produce more of the protein that gene encodes. If that increase in protein level is beneficial to the organism, evolution will tend to preserve the duplication [28]. This idea of increased gene *dosage* will help us understand how yeast cells responded to their genome duplication.

About 10 years after Ken Wolfe discovered the genome duplication, I was lucky enough to work with him on trying to understand some of the evolutionary changes it might have produced. As shown in **Figure 15.14**, one of the oddest features of *S. cerevisiae* is that after glucose has passed through the glycolysis pathway to produce pyruvate, yeast will *ferment* ethanol from that pyruvate [29, 30]. It will perform this conversion even though the process is energetically inefficient in the presence of molecular oxygen [17]. Of course, this ability is vital to the human uses of this microbe, since ethanol fermentation defines the production of beer and wine.

Oddly, however, when we refer back to introductory biology, we recall that most eukaryotes prefer instead to route pyruvate through the tricarboxylic acid (TCA) cycle and the electron transport chain (ETC). The reason most cells use the

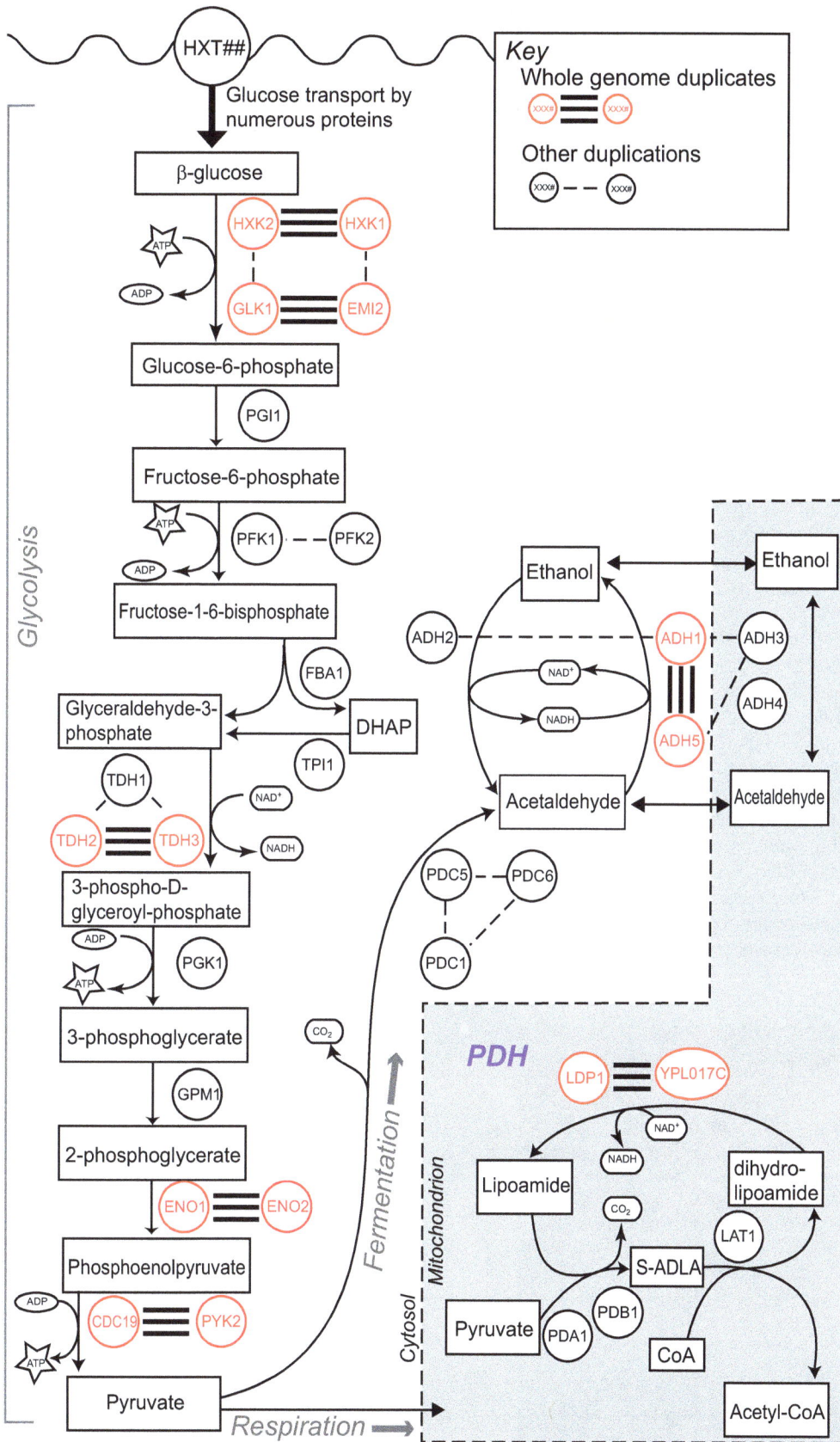

Figure 15.14 Glycolysis in *S. cerevisiae* and its connections to the first steps of ethanol fermentation and respiration. Baker's yeast has remnants of an ancient genome duplication, and some of the genes for glycolytic enzymes are still duplicated as a result (red pairs of genes). (Adapted from Conant GC & Wolfe KH (2007) *Molecular Systems Biology* 3:129.)

TCA and the ETC is their high energy yield [17]: the energy yield of glycolysis and fermentation is perhaps 10-fold less than that produced by the combination of the TCA cycle and the ETC [2]. Yet *S. cerevisiae* will, if sufficient glucose is available, instead always ferment that glucose into ethanol, despite having all the enzymatic machinery of the TCA and ETC cycles and the molecular oxygen needed to run them. Why is this?

Back in Chapter 6 we saw a full differential equation model of yeast glycolysis that was created by Teusink and colleagues [31]. Ken and I got to thinking about the potential effects of genome duplication on a model like this one (**Figure 15.15**). One of the more interesting effects of a genome duplication is that the volume of the cell possessing it tends to increase essentially in proportion to the amount of duplicated genetic material. In other words, a doubling in DNA content results in a roughly twofold increase in cell volume [32]. Doubling the cell volume and doubling the gene content should not, therefore, to a first approximation, change the *concentration* of any of the cellular enzymes. In this view (which we will shortly see is incomplete), genome duplication should not alter cellular metabolism.

However, we next thought about the effects of the loss of duplicated genes on genome size, cell volume, and metabolism. As duplicate genes are lost through drift, we expect both the genome size and therefore the cell volume to decline in proportion. However, they will not fully return to the ancestral condition due to the survival of some of the duplicated genes [33]. What would happen to the concentration of the various enzymes in the cell during this process? Well, any enzymes whose genes remained duplicated might now see their concentration *increase* relative to the other proteins in the cell, since they are being expressed from two genomic copies rather than one. In principle, at least, that concentration increase might then allow the pathways those enzymes catalyzed to run at higher flux (think back to Chapters 5 and 6).

As it turns out, nearly half the genes for the enzymes of glycolysis survived the genome duplication in duplicate (see Figure 15.14). This proportion was quite different from the overall proportion of only about 10% of the genes having surviving duplicates [34]. By the argument above, this excess of duplicated glycolysis enzymes could have resulted in an increase in the flux through glycolysis. As we recall from Chapter 5, not all enzymes in a biochemical pathway have an equivalent impact on pathway flux. We could thus test our hypothesis that glycolytic flux increased after genome duplication by asking whether the enzymes with surviving duplicates were also the enzymes the metabolic model of Teusink et al. [31] predicted to influence glycolytic flux the most. This was what we found (see Figure 15.15; also [33]).

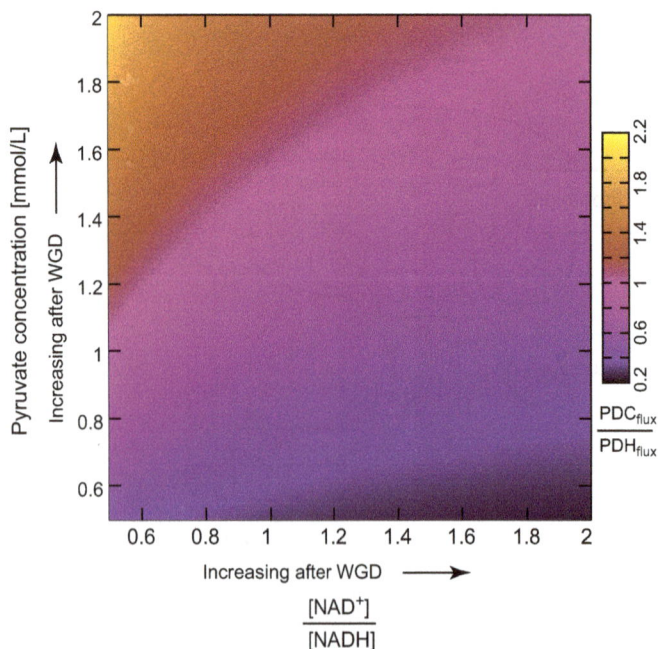

Figure 15.15 Models of metabolism suggest that genome duplication could increase flux through glycolysis due to the preserved glycolytic duplicate genes. These duplicate genes would tend to increase the NAD+/NADH ratio (x axis) and the pyruvate levels (y axis), the combination of which will tend to drive the flux through fermentation up relative to the flux through respiration, potentially showing that genome duplication (yellow shades on the graph) drove the appearance of aerobic ethanol production. (Adapted from Conant GC & Wolfe KH (2007) *Molecular Systems Biology* 3:129.)

So far, our argument was rather simple and not terribly informative. The insight came from thinking about the influence that the genome duplication might have had on the TCA cycle and particularly on the ETC. Glycolytic flux depends on the concentration of glucose and the dissolved enzymes of the pathway: increasing those enzyme concentrations could increase flux. Respiration, on the other hand, has very complex dependencies. Some of these requirements include [2]:

1. The need for molecular oxygen, the cellular concentration of which is not easy for the organism to change.
2. The need for the transport of pyruvate into the mitochondria.
3. The need for the movement of $FADH_2$ and NADH to the mitochondrial membrane so that they can initiate the ETC.
4. The need for the ETC itself, which occurs through enzymes embedded in the cristae (i.e., Figure 15.13).

Notice that, unlike glycolysis, all these processes are subject to surface area constraints because they either occur on a surface (the ETC) or require the moment of a metabolite across a surface. How do surface-bound processes respond to genome duplication? Well, if the volume of the cell increases due to genome duplication, its surface area to volume ratio necessarily decreases (even under the ¾ scaling we proposed above; see Figure 15.4). Hence, genome duplication could have had a paradoxical effect on yeast metabolism, allowing glycolytic flux to increase but impairing respiration due to a decreased relative surface area.

And indeed, when we ran Teusink and colleagues' model [31], we saw that the respiratory flux *decreased* relative to the fermentative flux after a hypothetical genome duplication. In other words, the genome duplication might have actually hurt the yeast cells by restricting respiration and the TCA cycle. However, yeast cells had long possessed an alternative to respiration: the fermentation of glucose to ethanol. All yeasts use this fermentative pathway when oxygen is not available [35], and it is not strongly surface-bound. Hence, fermentation was available to baker's yeast in the aftermath of the genome duplication as an alternative route to accept the excess glycolytic flux that could not be routed through respiration. In other words, we and others argued that genome duplication may have been the first step in the creation of a yeast that ferments sugars in the presence of oxygen [36–39], and thus the first step in the creation of a domestic microorganism for making wine and beer.

REFLECTIONS AND A FINAL PREVIEW

In this chapter we considered two features of an organism that might not have seemed obviously connected: its metabolism and its shape. Because metabolism occurs throughout the volume of the organism but requires that compounds enter and exit that organism over surfaces, we cannot treat large organisms simply as scaled-up versions of smaller ones. At first, we might have thought that this relationship would require that metabolism grow no faster than as a 2/3 power of the organism's mass. However, we realized that the sort of complex, self-similar structures seen in Figure 15.13 provided the chance for a partial escape from the constraints of 2/3 power scaling.

Taking a brief detour into using the tools of vectors and matrices to fit data to a scaling curve, we found that indeed there was evidence for a >2/3 scaling of metabolism with body size. We then saw how metabolic modeling and considerations of surface area and volume relationships allowed us to make a prediction about the origins of an important behavior in baker's yeast: the aerobic production of ethanol from sugar, a reaction vital to humans' production of beer and wine.

Beyond the intrinsic interest of this finding about yeast and fermentation, it speaks to the common theme of this book: that understanding, *in detail*, the function and evolution of living things is critically dependent on appropriate mathematical and computational models of those living things. To me, the excitement of the current moment in biology is just how many unexpected and subtle features of life we are uncovering using the combination of mathematical modeling and computational approaches. In the final chapter of the book we will consider this increasing importance of computation in understanding biology. In so doing, we will also discover a strange and deep connection between those computers and the nature of life itself.

REFERENCES

1. Raymond J & Segre D (2006) The effect of oxygen on biochemical networks and the evolution of complex life. *Science* 311(5768):1764–1767.

2. Mathews CK & Van Holde KE (1996) *Biochemistry*. 2nd Edition (The Benjamin/Cummings Publishing Company Inc., Menlo Park).

3. Marieb EN (1991) *Human anatomy and physiology* (The Benjamin/Cummings Publishing Company, Redwood City, CA), p. 1040.

4. White CR & Kearney MR (2013) Determinants of inter-specific variation in basal metabolic rate. *Journal of Comparative Physiology B* 183:1–26.

5. Brody S (1945) *Bioenergetics and growth with special reference to the efficiency complex in domestic animals* (Reinhold Publishing Corporation, New York), p. 1023.

6. Jürgens KD (2002) Etruscan shrew muscle: the consequences of being small. *Journal of Experimental Biology* 205(15): 2161–2166.

7. White CR & Seymour RS (2003) Mammalian basal metabolic rate is proportional to body mass$^{2/3}$. *Proceedings of the National Academy of Sciences, USA* 100(7):4046–4049.

8. Banavar JR, Moses ME, Brown JH, Damuth J, Rinaldo A, Sibly RM, & Maritan A (2010) A general basis for quarter-power scaling in animals. *Proceedings of the National Academy of Sciences, USA* 107(36):15816–15820.

9. Kleiber M (1932) Body size and metabolism. *Hilgardia* 6(11): 315–353.

10. Savage VM, Gillooly JF, Woodruff WH, West GB, Allen AP, Enquist BJ, & Brown JH (2004) The predominance of quarter-power scaling in biology. *Functional Ecology* 18(2): 257–282.

11. Lay DC (1994) *Linear algebra and its applications* (Addison-Wesley, Reading, MA), p. 445.

12. White CR & Seymour RS (2005) Allometric scaling of mammalian metabolism. *Journal of Experimental Biology* 208(9): 1611–1619.

13. Koch H (1904) Sur une courbe continue sans tangente, obtenue par une construction géométrique élémentaire. *Arkiv for Matematik, Astronomi och Fysik* 1:681–704.

14. Flake GW (1998) *The computational beauty of nature: computer explorations of fractals, chaos, complex systems, and adaptation* (The MIT Press, Cambridge, MA).

15. Mandelbrot B (1967) How long is the coast of britain? Statistical self-similarity and fractional dimension. *Science* 156(3775):636–638.

16. Kopylova V, Boronovskiy S, & Nartsissov YR (2018) Tree topology analysis of the arterial system model. *Journal of Physics: Conference Series*: 012027.

17. Lodish H, Baltimore D, Berk A, Zipursky SL, & Matsudaira P (1995) *Molecular cell biology*. 3rd Edition (Scientifc American Books, New York), p. 1344.

18. West GB, Brown JH, & Enquist BJ (1999) The fourth dimension of life: Fractal geometry and allometric scaling of organisms. *Science* 284:1677–1679.

19. West GB, Woodruff WH, & Brown JH (2002) Allometric scaling of metabolic rate from molecules and mitochondria to cells and mammals. *Proceedings of the National Academy of Sciences, USA* 99(Suppl 1):2473–2478.

20. Sicard D & Legras J-L (2011) Bread, beer and wine: yeast domestication in the Saccharomyces sensu stricto complex. *Comptes Rendus. Biologies* 334(3):229–236.

21. Goffeau A, Barrell BG, Bussey H, Davis RW, Dujon B, Feldmann H, Galibert F, Hoheisel JD, Jacq C, Johnston M, Louis EJ, Mewes HW, Murakami Y, Philippsen P, Tettelin H, & Oliver SG (1996) Life with 6000 genes. *Science* 274(5287):546,563–567.

22. Wolfe KH & Shields DC (1997) Molecular evidence for an ancient duplication of the entire yeast genome. *Nature* 387 (6634):708–713.

23. Van de Peer Y, Mizrachi E, & Marchal K (2017) The evolutionary significance of polyploidy. *Nature Reviews Genetics* 18(7): 411–424.

24. Holland PWH, Garciafernandez J, Williams NA, & Sidow A (1994) Gene duplications and the origins of vertebrate development. *Development*:125–133.

25. Scannell DR, Frank AC, Conant GC, Byrne KP, Woolfit M, & Wolfe KH (2007) Independent sorting-out of thousands of duplicated gene pairs in two yeast species descended from a whole-genome duplication. *Proceedings of the National Academy of Sciences, USA* 104:8397–8402.

26. Ohno S (1970) *Evolution by gene duplication* (Springer, New York), p. 160.

27. Hao Y, Fleming J, Petterson J, Lyons E, Edger PP, Pires JC, Thorne JL, & Conant GC (2021) Convergent evolution of polyploid genomes from across the eukaryotic tree of life. *submitted*.

28. Conant GC & Wolfe KH (2008) Turning a hobby into a job: how duplicated genes find new functions. *Nature Reviews Genetics* 9:938–950.

29. Johnston M & Kim J-H (2005) Glucose as a hormone: Receptor-mediated glucose sensing in the yeast Saccharomyces cerevisiae. *Biochemical Society Transactions* 33:247–252.

30. Geladé R, Van de Velde S, Van Dijck P, & Thevelein JM (2003) Multi-level response of the yeast genome to glucose. *Genome Biology* 4:233.

31. Teusink B, Passarge J, Reijenga CA, Esgalhado E, van der Weijden CC, Schepper M, Walsh MC, Bakker BM, van Dam K, Westerhoff HV, & Snoep JL (2000) Can yeast glycolysis be understood in terms of *in vitro* kinetics of the constituent enzymes? Testing biochemistry. *European Journal of Biochemistry* 267:5313–5329.

32. Epstein CJ (1967) Cell size, nuclear content, and the development of polyploidy in the mammalian liver. *Proceedings of the National Academy of Sciences* 57(2):327–334.

33. Conant GC & Wolfe KH (2007) Increased glycolytic flux as an outcome of whole-genome duplication in yeast. *Molecular Systems Biology* 3:129.

34. Byrne KP & Wolfe KH (2005) The Yeast Gene Order Browser: Combining curated homology and syntenic context reveals gene fate in polyploid species. *Genome Research* 15(10): 1456–1461.

35. Visser W, Scheffers WA, Batenburg-van der Vegte WH, & van Dijken JP (1990) Oxygen requirements of yeasts. *Applied and Environmental Microbiology* 56(12):3785–3792.

36. Blank LM, Lehmbeck F, & Sauer U (2005) Metabolic-flux and network analysis of fourteen hemiascomycetous yeasts. *FEMS Yeast Research* 5:545–558.

37. Piškur J, Rozpedowska E, Polakova S, Merico A, & Compagno C (2006) How did *Saccharomyces* evolve to become a good brewer? *Trends in Genetics* 22:183–186.

38. Merico A, Sulo P, Piškur J, & Compagno C (2007) Fermentative lifestyle in yeasts belonging to the *Saccharomyces* complex. *FEBS Journal* 274:976–989.

39. van Hoek MJ & Hogeweg P (2009) Metabolic adaptation after whole genome duplication. *Molecular Biology and Evolution* 26(11):2441–2453.

40. NIAID Visual and Medical Arts (8/28/2024) H5N1. in *NIAID BIOART Source*. https://bioart.niaid.nih.gov/bioart/187

41. NIAID Visual and Medical Arts (1/30/2025) Gram Negative Bacteria. in *NIAID BIOART Source*. https://bioart.niaid.nih.gov/bioart/179

42. NIAID Visual and Medical Arts (9/18/2024) Mast Cell. in *NIAID BIOART Source*. https://bioart.niaid.nih.gov/bioart/335

43. Shykind D (2012) Pellicle of Euglena (picture by David Shykind). https://commons.wikimedia.org/wiki/File:Euglena_pellicle_2.jpg

44. Kbradnam (2006) Caenorhabditis elegans. https://en.wikipedia.org/wiki/Caenorhabditis_elegans#/media/File:Adult_Caenorhabditis_elegans.jpg

45. Murphy M (2005) The "Grizzly Giant" Sequoia tree, 88 meters high and 2,800 years old. in *California*. https://commons.wikimedia.org/wiki/File:Grizzly_Giant_Mariposa_Grove.jpg

46. Trebol-a (2005) Suncus etruscus, the smallest mammal by weight of the world. *Portrait over hand for comparison*. https://commons.wikimedia.org/wiki/File:Suncus_etruscus.jpg

47. de Campos AM (2007) 3 first steps of the building of the Peano fractal curve. https://commons.wikimedia.org/w/index.php?curid=2304574

48. Dickau R (2008) First six stages of fractal Sierpinski arrowhead curve. https://commons.wikimedia.org/w/index.php?curid=4260404

49. Mannella CA, Lederer WJ, & Jafri MS (2013) The connection between inner membrane topology and mitochondrial function. *Journal of Molecular and Cellular Cardiology* 62:51–57.

50. Bussi Y, Shimoni E, Weiner A, Kapon R, Charuvi D, Nevo R, Efrati E, & Reich Z (2019) Fundamental helical geometry consolidates the plant photosynthetic membrane. *Proceedings of the National Academy of Sciences, USA* 116(44):22366–22375.

Bits

Life as an Information Transfer Process

16

> *"For now we see through a glass, darkly; but then face to face:*
> *now I know in part; but then shall I know even as also I am known."*
> —1 Corinthians 13:12

In this final chapter I will focus on two topics that have been woven throughout the book, those of *data* and *computation*.

Figure 16.1 illustrates the explosive growth of four different types of biological data over the past decades. As mentioned in Chapter 12, the increase is most impressive for DNA base pairs (notice the log scale for Figure 16.1A), but the growth has been significant across different fields of biology.

What this growth means for practicing biologists is at once obvious and yet hard to appreciate. In fact, Figure 16.1 is the central motivation for this book. The success of biology as a science will be defined, for the foreseeable future, by our ability to integrate mathematics, statistics, and computation into the development of quantitative models and theory that can accommodate the explosive growth in the quantities of data about life we are collecting.

In my lifetime as a practicing biologist, genetics and genomics have gone from fields that were data limited to ones where, in practical terms, we are overwhelmed by data. Hence, I chose to write this book to help train a new generation of biologists who can look at all facets of the field from a quantitative perspective, bringing the tools described in these chapters (and, of course, others you will teach yourselves) to the big problems in biology.

MEASURING DATA: BITS AND BYTES

The right-hand axis in Figure 16.1A reports the size of GenBank, not in terms of number of bases of DNA, but rather of the number of *megabytes* those bases require for their storage on the GenBank computers. In the modern era, we are so saturated with computer technology that we use terms such as *megabyte* and *gigabyte* without noticing. It is, however, worth stepping back and refreshing our intuition about what exactly these words measure and mean.

In Exercise 16.1, you and a partner will play a game where you attempt to reconstruct a DNA sequence known to your partner but not to you. The trick of the game is that you are only allowed to ask questions that can be answered with a "yes" or a "no."

Exercise 16.1

Form a team with another student. The first team member will go to the website http://qbio.statgen.ncsu.edu/RandDNA/ and generate a random DNA sequence of 10 bases. She or he will *not* show this sequence to the other player. Instead, the other player will try to reconstruct the sequence by asking the first player questions. You may ask any questions you wish, as long as they can be answered with a "yes" or a "no." Record how many questions you need to recover the sequence. Then switch sides and see if you can recover the next sequence with fewer questions.

DOI: 10.1201/9781003687504-16

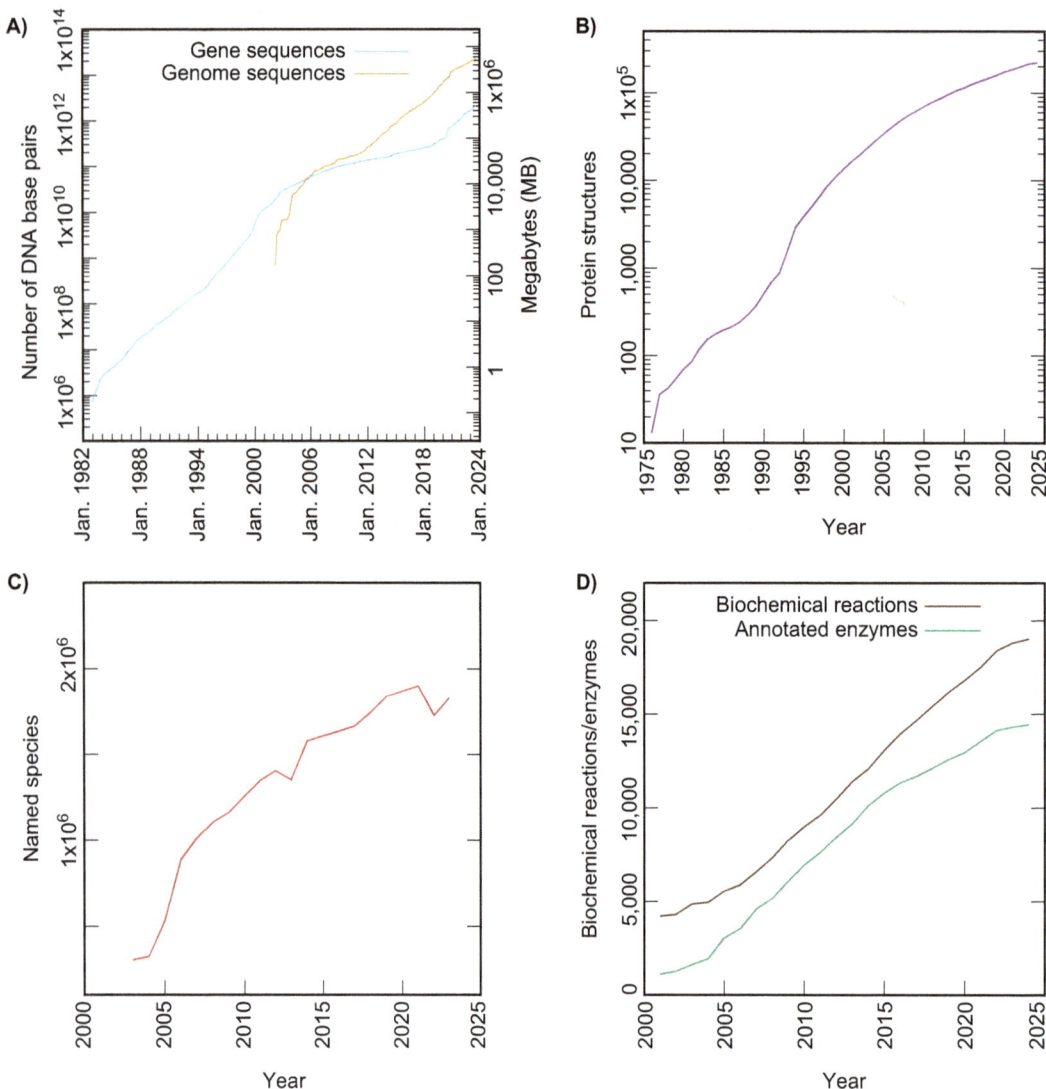

Figure 16.1 Growth of biological datasets in four domains over the past decades. (**A**) Growth in the Genbank public data repository [6] of DNA bases from individual gene sequences (*blue*) and from genome sequences (*tan*). Notice the logarithmic scale of the *y* – axis. (The data here are the same as in Figure 12.1B, with the addition of measurements of the database size on the right.) (**B**) Number of known protein structures from the Protein Data Bank (PDB; https://www.rcsb.org) [13]. (**C**) Number of named biological species from the Catalogue of Life (https://www.catalogueoflife.org) [17]. **D**) Number of distinct biochemical reactions and identified enzymes from the MetaCyc database (https://metacyc.org) [21].

Because the sequences generated in Exercise 16.1 are random, you might occasionally simply guess the full sequence with a single question. For instance: "Is the sequence TGTGAATGGA?" However, we can refine the game by requiring the players to submit their planned questions before the sequence is generated, or, better yet, to use their questions repeatedly against a whole series of random sequences.

For such a repeated game, we should make our questions independent of the base position: in other words, you would ask the same series of questions about each base. This approach is also sensible because we assume that the bases are independent: knowing base i tells us nothing about base $i+1$. (Real DNA sequences slightly violate this rule, but not to a degree that will change anything we are about to discuss.)

The most natural question scheme you might come up is illustrated in Figure 16.2A. Notice that since we know the answer falls in the set {A, C, G, T}, we need not ask the last question: three no's in a row means we have a T.

If we assume that the four bases are equally likely to appear, how many questions, on average, will we ask with the scheme above? Well, we will always ask

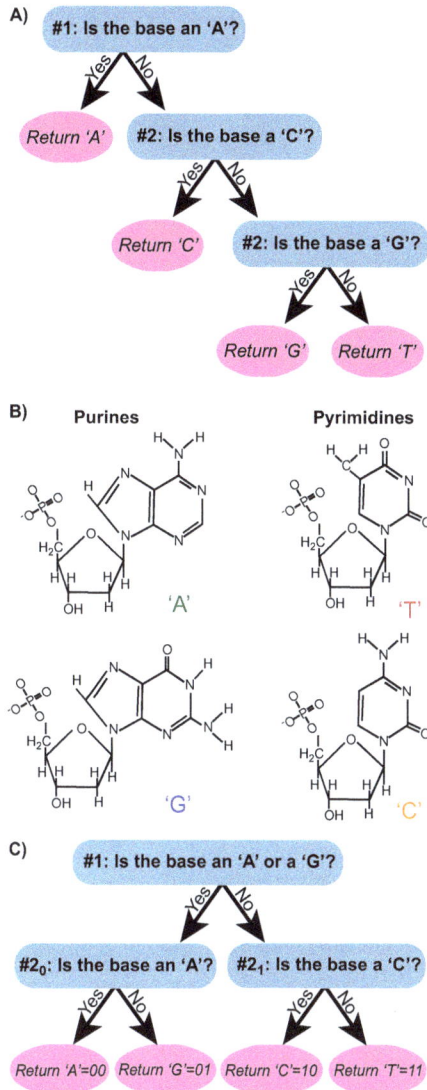

Figure 16.2 Recovering an unknown DNA base pair with "Yes" and "No" questions. (A) The most naïve approach serially asks questions to recover the base. (B) A reminder of the chemical structures of the four DNA nucleotides, split into the chemically similar purines and pyrimidines. (C) Using this natural division, we can reduce our questioning scheme to a pair of questions.

Question 1. Seventy-five percent of the time we will be wrong (we receive the answer "no" because the base is not an A), and so we will ask Question 2. Of the times we ask Question 2, two out of three times we will be wrong (the base is not a C), and we will need to ask Question 3. So, on average, we ask:

$$Questions = 1 + \frac{3}{4} + \frac{1}{2} = 2.25 \tag{16.1}$$

Notice that we use 1/2 not 2/3 in Equation 16.1 to express the fraction of total cases in which we need to ask Question 3, giving the average number of questions needed as 2¼.

However, if we look at our question scheme from Figure 16.2A, we might wonder if there is some wastefulness there. In particular, for both Questions 1 and 2, the answers are not balanced: We will answer "no" more often than "yes." So, the point arises: Can we ask fewer questions than indicated by Equation 16.1?

In **Figure 16.2B**, we remind ourselves that the four DNA bases divide themselves naturally into two groups: the purines and the pyrimidines. What if we were to structure our questions around this division? As shown in **Figure 16.2C**, that approach reduces the number of questions to identify a base pair to precisely two in every case.

Although the DNA bases have a natural division into two groups, this approach of using successive reductions of an "alphabet" by half with yes or no questions can be used for any symbolic alphabet. The average number of questions Q needed to recover a single letter from an alphabet of size a is:

$$Q = \log_2(a) \tag{16.2}$$

We can see that for the $a = 4$ of DNA, the number of questions is indeed two: for the English alphabet of 26 letters (ignoring punctuation and capitalization), we would need $\log_2(26) = 4.7$. Hence, for English letters, we would build a tree like that in Figure 16.2C, except that for some branches we will need to descend four questions and some five.

In 1948, Claude Shannon wrote one of the most important papers of 20th-century science: "A Mathematical Theory of Communication" [22]. In it he laid out the argument above, using the term *bit* to represent the answer to one "yes" or "no" question.

What made Shannon's paper so influential is that he formalizes the idea that *information* can be measured. The use of yes or no questions may seem odd at first, but it actually allows us to make some very interesting links across scientific disciplines. First, it naturally links the question scheme to a mapping of the alphabet into a *binary* coding scheme (see Figure 16.2C). Since computers represent numbers (and other data) in binary format, the concept of a *bit* represents the size of a message stored in this binary format on a computer.

Looking back at Figure 16.1A, we can now almost understand the measurements of sequence database sizes. The only further definition we need is that 8 bits are conventionally referred to as a *byte*. Hence, a DNA sequence of n bases can be stored in $n / 4$ bytes (i.e., 2 bits per base).

Information

The presentation so far has overlooked a key nuance pointed out by Shannon: What if the different letters of the alphabet are not seen in the message at equal frequencies? For DNA, the frequency of the four DNA bases does in fact vary both within different parts of the same genome and between different genomes [23, 24]. The case is even more obvious for the English alphabet. As shown in **Figure 16.3**, the 26 letters of the alphabet appear in English words at very different frequencies [5], and we should ask how such differences would alter our computation of the information in an English message. (Sherlock Holmes fans among you will remember that it is this unevenness in letter frequencies that allows Holmes to decipher the code embedded in the drawings of "The Adventure of the Dancing Men.")

Shannon's answer to this question was to scale the contribution of the different states (e.g., letters) by their probability. He defined the per-letter information content I of a message composed from a set of a discrete letters to be:

$$I = -\sum_{i=1}^{a} p_i \log_2(p_i) \tag{16.3}$$

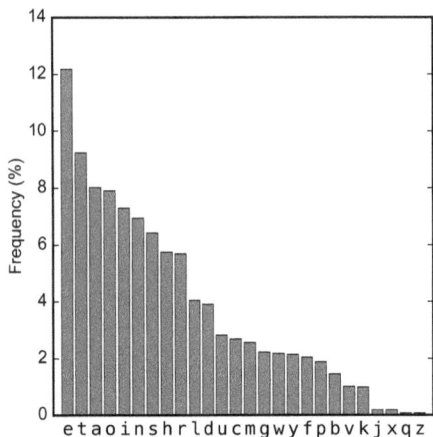

Figure 16.3 Frequencies of different letters in English words, showing that the letters are not equally frequent. Data from [5].

Here p_i is the frequency or probability of that letter. If we apply Equation 16.3 to our example of the DNA bases, what do we find? Assuming that the base frequencies are equal, we have $p_i = 1/4$, which yields the quantity $-1/4 \cdot \log_2(1/4)$ four times, which is just $-\log_2(1/4) = 2$. We can then notice that, as the base frequencies become unequal, I decreases, meaning DNA has a maximal information content when the base frequencies are equal. On the other hand, in the limit where $p_1 = 1$ and $p_2..p_4 = 0$, we find that $\log_2(1) = 0$. At that point, there is no information in the message because it is just a stream of copies of a single letter. This result is general: the more uneven the frequencies of the letters in our alphabet, the less information is contained in a sequence of them.

RELATIONSHIP BETWEEN INFORMATION AND ENTROPY

Equation 16.3 is, curiously enough, referred to as Shannon's *entropy*. We very briefly encountered entropy in Chapter 5, where we noted that a chemical reaction's equilibrium was driven by both the difference in chemical energy between the reactants and the products, ΔH, and by the entropy or *disorder* present in the system (ΔS):

$$\Delta G = \Delta H - T\Delta S \tag{16.4}$$

In that chapter, we did not pause to define or measure entropy. However, if we now take a toy problem, we can see something interesting.

In **Figure 16.4A**, we remind ourselves that the amino acids used to make proteins can exist in two different *stereoisomers*, which we can essentially think of as molecules that are mirror images of each other [3]. These isomers are chemically and energetically identical. However, with very few exceptions, life on Earth only uses the L forms of the amino acids in its proteins [3]. Nonetheless, enzymes called racemases can reduce a population of pure L amino acids to a mixture of 50% L and 50% D forms [25].

We will consider the reaction:

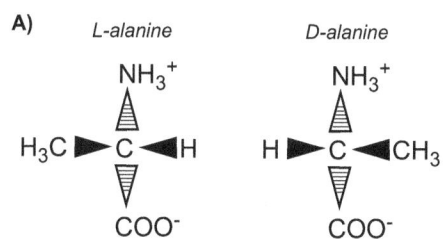

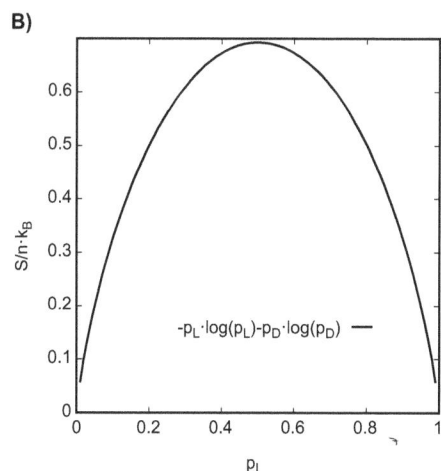

$$L-\text{alanine} \rightleftharpoons D-\text{alanine} \tag{16.5}$$

Figure 16.4 **Entropy in a simple chemical reaction. (A)** The L and D forms of the amino acid alanine. The darker triangles to the left and right indicate the CH_3 and H atoms coming *out* of the page toward us, the top and bottom triangles showing that these atoms project *down* into the page. Adapted from [3]. **(B)** Relative entropy (y) of systems with differing proportions of *L*- and *D*-alanine (p_L, x – axis). See Equation 16.16.

and assume we start with n_L molecules of L-alanine and no D-alanine ($n_D = 0$). Because L- and D-alanine are energetically identical, ΔH must be zero for the reaction of 16.5. As a result, any interesting behavior in this reaction is driven by its entropy.

We can approach the question of the entropy in an equation like Equation 16.5 by considering the set of microscopic versions of the chemical system that would be identical at the macroscopic level. In this case, we will cheat slightly and define our macroscopic variables as just the number of molecules of L- and of D-alanine because we can think of those numbers as easily convertible to the observed concentrations of the two compounds. For a fixed combination of $n_L + n_D = n$, a typical way to express the entropy in Equation 16.4 would be [26]:

$$S = k_B \cdot \log(W) \tag{16.6}$$

where k_b is a physical constant called the Boltzmann constant that we will not worry about. Instead, we will consider W, which is the number of microscopic combinations of the system that would have the same values of n_L and n_D. What is W? As it happens, we have already computed a version of it, although we might not have realized it. Because all the L- and all the D-alanine molecules are identical and hence interchangeable, they behave like the heads and tails of the coins we analyzed in Chapter 8. Using the equations we developed there, we wrote:

$$W = \frac{n!}{n_L! \cdot n_D!} \tag{16.7}$$

In other words, starting with the $n!$ total states of the system, we apply the approach from Chapter 8 for removing the $n_L!$ and $n_D!$ identical states, just as we did with heads and tails in that chapter. Making the corresponding substitution into Equation 16.6, we have:

$$S = k_B \cdot \log\left(\frac{n!}{n_L! \cdot n_D!}\right) \tag{16.8}$$

or:

$$S = k_B \cdot \log(n!) - k_B \cdot \log(n_L! \cdot n_D!) \tag{16.9}$$

These terms in $\log(n!)$ are not particularly easy to work with. However, because we expect that n, n_L, and n_D are all very large, we can use an approximation called Stirling's rule [26] to simplify them for this analysis (Exercise 16.2). Stirling's rule states that, for large n, we can write:

$$\log(n!) \approx n \cdot \log(n) - n \tag{16.10}$$

which, if we also recall $\log(a \cdot b) = \log(a) + \log(b)$, allows us to rewrite Equation 16.9 as:

$$S = k_B \cdot \left(n \cdot \log(n) - n - \left(n_L \cdot \log(n_L) - n_L + n_D \cdot \log(n_D) - n_D\right)\right) \tag{16.11}$$

Because $n = n_L + n_D$, we can cancel the linear terms in $-n$, n_L and n_D:

$$S = k_B \cdot \left(n \cdot \log(n) - \left(n_L \cdot \log(n_L) + n_D \cdot \log(n_D)\right)\right) \tag{16.12}$$

We can represent the proportion of all the molecules that are in state $i = L$ or D as:

$$p_i = \frac{n_i}{n} \tag{16.13}$$

Making the substitution $n_i = n \cdot p_i$ for both n_D and n_L gives us:

$$S = k_B \cdot \left(n \cdot \log(n) - \left(n \cdot p_L \cdot \log(n \cdot p_L) + n \cdot p_D \cdot \log(n \cdot p_D) \right) \right) \qquad (16.14)$$

If we expand the terms in $\log(n \cdot p_i)$ and then group the common terms in $n \cdot \log(n)$, we find:

$$S = k_B \cdot \left(n \cdot \log(n) - \left(n \cdot (p_L + p_D) \cdot \log(n) + n \cdot p_L \cdot \log(p_L) + n \cdot p_D \cdot \log(p_D) \right) \right) \quad (16.15)$$

We can pull an n out of all these terms to give:

$$S = k_B \cdot n \left(\log(n) - \left((p_L + p_D) \cdot \log(n) + p_L \cdot \log(p_L) + p_D \cdot \log(p_D) \right) \right)$$

But since $p_L + p_D = 1$, we can cancel the first and second terms, giving us:

$$S = -n \cdot k_B \cdot \left(p_L \cdot \log(p_L) + p_D \cdot \log(p_D) \right) \qquad (16.16)$$

We can use Equation 16.16 in a few ways. First, we notice that, in **Figure 16.4B**, the entropy of a mixture of L- and D-alanine, computed using Equation 16.16, is maximal when the number of molecules of each is equal ($p_L = p_D = \frac{1}{2}$). Again, because we know that the enthalpy of L- and D-alanine are the same, the lowest-energy state of the reaction in Equation 16.5 must be driven by the entropy. As a result, the lowest-energy state for the reaction occurs when S is maximized at $p_L = \frac{1}{2}$.

What is more interesting about Equation 16.16 is what happens if we extend the p_is to more than two states. Then we can write:

$$S = -n \cdot k_B \cdot \sum_i p_i \cdot \log(p_i) \qquad (16.17)$$

But this expression differs only by constants and the logarithmic base from Equation 16.3. In other words, there is an underlying linkage between computing the number of possible messages we can produce with an alphabet (the Shannon entropy) and the possible amount of disorder in a chemical system (the chemical entropy).

Interpretation of Entropy and Information

What should we make of this linkage between chemical entropy and information? One useful approach is to conceptualize information as *the amount that our uncertainty about the system has decreased as a result of receiving a message*. If we think about our chemical example for a moment, our uncertainty about the microscopic state of the system is maximal when $p_L = \frac{1}{2}$. Under those circumstances, if we are told the precise microscopic state of the system, that information excludes all the other $n!/\left((n/2)! \right)^2 - 1$ microstates that also have the macroscopic state of $p_L = \frac{1}{2}$. In the same way, our uncertainty about a message in an arbitrary alphabet decreases as we receive letters most rapidly if the letters of that alphabet are equally frequent. We will next see how this concept of measuring information content has applications far afield from both computation and chemistry.

APPLICATION OF SHANNON'S ENTROPY: ECOLOGICAL DIVERSITY

We have just linked the idea of measuring information in a message to that of the chemical disorder of a system. But the utility of Shannon's equation does not end there. In fact, we can use it unaltered in the apparently unrelated field of ecology.

In Chapter 12, I commented on the increasing ease with which it has become possible to collect DNA sequence data. One surprising result of that change has been a revolution in our understanding of the microbial world. Most of the microbes on our planet cannot be grown in pure cultures in a laboratory, meaning that they were almost impossible to identify and study with conventional techniques in microbiology [27]. The advent of inexpensive DNA sequencing allowed researchers to sidestep this barrier by extracting and sequencing microbial DNA directly from an environment [27, 28]. When biologists performed such extractions across a variety of habitats, they found enormous numbers of new kinds of microbes that had never been seen before [28, 29].

The most common approach to finding these new microbial taxa has been to sequence one of the genes than encode the ribosome (the macromolecular machine that we saw synthesizing proteins in Chapter 7). This 16S rDNA gene is found in the genome of every microbe, but, as we saw in Chapter 12, it varies a bit from microbial taxon to microbial taxon. Hence, we can sequence all the copies of this gene in an environment and compare them to each other. If we count the frequency of each different type of the 16S rDNA gene, we can get an unbiased assessment of the taxa present in that environment [27]. A cartoon of how such sampling might work is shown in Figure 16.5: our goal would be to determine all the different types of bacteria present and how common each type is.

What do we see as we sequence more and more of these 16S genes? Most often, we find a pattern like that in Figure 16.6A, which was made with some data from collaborators of mine [9]. At first, every time we look at a new 16S rDNA gene, we find that it comes from a taxon we have not seen before. However, as we sequence more copies of the gene, we start to see new individuals from taxa that we have already seen. The *collectors' curve* that results from plotting this pattern rises quickly at first but eventually starts to level off (see Figure 16.6A) as most of the new genes we sequence start to come from taxa we have already seen. The shape of the collectors' curve, then, allows us to infer whether or not we would be likely to see new taxa were we to sequence more 16S rDNA genes from that environment. If you want a macroscopic example, this analysis would correspond to walking into a forest and, at each new tree you encounter, either adding a tick next to the name of a type of tree you have already seen or adding a new type of tree to your list with a single tick next to it.

The other characteristic of these sequencing experiments is that the frequencies of the taxa are generally quite unbalanced (Figure 16.6B). A few taxa will represent most of the individuals in the environment, with a long tail of rare taxa also present. (By the way, if you are wondering why I keep writing "taxa" rather than "species," keep in mind that the standard definition of species as "a population of mutually-fertile, obligately sexually reproducing individuals," does not apply terribly well to prokaryotes [30].)

The combination of features in Figure 16.6A, B makes the obvious approaches for describing the diversity of taxa in an ecosystem problematic. For instance, from Figure 16.6A it is clear that simply reporting the number of taxa found is misleading because if one researcher sampled more individuals than another did, she or he will almost certainly find more taxa present. Likewise, there could be ecosystems with many taxa distributed rather evenly, with about the same number of individuals present from each taxa. On other hand, in the more common case like that of Figure 16.6B, a few taxa dominate. Even if we were sure that we had counted every

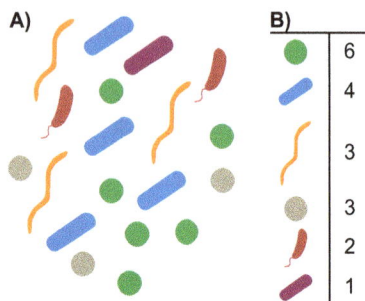

Figure 16.5 Sampling a microbial ecosystem. (A) Different kinds of microbes live in an environment. Here we illustrate these differences with shape and color, although in real ecosystems the bacteria can be very difficult to distinguish under a microscope [7]. **(B)** A table of the sorted counts of each type from panel **A.**

A)

B)

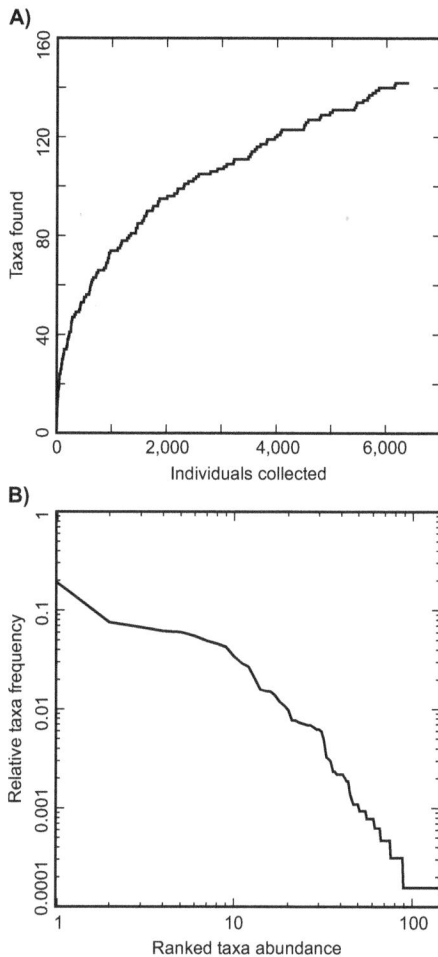

Figure 16.6 Quantifying the microbial individuals present in an environment by sequencing the 16S rDNA genes present there. (**A**) Number of *distinct* taxa of microbes identified (e.g., unique 16S rDNA genes) in the rumen of a sheep (y−axis) versus the total number of 16S rDNA genes sequenced ("Individuals collected," x−axis). (**B**) Comparison of the ranked abundance of the taxa in **A** (x−axis) versus their frequency (y−axis). Data taken from individual 1101 in [9].

last taxon in two ecosystems and found those numbers to be equal, the two ecosystems would still be quite different. Ecologists call the first concept, the number of taxa present, *richness*, and the second, how they are distributed, *evenness* [31].

Could we propose one measure of an ecosystem diversity that would capture *both* richness and evenness? As a matter of fact, probably the most common such measure is the *Shannon diversity*. That diversity is just the application of Equation 16.3 to data such as that in Figure 16.6. In this view, each taxon i is represented by p_i—its proportional abundance in the sample [31]. The Shannon entropy (or diversity) of the data in Figure 16.6, is, for reference, 4.71.

Utility of Shannon Diversity

How might one apply Equation 16.3 in an analysis? In some work I did with collaborators, we were comparing sheep that had gained weight well versus those that had not. We found that the animals that gained weight did not differ significantly from the other animals when we looked at the taxonomic diversity of their microbial ecosystems (i.e., comparing data like those in Figure 16.6 between the two groups). However, when we used Shannon entropy (Equation 16.3) to consider the diversity of the *metabolic networks* from this ecosystem, we saw something interesting. Thinking back to Chapter 14, these metabolic networks have nodes representing reactions connected by shared metabolites see (Figure 14.6). We can simply count, for each such reaction, how often we see a gene associated with that reaction in our DNA sequence data from that ecosystem (which, again, are the gastrointestinal tracts of the sheep [9]). We then apply Equation 16.3 to the proportional abundances of the enzymes. When we did this, we found that the animals that gained weight had a significantly higher Shannon diversity among the metabolite reactions of their microbes than did the other animals [32].

In other words, these animals had more different types of bacterial enzymes in their gastrointestinal tracts, and at more even abundance, than did the animals that gained weight less well.

This single result speaks to a more general idea: perhaps ecosystems with higher diversity (higher I) are more metabolically efficient and stable [33, 34]. To understand this proposal, we can first argue that ecosystems have many niches that are characterized by different metabolic potentials, such as the carbon and nitrogen sources available, the temperature and pH, and so forth. In such a complex ecosystem, we would need a good deal of metabolic and taxonomic diversity to fill it in such a way that each of those niches is used efficiently. This idea is actually supported by data from humans, who play host to an enormous community of microbes in their intestinal tracts [35]. These microbes help metabolize the complex human diet into compounds our bodies can use. Researchers have found that individuals who have less diversity in this microbial ecosystem are more prone to bowel diseases [36, 37]. There are also, unfortunately, other implications of this idea: as we humans continue to stress the ecosystems of our planet, we should be aware that every aspect of biological diversity is potentially important, making the extinction of any species a matter of concern to all of us.

LIFE AS AN INFORMATION TRANSFER PROCESS

The previous section argued for an appreciation of entropy and information at scales from the intestinal to the planetary. But this idea of information reaches down to smaller scales and into the very heart of what it means to be alive. We will next discuss how information is a key concept for discussing genetics and inheritance.

Tracking Information through the Central Dogma

We will start by returning to the central dogma discussed in Chapter 7. **Figure 16.7** is a recapitulation of two figures from that chapter, illustrating both the central dogma and the universal genetic code. Lest we think we know all there is to know about these two ideas, let us see if the concept of informational entropy adds anything to our understanding.

We have already calculated that one base of DNA has effectively 2 bits of information. Hence, a codon of three bases should convey roughly 6 bits. Now, with the exception of the three stop codons, each codon encodes a single amino acid. What is the information content of one amino acid? We know Shannon's entropy is maximal under the assumption that all letters in an alphabet are equally frequent. In fact, the 20 amino acids are *not* equally frequent in proteins, but even if they were, we can compute an informational entropy for one amino acid as:

$$I = -\sum_{i=1}^{20} \frac{1}{20} \log_2\left(\frac{1}{20}\right) = -\log_2\left(\frac{1}{20}\right) = 4.32 \tag{16.18}$$

In other words, the central dogma maps codons with 6 bits of information onto amino acids with (at most) slightly more than 4 bits. Where is the excess information? It is in the degeneracy of the genetic code. In other words, information theory would predict that the genetic code is degenerate without ever having seen it, because there are fewer than 64 amino acids. (Of course, entropy could not predict the exact structure of that degeneracy.)

DNA Synthesis, Entropy, and Energy

In their 1953 paper proposing a chemical structure for DNA [38] based on X-ray diffraction data from Rosalind Franklin [39], Watson and Crick wrote:

> It has not escaped our notice that the specific pairing we have postulated immediately suggests a possible copying mechanism for the genetic material.

A)

B)

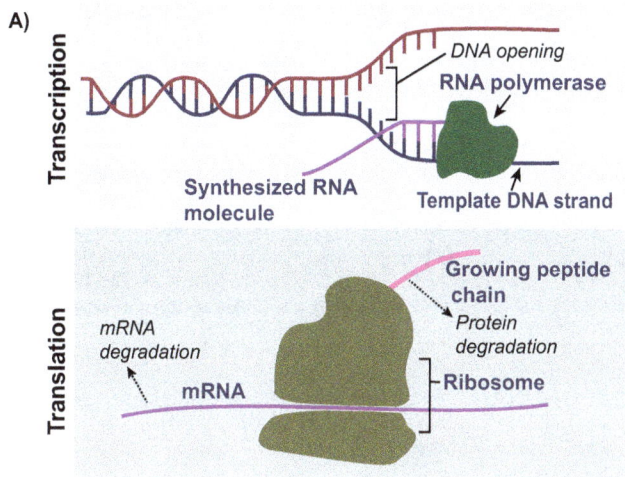

Figure 16.7 Information and the central dogma. (**A**) Figure 7.3 from Chapter 7, illustrating the central dogma of molecular biology: DNA codes for RNA codes for protein. (**B**) Figure 7.4 from Chapter 7: the universal genetic code.

In Figure 16.8, I show an illustration of this process, where the DNA polymerase adds a single new base of the growing DNA strand, using the other strand as a template. We could consider this process from a number of perspectives, but right now I would like to draw your attention back to Equation 16.4. What is ΔS for the synthesis of DNA? In fact, $\Delta S \ll 0$, meaning that the disorder of the system has greatly *decreased* through the synthesis of the DNA molecule. We can see this fact rather clearly when we realize that we have gone from a state with thousands to millions of free dNTP molecules diffusing around to a state with a single, highly ordered DNA molecule. If ΔS is negative, then $-T\Delta S$ is necessarily *positive*. In other words, considered solely from an entropy perspective, DNA synthesis is highly unfavorable. So how is it that cells can carry out such synthesis? Of course, the answer must lie in the ΔH term. Looking at Figure 16.8, it is clear where that ΔH contribution is coming from: we are using highly energetic dNTPs for synthesis, and it is their energy that allows us to force forward a reaction with such a large reduction in entropy. If this idea seems surprising to you, recall that dATP is just the DNA form of the ubiquitous cellular energy currency ATP [40]. From a larger perspective, the central importance of DNA to the organism is clear from the fact that it is willing to spend this currency so extravagantly in the service of DNA production.

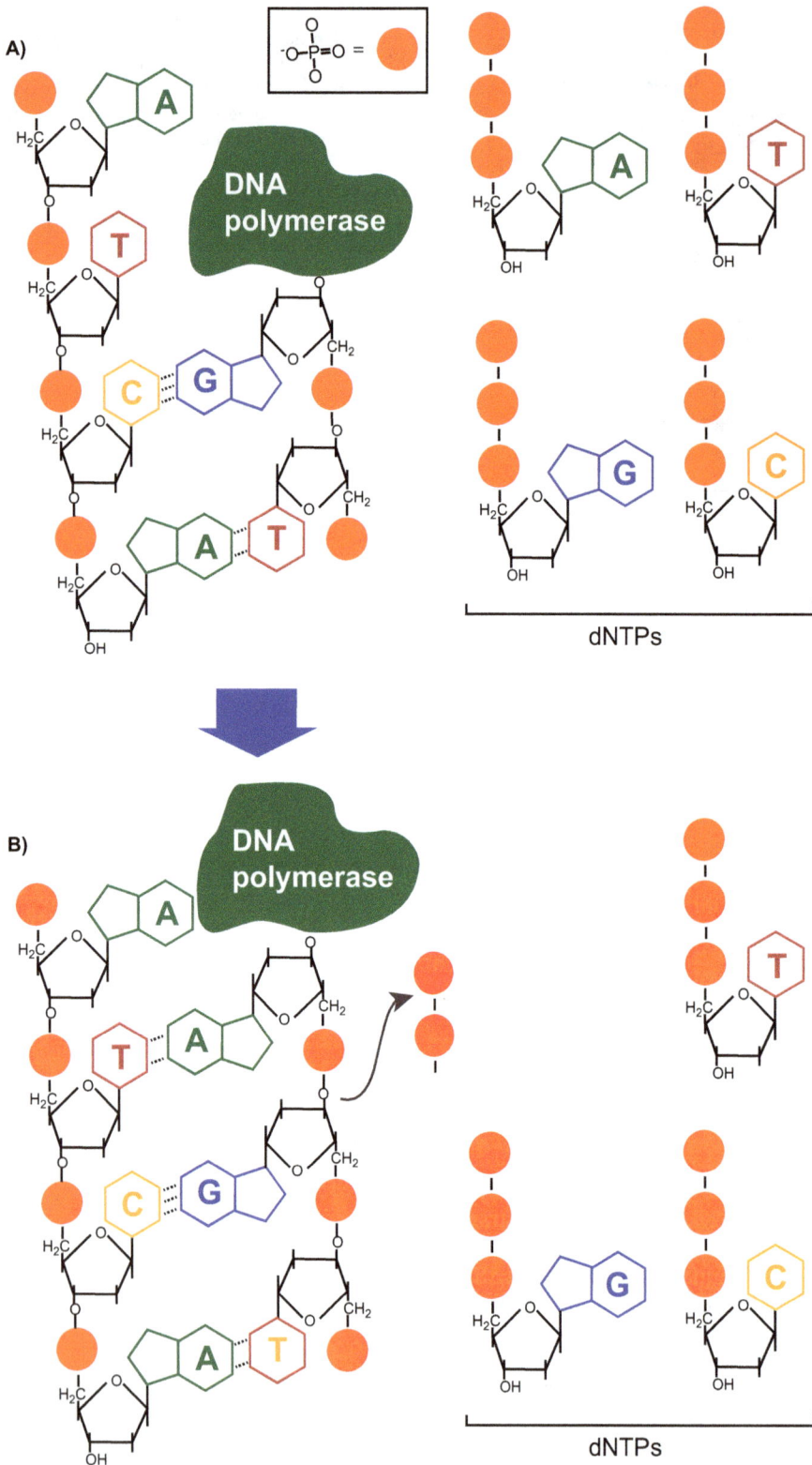

Figure 16.8 Using a DNA *template* strand to direct the synthesis of a new DNA molecule. The bases are simplified relative to Figure 16.2, representing the phosphate groups as colored diamonds. (**A**) Synthesis occurs by adding a new base to the OH group attached to the #3 carbon of the ribose sugar. (**B**) The DNA polymerase identifies the correct new base to add, using high-energy dNTPs (dATP, dGTP, etc.) for synthesis. The breaking of this bond releases the two excess phosphate groups.

Efficiency of Information Storage in Living Things

Stepping still further back, we can think about the problem of storing biological information in living systems. **Figure 16.9B** compares the information content of the human genome to a few other datasets. Three key points are clear. First, there are living vertebrate animals with genomes both 10 times smaller and 10 times larger than that of humans [1, 2, 8]. Second, while the human genome is relatively

A)

B)

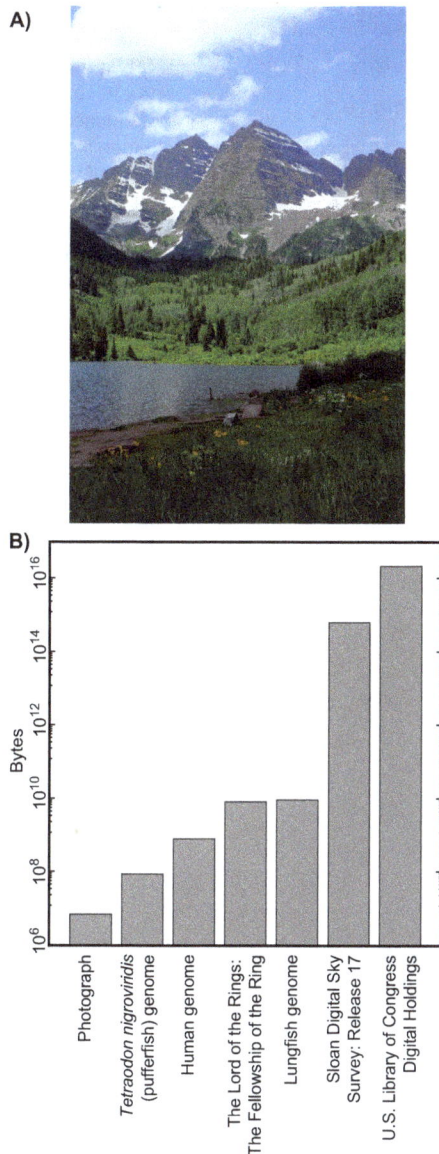

Figure 16.9 **Information content of various items.** (**A**) For reference, a typical digital photograph occupies about 7 million bytes (7 megabytes) Photo by the author. (**B**) Information content of some selected biological and nonbiological datasets. Note the log scale. Sources: *Tetraodon nigroviridis* genome: [1]; human genome: [2]; lungfish genome: [8]; Sloan Digital Sky Survey DSS: [12]; Library of Congress digital holdings: [15].

large compared to a single photo, even when compressed to store on a mobile phone, a digital movie of a few hours has 10 times the information content of the human genome. Third and finally, human datasets from astronomy or other areas are vastly greater in their information content.

Humans are prone to overstating how special they are, but I think most of us would agree than a human being is more complex that the series of moving images making up a motion picture. Yet the information content of our genomes is 10 times smaller than that of a movie. It must be, therefore, that our discussion so far has missed a key component: the *efficiency* with which information is used. As it happens, life is fantastically efficient in its storage of information. Humans have been slow to appreciate the value of being able to compress information so effectively. But Richard Feynman, in his 1959 lecture "There's plenty of room at the bottom" [41], stated the problem well:

> In the year 2000, when they look back at this age, they will wonder why it was not until the year 1960 that anybody began seriously to move in this direction.
> Why cannot we write the entire 24 volumes of the Encyclopedia Brittanica on the head of a pin?

Figure 16.10A, B contrast the information density for a modern piece of computing hardware with the information density of an average human cell. Recall that we saw

A)

Volume: 123 cm^3
Capacity: 4×10^{12} bytes
Information density: 3.3×10^8 bytes/cm^3

B)

Volume: 2×10^{-9} cm^3
Capacity: 1.6×10^9 bytes
Information density: 8×10^{17} bytes/cm^3

C)

D)

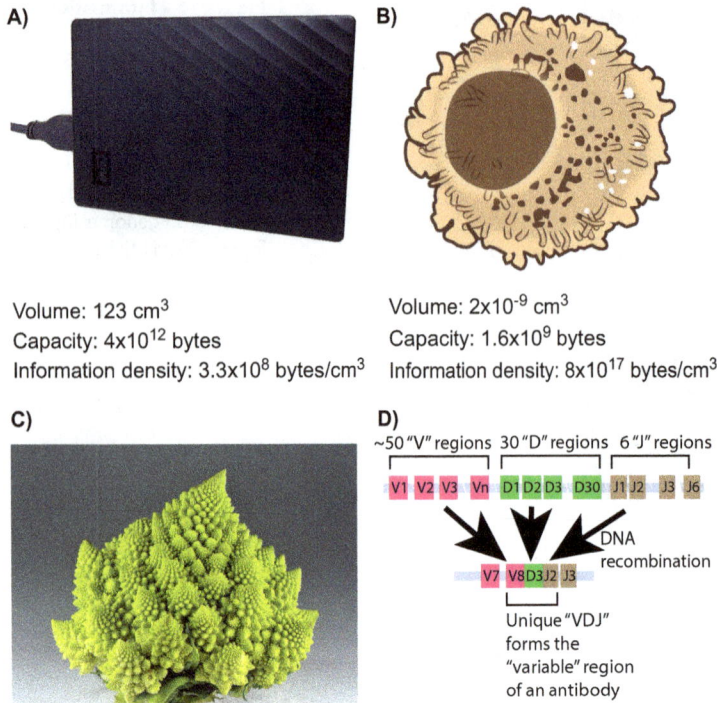

Figure 16.10 **Information compression with human technology and in biological systems.** (**A**) A typical 4-terabyte (4×10^{12} bytes) external hard drive. Its physical dimensions and capacity give us an information density of 100s of megabytes per cm^3. (**B**) A typical mammalian cell [4] has a volume of 2,000 µm^3 [10] yet contains two copies of the entire genome. As a result, its information density is at least a billion times greater than that of the hard drive in **A**. Obviously, the disk in **A** does not represent the most efficient information storage humans have created, but since we have used the volume of the entire cell in **B** rather than just the nucleus, the difference in density is still enormous. (**C**) Fractal patterns of growth can give rise to complex forms using relatively little genetic information [14]. (**D**) Formation of distinct antibody genes in each antibody-producing cell line through VDJ recombination. Each antibody cell line picks a distinct V, D, and J region to form the variable part of the antibody it will produce. The combination of 50 V, 30 D, and 6 J regions allows for up to $50 \times 30 \times 6 = 9,000$ different products from this single genetic region. (B, Mast cell art courtesy of NIAID. C, Courtesy of Ivan Leidus, on Wikimedia, published under CC BY-SA 4.0 license. D, Adapted from Roitt I, Brostoff J & Male D (1998) *Immunology*, 5th Edition. Mosby International Ltd., London.)

in Chapter 7 that a typical mammalian cell might be 2,000 µm^3 in volume [10]. And yet every one of those cells in your body (excepting red blood cells [42]) manages to compress two complete copies of the genome into that tiny volume. That information density is a factor of more than a billion greater than that of the hard disk drive in Figure 16.10A. So while our technology is still some distance from Feynman's goal of using a handful of atoms to store a single bit of information, life has come rather closer, with the each DNA base in Figure 16.2 consisting of a few dozen atoms. But even this molecular efficiency is not the extent of life's compression ability. The DNA content of a single one of your cells would, if fully stretched out, have a length of about 2 meters [43]. Your cells compress these very long molecules in several ways. The DNA is *supercoiled*, which is to say the DNA helix is slightly unwound relative to its relaxed state. This unwinding, in turn, causes the DNA to wrap around itself, just as a rubber band or old-style telephone cord would do when similarly unwound [43]. The DNA is further wrapped around a group of proteins called *histones*: these ancient and highly conserved proteins [44] coil the DNA around themselves, reducing the one-dimensional length of the DNA by packing it into three dimensions as we saw in Chapter 15 [43]. The net result is to fit 2 meters of DNA into a cell nucleus with a diameter more than a thousandfold smaller.

Information Efficiency

Yet the biggest difference between human information storage technology and that in biological system is probably not this high physical information density. Instead, Figure 16.9B suggests a second dimension to life's efficiency. While the information content of the human genome is a factor of 10 less than that of a digital movie, as mentioned, animals like the pufferfish *Tetraodon nigroviridis* have genomes a tenth the size of that of humans [1]. Despite this, these animals possess the complex cell types and organs systems of all vertebrates. In fact, as we discussed in Chapter 11, we have good reasons to suspect that much of the information in the human genome is in fact evolutionary detritus that does not directly describe the operation of a human organism [45, 46]. Recall that 33% of the human genome consists of just two types of repetitive elements (called LINEs and SINEs; see Figure 11.4 [2]).

Getting back to our examples from Figure 16.9, I happen to think our human chauvinism is not wrong in this case. A human body is indeed a more complex thing than a movie, and yet the information needed to build a new one is encoded in a fraction of the 800 megabytes of our genome. That genetic information is thus used exquisitely well. **Figure 16.10C, D** presents just two examples of how cleverly life uses its information. In Figure 16.10C, we see again a fractal-like pattern, but this time from a plant. As complex as this pattern appears, it is at least possible to "code" it with very simple rules similar to those we used in the Koch snowflake of the preceding chapter [47]. And although the details are not fully worked out, the developmental genetics of plants appear to encode rules for such fractal patterns with only a handful of genes, expressed with very clever timing [14]. The second example is perhaps even more interesting. As large, multicellular animals, we are constantly challenged by the kinds of infectious agents we considered in Chapter 2. In that chapter we assumed that once we had been infected with a particular disease and recovered, we would be immune to it thereafter. However, we did not consider how this immunity was achieved. We know that the immunity is not hard-coded into the genome because it requires an initial infection to be activated. And if you think about the problem of infectious disease broadly, it is apparent that the genome, however large it is, could never encode immunity to all possible infectious agents in the world. The problem is especially serious given that those agents evolve so quickly that they could easily evade any fixed genomic protection in the host [48]. Instead, your genome can create an enormous range of functional molecules like antibody proteins by using *mix-and-match* genetic elements that can be combined in many, many combinations [18]. Figure 16.10D illustrates just a part of the antibody production process. Antibodies recognize infectious agents and are manufactured in special cells called B cells. Each progenitor B cell differs in the combination of elements (V, D, and J in Figure 16.10D) that it uses as the constituent parts of its particular antibody protein. Most of the B cells produce antibodies that will not react to a particular pathogen. Just by chance, however, some of them do. That recognition causes those B cells to divide and reproduce themselves during an infection. In so doing, they produce the antibodies that contain the infection. After recovery, a number of these responsive cells, now called *memory B cells*, are kept in circulation by the immune system. If the pathogen tries to reinfect you, those memory B cells recognize it immediately and start to reproduce again, usually strangling the infection before you are aware of it. This system explains both the immunity of Chapter 2 and how a vaccination can train your immune system to recognize a pathogen before it is ever exposed it, giving you the protection of immunity without first becoming infected.

A FINAL REFLECTION: THE GENOTYPE-TO-PHENOTYPE MAP IS CENTRAL TO THE UNSOLVED PROBLEMS IN BIOLOGY

The example of the development of the cells of the immune system from the relatively sparse information of the genome is one of the cases where we understand fairly well how the DNA bases in the genome are used for building a part of the organism. But the question is a bigger one; indeed, understanding how the information in the genome encodes the form and capabilities of the organism is perhaps the central unsolved problem in biology. It is called the problem of the *genotype-to-phenotype map* [49]. Without naming it, we have been dancing around it at many different points in this book. As **Figure 16.11** suggests, it is a tremendously difficult problem because it is not a single problem: we will need dozens of different kinds of models, coupled with many terabytes of experimental data, to advance against it. Yet, as Figure 16.11 also suggests, it is critical for biology at every scale: the ecological interactions of Chapter 3 are driven by the phenotypes

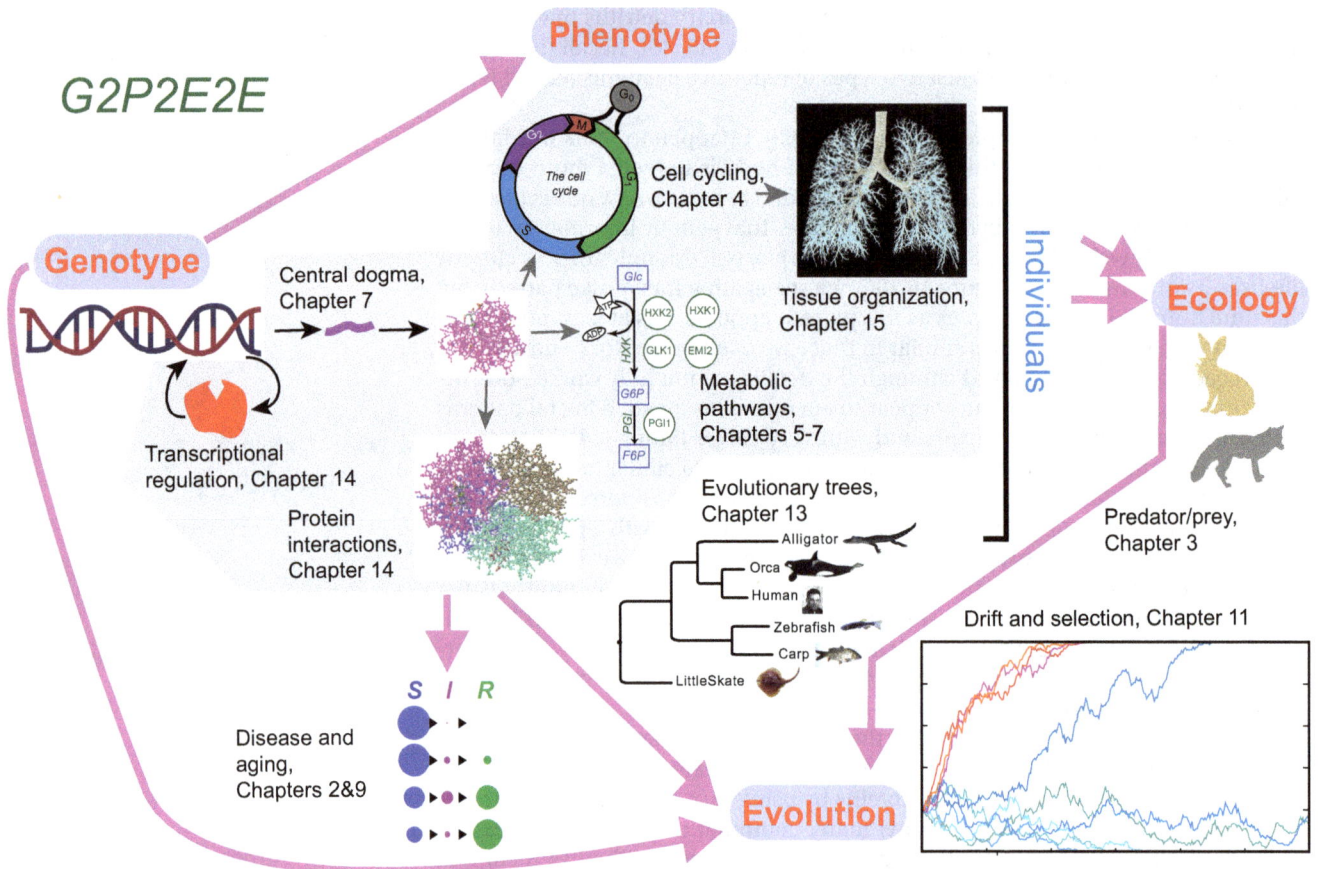

Figure 16.11 *Genotype to phenotype to ecology to evolution* (**G2P2E2E**). The need for a *genotype-to-phenotype* map is central to biology and to the topics covered in this book. The phenotype encompasses many topics in this book and, of course, beyond. This phenotype then underlies the actions of organisms in an ecosystem and their evolutionary dynamics. Proteins shown are myoglobin (PDB ID 3RGK [11]) and hemoglobin PDB structure (PDB ID 4MQJ [13]). (Fox and hare art courtesy of NIAID. Branching lung image courtesy of Pacific Northwest National Laboratory. Carp image courtesy of George Chernilevsky, on Wikimedia, published under CC BY-SA 3.0 license. Skate image courtesy of MDI Biological Laboratory, Maine, on Wikimedia, released into the public domain.)

of the organisms involved, while the natural selection we modeled in Chapter 11 acts not on the genotype but on the phenotype. As a result, we cannot fully understand the genetic trajectories of genomes over millions of years until we understand how those genes encode the form and function of the organisms carrying them.

Thus, this book ends, I think, both with ambiguity and with hope. Ambiguity because, for all the tools and models we have considered, we still have only a partial outline of how we can develop a quantitative understanding of biological systems at scales ranging from the atomic to the planetary. The hope is that this book will furnish a small part of the toolkit by which you, the reader, will help solve this extraordinary set of scientific mysteries.

Exercise 16.2

Test the validity of Stirling's approximation to $\log(n!)$ from Equation 16.10 for a few increasingly large values of n.

REFERENCES

1. Jaillon O, Aury JM, Brunet F, Petit JL, Stange-Thomann N et al., (2004) Genome duplication in the teleost fish Tetraodon nigroviridis reveals the early vertebrate proto-karyotype. *Nature* 431(7011):946–957.

2. Lander ES, Linton LM, Birren B, Nusbaum C, Zody MC et al., (2001) Initial sequencing and analysis of the human genome. *Nature* 409(6822):860–921.

3. Mathews CK & Van Holde KE (1996) *Biochemistry*. 2nd Edition (The Benjamin/Cummings Publishing Company Inc., Menlo Park).

4. NIAID Visual and Medical Arts (9/18/2024) Mast Cell. in *NIAID BIOART Source*. https://bioart.niaid.nih.gov/bioart/335

5. Ridley DR, Ridley LG, & Walker CB (1999) English letter frequencies as found in Whissell's parsimonious sampling of English words. *Perceptual and Motor Skills* 88(2):607–614.

6. Sayers EW, Cavanaugh M, Clark K, Pruitt KD, Schoch CL, Sherry ST, & Karsch-Mizrachi I (2022) GenBank. *Nucleic Acids Research* 50(D1):D161–D164.

7. Xu J (2006) Invited review: microbial ecology in the age of genomics and metagenomics: concepts, tools, and recent advances. *Molecular Ecology* 15(7):1713–1731.

8. Meyer A, Schloissnig S, Franchini P, Du K, Woltering JM, Irisarri I, Wong WY, Nowoshilow S, Kneitz S, & Kawaguchi A (2021) Giant lungfish genome elucidates the conquest of land by vertebrates. *Nature* 590(7845):284–289.

9. Wolff SM, Ellison MJ, Hao Y, Cockrum RR, Austin KJ, Baraboo M, Burch K, Lee HJ, Maurer T, Patil R, Ravelo A, Taxis TM, Truong H, Lamberson WR, Cammack KM, & Conant GC (2017) Diet shifts provoke complex and variable changes in the metabolic networks of the ruminal microbiome. *Microbiome* 5:60.

10. Milo R & Phillips R (2015) *Cell biology by the numbers* (Garland Science).

11. Hubbard SR, Hendrickson WA, Lambright DG, & Boxer SG (1990) X-ray crystal structure of a recombinant human myoglobin mutant at 2.8 A resolution. *Journal of Molecular Biology* 213(2):215–218.

12. Abdurro'uf N, Accetta K, Aerts C, Silva Aguirre V, Ahumada R, Ajgaonkar N, Filiz Ak N, Alam S, Allende Prieto C, & Almeida A (2022) The seventeenth data release of the sloan digital sky surveys: complete release of MaNGA, MaStar, and APOGEE-2 data. *The Astrophysical Journal* 259(2).

13. Berman HM, Westbrook J, Feng Z, Gilliland G, Bhat TN, Weissig H, Shindyalov IN, & Bourne PE (2000) The protein data bank. *Nucleic Acids Research* 28(1):235–242.

14. Azpeitia E, Tichtinsky G, Le Masson M, Serrano-Mislata A, Lucas J, Gregis V, Gimenez C, Prunet N, Farcot E, & Kater MM (2021) Cauliflower fractal forms arise from perturbations of floral gene networks. *Science* 373(6551):192–197.

15. Webmaster L (2022) Library of congress: frequently asked questions. https://www.loc.gov/programs/digital-collections-management/about-this-program/frequently-asked-questions/#:~:text=The%20Library%20of%20Congress'%20digital%20collections%20are%20dynamic%3B%20they%20continually,comprising%20914%20million%20unique%20files.

16. Leidus I (2021) Romanesco broccoli (Brassica oleracea). https://en.wikipedia.org/wiki/File:Romanesco_broccoli_(Brassica_oleracea).jpg

17. Hobern D, Barik SK, Christidis L, Garnett S, Kirk P, Orrell TM, Pape T, Pyle RL, Thiele KR, & Zachos FE (2021) Towards a global list of accepted species VI: The Catalogue of Life checklist. *Organisms Diversity & Evolution* 21(4):677–690.

18. Roitt I, Brostoff J, & Male D (1998) *Immunology*. 5th Edition (Mosby International Ltd., London), p. 423.

19. NIAID Visual and Medical Arts (9/5/2024) Fox Silhouette. in *NIAID BIOART Source*. https://bioart.niaid.nih.gov/bioart/164

20. NIAID Visual and Medical Arts (9/16/2024) Hare Silhouette. in *NIAID BIOART Source*. https://bioart.niaid.nih.gov/bioart/195

21. Caspi R, Billington R, Keseler IM, Kothari A, Krummenacker M, Midford PE, Ong WK, Paley S, Subhraveti P, & Karp PD (2020) The MetaCyc database of metabolic pathways and enzymes-a 2019 update. *Nucleic Acids Research* 48(D1):D445–D453.

22. Shannon CE (1948) A mathematical theory of communication. *The Bell System Technical Journal* 27(3):379–423.

23. Eyre-Walker A & Hurst LD (2001) The evolution of isochores. *Nature Reviews Genetics* 2(7):549–555.

24. Mooers AØ & Holmes EC (2000) The evolution of base composition and phylogenetic inference. *Trends in Ecology & Evolution* 15(9):365–369.

25. Soda K & Osumi T (1969) Crystalline amino acid racemase with low substrate specificity. *Biochem Biophys Res Commun* 35(3):363–368.

26. Levine IN (1995) *Physical chemistry*. 4th Edition (McGraw-Hill, Inc, New York), p. 901.

27. Tringe SG & Rubin EM (2005) Metagenomics: DNA sequencing of environmental samples. *Nature Reviews Genetics* 6(11):805–814.

28. Venter JC, Remington K, Heidelberg JF, Halpern AL, Rusch D, Eisen JA, Wu D, Paulsen I, Nelson KE, Nelson W, Fouts DE, Levy S, Knap AH, Lomas MW, Nealson K, White O, Peterson J, Hoffman J, Parsons R, Baden-Tillson H, Pfannkoch C, Rogers YH, & Smith HO (2004) Environmental genome shotgun sequencing of the Sargasso Sea. *Science* 304(5667):66–74.

29. Tringe SG, von Mering C, Kobayashi A, Salamov AA, Chen K, Chang HW, Podar M, Short JM, Mathur EJ, Detter JC, Bork P, Hugenholtz P, & Rubin EM (2005) Comparative metagenomics of microbial communities. *Science* 308(5721):554–557.

30. Bobay L-M (2020) The prokaryotic species concept and challenges. *The pangenome: diversity, dynamics and evolution of genomes*:21–49.

31. Peet RK (1974) The measurement of species diversity. *Annual Review of Ecology and Systematics* 5(1):285–307.

32. Patil RD, Ellison MJ, Wolff SM, Shearer C, Wright AM, Cockrum RR, Austin KJ, Lamberson WR, Cammack KM, & Conant GC (2018) Poor feed efficiency in sheep is associated with several structural abnormalities in the community metabolic network of their ruminal microbes. *Journal of Animal Science* 96:2113–2124.

33. Cardinale BJ, Palmer MA, & Collins SL (2002) Species diversity enhances ecosystem functioning through interspecific facilitation. *Nature* 415(6870):426–429.

34. Cotillard A, Kennedy SP, Kong LC, Prifti E, Pons N, Le Chatelier E, Almeida M, Quinquis B, Levenez F, & Galleron N (2013) Dietary intervention impact on gut microbial gene richness. *Nature* 500(7464):585–588.

35. Ley RE, Lozupone CA, Hamady M, Knight R, & Gordon JI (2008) Worlds within worlds: evolution of the vertebrate gut microbiota. *Nature Reviews Microbiology* 6(10):776–788.

36. Sha S, Xu B, Wang X, Zhang Y, Wang H, Kong X, Zhu H, & Wu K (2013) The biodiversity and composition of the dominant fecal microbiota in patients with inflammatory bowel disease. *Diagnostic Microbiology and Infectious Disease* 75(3):245–251.

37. Carroll IM, Ringel-Kulka T, Siddle JP, & Ringel Y (2012) Alterations in composition and diversity the intestinal microbiota in patients with diarrhea-predominant irritable bowel syndrome. *Neurogastroenterol Motil* 24(6):521–530, e248.

38. Watson JD & Crick FHC (1953) Molecular structure of nucleic acids. *Nature* 171:737.

39. Franklin RE & Gosling RG (1953) Molecular configuration in sodium thymonucleate. *Nature* 171(4356):740–741.

40. Lodish H, Baltimore D, Berk A, Zipursky SL, & Matsudaira P (1995) *Molecular cell biology*. 3rd Edition (Scientifc American Books, New York), p. 1344.

41. Feynman R (2018) There's plenty of room at the bottom. *Feynman and computation* (CRC Press), pp. 63–76.

42. Marieb EN (1991) *Human anatomy and physiology* (The Benjamin/Cummings Publishing Company, Redwood City, CA), p. 1040.

43. Russell PJ (1994) *Fundamentals of genetics* (Harper Collins, New York).

44. Freeman S & Herron JC (1998) *Evolutionary analysis* (Prentice-Hall, Inc, Upper Saddle River, NJ), p. 786.

45. Doolittle WF (2013) Is junk DNA bunk? A critique of ENCODE. *Proceedings of the National Academy of Sciences* 110(14): 5294–5300.

46. Lynch M & Conery JS (2003) The origins of genome complexity. *Science* 302:1401–1404.

47. Flake GW (1998) *The computational beauty of nature: Computer explorations of fractals, chaos, complex systems, and adaptation* (The MIT Press, Cambridge, MA).

48. Holmes EC (2008) Evolutionary history and phylogeography of human viruses. *Annu. Rev. Microbiol.* 62:307–328.

49. Pigliucci M (2010) Genotype-phenotype mapping and the end of the 'genes as blueprint' metaphor. *Philosophical Transactions of the Royal Society B: Biological Sciences* 365(1540):557–566.

Glossary

Activation energy

In a biochemical reaction, the input energy to the reactants required to allow the conversion of those reactants to products of lower free energy.

Adaptive therapy

A cancer treatment where the use of drugs is modulated with periods when the tumor is allowed to grow untreated by the drug to reduce the proportion of tumor cells resistant to that drug treatment. Removing the drug reduces the population of resistant cells under the assumption that resistance is costly, meaning such cells are at a competitive disadvantage in the absence of the drug.

Agent-based model

A modeling approach that represents the individual elements ("agents") in the model explicitly in the computer and uses the computer to simulate the behavior and interactions between those agents.

Algorithm

A set of described steps, such as computations, that, when followed, produce a defined outcome.

Alignment

For two strings of differing length, an arrangement of the two into a table of two rows and an equal number of columns for each sequence. In each row, the corresponding string is listed in order, except that some cells in the table may have gap characters rather than characters from the string. No column of an alignment may have both rows possessing a gap.

Allele

For a position in the genome where individuals in a population have genetic differences, *allele* is the name given to each of these different genetic variants.

Antiderivative

For a given function, its antiderivative is a function that, when differentiated, gives back the original function.

Basic reproduction rate

In various forms of the SIR model, the basic reproduction rate (often written R_0) is the number of new infections spawned through direct infection from a single infected individual introduced into the population. When R_0 is less than 1.0, the population will not produce an epidemic from the introduction of the disease.

Biomass reaction

A pseudo-biochemical reaction in a constraint-based metabolic model that describes the metabolites that comprise a cell and the relative abundance of those metabolites in that cell.

Byte

A collection of eight binary digits (bits).

Carrying capacity

In an ecological model, the carrying capacity describes the maximum number of individuals the environment is assumed to be able to support.

Catalyst

A compound in a (bio)chemical reaction that is not consumed in that reaction but reduces the activation energy of the reaction such that the rate proceeds in the direction determined by the free energy but at a faster rate.

Central dogma of molecular biology

The shorthand claim that information, and hence biological function, in most systems is encoded by DNA, transcribed into RNA, and then translated into functional proteins.

Chaos

In dynamical systems, a model is said to show chaotic behavior if it lacks any random component but is neither periodic nor converging to a steady state in its long-run behavior. Some chaotic systems show strong dependence on initial conditions, such that initially very similar systems may greatly diverge over time.

Clustering coefficient

A measure of how likely the neighbors of a node in a network are also to be each other's neighbors.

Codon

A group of three RNA bases that are translated by the ribosome and tRNAs into a single amino acid in a growing peptide chain.

Constraint-based models

Models of biochemical networks that represent reactions with linear equations under the assumption that metabolite concentrations in the model are held in steady state.

Continuous random variable

A random variable where the outputs are real numbers over some range.

Cumulative distribution function

A function that, when evaluated between two points, gives the probability of the outcome being in that range. Hence, it is the result of integrating the probability density function between those endpoints.

Derivative

A function that, for every input point, gives the instantaneous rate of change of another function at that point.

Differential equation

An equation that gives the instantaneous rate of change of a (generally unknown) state function in terms of its functional inputs or of other state variables.

Diploid

An organism, usually sexually reproducing, that carries two copies of all of its genes in its cells.

Discrete random variable

A random variable that produces integer outputs.

Dominance

In genetics, it is often observed that, at loci that are polymorphic in the population, there is a (dominant) allele that gives the same observed phenotype whether it is present in one copy (heterozygous) or two copies (homozygous), whereas the other phenotype is seen only in the case of individuals possessing two copies of the other allele.

Dynamical system

A system of related state functions where the values of at least some of the state functions cause other state functions in the system to change. As a result, the overall state of the system changes through time.

Effective population size

For a given population, the size of an idealized population that experiences the same rate of allele frequency change in time as does the original population itself.

Emergent property

A behavior in a model or dynamical system that emerges from the individual components of the system and their interactions with each other that nonetheless cannot be observed or modeled when considering those parts of the system in isolation from each other.

Entropy

The degree of disorder or information in a system.

Exponential process

A process where the rate of change of that process at any point is directly proportional to the value of the function at that point.

Fitness

Measures the reproductive success of carriers of a particular genetic variant relative to the reproductive success of the population as a whole.

Fixation

When a previously variable genetic location in a population loses that variability, such that all members of the population now possess the same genetic sequence at that position.

Floating point number

A computer representation of a real number.

Flux

The rate of product production per unit time in a biochemical reaction or pathway.

Flux-balance analysis (FBA)

Use of constraint-based models to find a set of biochemical pathway fluxes that satisfy some set of nutrient limitations. The analysis usually maximizes the flux through a particular reaction, typically the biomass reaction.

Fractal dimension

A way to measure the degree to which complex curves have shapes falling between conventional integer dimensions.

Function

A mathematical "machine" that uses one or more input numbers to compute another, output number.

Genetic code

The table of rules by which the 4-state DNA or RNA code is translated three bases at a time (i.e., a codon) into the 20-state amino acid code.

Genetic drift

For a genetically variable position in a population, descriptive of random, generation-to-generation changes in the frequencies of the alleles at that position.

Genotype

The particular combination of alleles possessed by a particular individual in a population.

Herd immunity

In a population that has achieved herd immunity, either through an epidemic or through vaccination, the number of susceptible individuals is so low that the basic reproduction rate of a disease, or R_0, is less than one, meaning that no epidemic can occur and that the remaining susceptible individuals are protected by the overall immunity of the population or *herd*.

Heterozygous

An individual that has two copies of a particular genetic location and for which the alleles at those two copies differs.

Homology

Describes two structures or sequences that derive from a single common ancestor in their evolutionary history.

Homozygous

An individual that has two copies of a particular genetic location but for which the individual has the same allele at both positions.

Hybrid

An individual or population that results from matings between two populations or species that have been genetically isolated from some period of time.

Initial conditions

In a dynamical system, the values of the state functions at the start of the model (i.e., at time 0).

Integration

For a given function, one can approximate the area between the function and the x-axis with a series of small rectangles of width Δx rising from the axis to touch the function curve at one point. Integration is the result of letting the interval width Δx become arbitrarily small, such that the approximation to the area under the curve becomes exact. If this process is carried out over the entire range of that function, it is equivalent to obtaining the antiderivative of that function.

Linear algebra

Studying systems of linear equations where the number of dependent variables is generally three or greater. Often these equations are represented as matrices.

Linear equations

Equations where the dependent variables contribute to the value of the independent variable only through the multiplication of those individual dependent variables by constant terms.

Matrix

A representation of a set of linear equations where only the coefficients of the dependent variables (the xs) are listed and the variable xs are represented in a column vector to the right of that matrix. In other applications, the term can be used simply to describe a two-dimensional table of numbers.

Metabolic channeling

A variety of strategies by which cells spatially organize metabolic pathways to avoid having those pathways be diffusion limited in their flux.

Model

A mathematical, computational, verbal, or statistical representation of a phenomenon in the outside world that allows some combination of prediction, data analysis, theory testing, and description.

Network

A large system of interacting entities represented as nodes and interactions.

Network motif

Given a defined number of nodes from a network, a particular pattern of connections between that fixed set of nodes.

Neural network

A computational representation of conceptual neurons that channel information from one to the next, often used to represent arbitrary computations.

Neutral alleles

Genetic variants in a population that do not measurably differ in their fitness.

Node

An individual entity in a network.

Node degree

The number of edges connecting to a node in a network.

Parameter

In a model, a tunable numerical value that can change the model's behavior but that is not a state function of the model and hence does not change over time in the model.

Partial differential equation

A differential equation describing a state function of several dependent variables in terms of its derivatives with respect to one or more of those dependent variables.

Path length

The number of other nodes in a network one must, by way of interactions, pass through to reach one node from another.

Phenotype

In the context of genetic variation, an observed form or behavior in an organism that is known to vary as a result of genetic differences.

Phylogeny

A branching tree representing the evolutionary relationships between biological sequences or species.

Polymerase

An enzyme that is able to use a DNA or RNA template to synthesize a new DNA or RNA strand from free nucleotides.

Probability

The odds, chance, or frequency with which a random event occurs.

Probability density function

A continuous function giving the rate of change of the probability of an outcome. When integrated over a range, this function gives the probability of the outcome being in that range.

Promoter

A region of a gene, typically located before the coding region of that gene, that contains DNA sequences that interact with other factors to induce or repress that gene's transcription.

Protein interaction

A physical interaction between two different protein peptide chains that can be transient or long-lasting.

Random variable

A mathematical process that produces outputs over a defined range at a defined relative proportion but where the individual values cannot be predicted beyond the information of those ranges and proportions.

Recessive

In genetics, it is often observed that, at loci that are polymorphic in the population, there is a dominant allele that gives the same observed phenotype whether it is present in one copy (heterozygous) or two copies (homozygous), whereas the other phenotype is seen only in the case of individuals possessing two copies of the other, *recessive* allele.

Robustness

A system is robust if it maintains its behavior in the presence of perturbations to its environment or internal state.

Small-world network

A network characterized by possessing both short average path lengths and high clustering coefficients.

State function

In a dynamical system, one of the functions being modeled as potentially changing in time and hence for which a differential equation is available.

Steady state

In dynamical systems, the state that is reached when all the functions achieve values that remain constant for all times t greater than some transition time t_i.

Stochastic

A process or model that incorporates some aspect of randomness (quantum mechanical or historical) into it.

Susceptible

An individual who has not yet become infected with a communicable disease and hence may become infected if they come in contact with an infected person.

Synonymous difference

Two alleles of a gene where the codon at that position differs between the alleles but, because of the redundancy of the genetic code, both alleles code for the same amino acid at that position.

Transcription

The production of an RNA molecule from a DNA template by an RNA polymerase.

Transcription factor

A protein that can induce or repress the transcription of its own gene or another gene.

Transition probability matrix

A matrix whose row and column indices are states in a discrete state model and where the entries of the matrix give the probability of moving from the row state to the column state.

Translation

The production of a polypeptide protein molecule from an RNA molecule by means of the ribosome and tRNA molecules.

Vaccine

A compound that, when administered, trains the immune system to recognize and respond to an infectious disease without the patient having been infected with that disease prior to the vaccine administration.

Variance

A measure of the degree of dispersion seen in realizations of a random variable. Typically measured as the squared difference between the outcomes and the mean outcome, taken over the range of the random process.

Vector

A list of numbers of a defined length, written in either a column or a row. The number of rows in a column vector is the dimension of that vector.

Index

Note: Pages in *italics* refer to figures and pages in **bold** refer to tables.

For Product Safety Concerns and Information please contact our EU
representative GPSR@taylorandfrancis.com
Taylor & Francis Verlag GmbH, Kaufingerstraße 24, 80331 München, Germany

* 9 7 8 1 0 4 1 1 7 0 1 5 0 *